RELIABILITY ENGINEERING FOR NUCLEAR AND OTHER HIGH TECHNOLOGY SYSTEMS

A Practical Guide

RELIABILITY ENGINEERING FOR NUCLEAR AND OTHER HIGH TECHNOLOGY SYSTEMS

A Practical Guide

ARMAND A. LAKNER

President, RSE Associates, Washington DC, USA
(formerly United States Nuclear Regulatory Commission,
Washington DC)

and

RONALD T. ANDERSON

President, Reliability Technology Associates, Chicago, USA

CRC Press
Taylor & Francis Group
Boca Raton London New York

CRC Press is an imprint of the
Taylor & Francis Group, an **informa** business

First published 1985 by Elsevier Applied Science Publishers
Taylor & Francis Group
6000 Broken Sound Parkway NW, Suite 300
Boca Raton, FL 33487-2742

Reissued 2018 by CRC Press

© 1985 by Taylor & Francis
CRC Press is an imprint of Taylor & Francis Group, an Informa business

No claim to original U.S. Government works

A Library of Congress record exists under LC control number: 85202540

Publisher's Note
The publisher has gone to great lengths to ensure the quality of this reprint but points out that some imperfections in the original copies may be apparent.

Disclaimer
The publisher has made every effort to trace copyright holders and welcomes correspondence from those they have been unable to contact.

ISBN 13: 978-1-138-10556-0 (hbk)
ISBN 13: 978-1-138-56173-1 (pbk)
ISBN 13: 978-0-203-71048-7 (ebk)

Visit the Taylor & Francis Web site at http://www.taylorandfrancis.com and the CRC Press Web site at http://www.crcpress.com

Foreword

This book presents a much needed practical methodology for the establishment of cost-effective reliability programs in nuclear or other high technology industries. Thanks to the high competence and practical experience of the authors in the field of reliability, it vividly illustrates the applicability of proven, cost-effective reliability techniques applied in the American space and military programs as hybridized with the avant-garde approach used by nuclear authorities, utilities and researchers in the United Kingdom and France. This emerged method will support a diligent effort in the enhancement of nuclear safety and protection of the health of the general public.

The methodology developed in this book exemplifies the total integrated reliability program approach in the design, procurement, manufacturing, test, installation and operational phases of an equipment life cycle. It is based on lessons learned in space and military programs with certain methodological modifications to enhance practicality.

The techniques described here are applicable to college instruction, plant upper and middle management personnel, as well as to regulating agencies with equal benefits; it provides a very pragmatic and cost-efficient approach to the reliability engineering discipline.

ANNICK CARNINO
Electricité de France,
Direction Générale, Paris, France

Preface

The realization of reliable systems, especially electrical power and control equipment used in nuclear power plants or other high technologies, in a cost-effective manner involves performing complex engineering efforts that begin with early cost–benefit studies during conceptual planning, continue through design and development and extend through power plant construction, acceptance and actual operation. This book addresses nuclear and other high technology system reliability from the standpoint of the operating utilities as well as from the standpoint of the system designers and hardware suppliers or vendors. It focuses on cost-effective, space technology-proven reliability techniques applicable to the procurement, installation and testing of electronic power control systems and instrumentation. The planning of a reliability program is discussed as well as its scope. It addresses the important question of determining the extent and type of reliability engineering effort that is necessary and cost-effective in order to avert a 'Salem-like' accident. In that classic case, a double failure of circuit breakers caused, on 21 February 1983, the malfunction of the automatic reactor trip system, a very serious and severe consequence scenario in a nuclear power plant operation.

This book emphasizes, first, the importance of a clearly defined cost-effective procurement specification that reflects the functions and needs that a system must meet and, second, the execution of effective, well-timed proven reliability engineering tasks during planning, design, development and plant construction to achieve and maintain the specified system reliability requirements during the system useful operational lifetime. A quantitative approach is described that recognizes that reliability, like other system parameters, can be specified in procurement, predicted in design, measured during test and controlled during construction and operation. It shows how the exercise of a very deliberate reliability engineering program, throughout the hardware system life cycle, can provide high reliability cost-effectively.

This book is written for the reliability instructor, program manager, system engineer, design engineer, reliability engineer, nuclear regulator, probability risk assessment (PRA) analyst, general manager and others who are involved in system hardware acquisition, design and operation and are concerned with plant safety and operational cost-effectiveness.

This book provides criteria, guidelines and comprehensive engineering data affecting reliability; it covers the key aspects of system reliability as it relates to conceptual planning, cost tradeoff decisions, specification, contractor selection, design, test and plant acceptance and operation. It treats reliability as an integrated methodology, explicitly describing life cycle management techniques as well as the basic elements of a total hardware development program, including: reliability parameters and design improvement attributes, reliability testing, reliability engineering and control. It describes how these elements can be defined during procurement, and implemented during design and development to yield reliable equipment.

This book is organized into six chapters, each covering an essential aspect of the procurement and development of reliable equipment. The chapters are organized in accordance with the planning and implementation required to obtain reliable equipment, from initial optimization tradeoff studies to ensuring that these plans are met during design, production and operation. In preparing this book each chapter was intended to present a complete, self-contained treatment of the pertinent subject matter following the planning and implementation process. Consequently certain specific topics and points are dealt with more than once in order to provide this completeness.

Chapter 1 provides general information related to planning and developing reliable equipment from an acquisition standpoint and discusses the reliability engineering tasks performed during this phase. Chapter 2 discusses the theory and the basic engineering foundations of reliability. It describes mortality concepts, degradation factors, reliability of non-maintainable equipment, reliability of maintainable equipment, availability analysis and reliability improvement. Chapter 3 provides specific information on how to perform reliability cost tradeoffs and describes reliability improvement attributes, cost-estimating relationships (CERs), reliability versus cost ratios and value impact studies that depict the cost of various reliability programs and test efforts. It describes a matrix scheme that facilitates the tailoring of standard reliability programs to meet specific needs. Chapter 4 describes how to effectively evaluate the adequacy of proposed hardware development programs for nuclear or other high technology systems. Evaluation guidelines are presented covering basic reliability engineering tasks, including prediction; failure mode, effects and criticality analysis (FMECA); design review; reliability testing and failure analysis. Chapter 5 describes the design process from a reliability standpoint and provides data and information that will enable the design engineer to incorporate reliability into his design. Reliability attributes, such as derating, part selection and the application of controlled screening and burn-in, are described. Chapter 6 provides basic engineering techniques for implementing a reliability engineering program during system hardware acquisition. Included are procedures covering reliability prediction, maintenance planning, FMECA, fault tree analysis (FTA), design review, reliability testing, degradation analysis and failure reporting, analysis and corrective action (FRACA).

ARMAND A. LAKNER
RONALD T. ANDERSON

Acknowledgements

The authors wish to thank a number of people for their help and advice in the preparation of this book, in particular, Dr Daniel Henry III and Mr Sidney Bass for their valuable assistance, constructive criticism, guidance and suggestions for improvement. Also, special thanks are given to Mme Annick Carnino of Electricité de France, Mr Peter Massoglia of the US Federal Aviation Administration and Mr Harold Lauffenburger of the IIT Research Institute for their encouragement in writing this book.

The authors are indebted to Stephen Lakner MD, of the George Washington University Hospital, Washington DC for his encouragement, support and advice principally addressed to the aspect of public health and safety, as well as Mr Jacob Pankowski, of McKenna, Conner and Cuneo, Washington DC, for his valuable legal and organizational counselling.

The book evolved from work performed by the authors over the past 15 years for US Government agencies such as the Nuclear Regulatory Commission, the Federal Aviation Administration, the National Aeronautics and Space Administration and the Department of Defense and for a number of major industrial organizations. This work has focused on the development and application of practical reliability engineering methods to the acquisition and operation of large, complex, high technology systems. The authors are grateful for the permission which has been given to reproduce selected tables and figures prepared for these organizations and previously presented in various government publications.

Mr George B. Olley of Elsevier Applied Science Publishers provided editorial comments and suggestions which added greatly to the quality of the final manuscript.

Acknowledgement is also given to Ms Karen Anderson for checking the entire manuscript and to Ms Laurie Brauer for preparing graphical and tabular material.

Finally, the authors would like to acknowledge Ms Janet Anderson and Ms Betty Kinney for their rapid and accurate typing of all draft material as well as for their efficient typing of the final manuscript.

A.A.L.
R.T.A.

Contents

Foreword v

Preface vii

Acknowledgements xi

Chapter 1   Introduction 1
             1.1 Overview of the Nuclear Industry 3
             1.2 Reliability Assurance Program Elements . . . 11

Chapter 2   Reliability Analysis 21
             2.1 Basic Reliability Concepts 25
             2.2 Basic Maintainability Concepts 31
             2.3 Availability Analysis and Improvement . . . 36
             2.4 Reliability Degradation and Control 51
             2.5 Software Reliability Analysis 59
             2.6 Reliability Analysis Techniques 70

Chapter 3   Specifying Safe, Reliable and Cost-Effective Equipment . 80
             3.1 Planning and Implementing a Life Cycle Reliability
                  Program 80
             3.2 Reliability Management 91
                  3.2.1 Element number 1: Reliability program
                         planning 101
                  3.2.2 Element number 2: Organization . . . 106
                  3.2.3 Element number 3: Program review . . 108
                  3.2.4 Element number 4: Configuration control and
                         management 108
                  3.2.5 Element number 5: Supplier audit and
                         surveillance 109
                  3.2.6 Element number 6: Reliability critical items list
                         (and control) 110
                  3.2.7 Element number 7: Reliability training and
                         indoctrination 111

3.3 Design Assurance Elements 112
    3.3.1 Element number 8: Tradeoff analysis . . 115
    3.3.2 Element number 9: Reliability and maintainability specification 117
    3.3.3 Element number 10: Reliability mathematical modeling 119
    3.3.4 Element number 11: Reliability prediction and allocation 120
    3.3.5 Element number 12: Maintainability prediction and allocation 123
    3.3.6 Element number 13: Diagnostics and modularity 124
    3.3.7 Element number 14: Human factors reliability analysis 125
    3.3.8 Element number 15: Failure mode and effects analysis (FMEA), failure mode, effects and criticality analysis (FMECA) and fault tree analysis (FTA) 126
    3.3.9 Element number 16: Common mode failure (CMF)/common cause failure (CCF) analyses 129
    3.3.10 Element number 17: Single failure analysis . 130
    3.3.11 Element number 18: Life cycle cost (LCC) assessment and analysis 130
    3.3.12 Element number 19: Reliability growth (and tracking) 131
    3.3.13 Element number 20: Reliability demonstration 133
    3.3.14 Element number 21: Design reviews . . 135
3.4 Component Availability 137
    3.4.1 Element number 22: Component derating . 138
    3.4.2 Element number 23: Component quality level 139
    3.4.3 Element number 24: Component failure rate modeling and estimation 140
    3.4.4 Element number 25: Component testing methods and compliance 141
    3.4.5 Element number 26: Component selection, specification and quality assurance . . . 143
    3.4.6 Element number 27: Effects of storage/dormant periods and shelf life 144
    3.4.7 Element number 28: Maintenance planning (RCM) 145
    3.4.8 Element number 29: Spare parts/logistics analysis 146
3.5 Reliability Degradation Control 147
    3.5.1 Element number 30: Controlled screening and burn-in 148
    3.5.2 Element number 31: Reliability acceptance . 150
    3.5.3 Element number 32: Degradation control . 151
3.6 Operational Reliability Assurance 153

3.6.1 Element number 33: Cyclic inspection . . 153
3.6.2 Element number 34: Reliability and maintainability (R & M) assessment 154
3.6.3 Element number 35: Probabilistic risk
analysis/reliability analysis 157
3.6.4 Element number 36: Reliability (or availability)
improvement 158
3.7 Experience Feedback 159
3.7.1 Element number 37: Failure reporting analysis
and corrective action (FRACA) . . . 160
3.7.2 Element number 38: Data recording and
feedback 162
3.8 Determining Reliability Specification Requirements . 163
3.9 Achieving Reliability in Commercial Equipment . 177

Chapter 4 **Evaluating Reliability Programs for Nuclear Power Plants** . 180
4.1 Reliability Program Audit 182
4.2 Reliability Program Monitoring Guidelines . . 193

Chapter 5 **Designing for Reliability** 207
5.1 Part Selection 209
5.1.1 Microcircuit selection guidelines . . . 211
5.1.2 Semiconductor selection guidelines . . 214
5.1.3 Resistor selection guidelines 216
5.1.4 Capacitor selection guidelines . . . 219
5.2 Application of Screening During Production (or
Maintenance) 222
5.3 Derating 240
5.4 Diagnostics and Modularity 244
5.5 Other Reliability Improvement Attributes . . . 248

Chapter 6 **Reliability Engineering and Control** 254
6.1 Reliability Allocation, Prediction and Assessment . 256
6.1.1 Part count prediction 260
6.1.2 Stress–strength prediction 263
6.2 Maintainability Prediction/Maintenance Planning
(Reliability Centered Maintenance) 267
6.3 Failure Mode Analysis Concepts 282
6.4 Part Control and Standardization 294
6.5 Production Reliability Assurance 297
6.6 Failure Reporting, Analysis and Corrective Action
(FRACA) 304
6.6.1 Failure report (FR) 311
6.6.2 Failure analysis report (FAR) form . . 313
6.6.3 Corrective action request (CAR) form . . 314
6.7 Design Review 314
6.8 Reliability Testing 317
6.9 Software Reliability Engineering and Control . . 326

Appendix A: Definitions 339

Appendix B: Bibliography 367
    Standards and Specifications 367
    Handbooks, Guidebooks and Notebooks 377
    Other Publications 386

Index 413

Chapter 1

Introduction

The systems and equipment used in nuclear power plants are sophisticated, complex, difficult to maintain and costly. They must operate for long periods of time without a serious failure and must have a very long total life. Furthermore, the physical environment in nuclear plants is very severe and can have a serious detrimental effect on the equipment, especially on complex mechanical and electronic components. High temperatures, vibration, high humidity, corrosive fluids and gases all take a toll on electromechanical devices and electronic components. Also, a large portion of the equipment must operate remotely after startup and depend on human operators for operation and control. These and other application factors make it essential that special attention be applied during acquisition to ensure the actual operational reliability of the equipment used in nuclear power plants.

There is great incentive for achieving high reliability. First, and most importantly, the safety requirements of nuclear power plants are of paramount concern. Also, the very large cost of designing and constructing a nuclear power plant and the large cost associated with plant downtime (as much as $800 000 per day in power replacement cost alone) provide a strong economic incentive toward designing-in, or improving equipment reliability. Furthermore, failures are highly visible and, if they are serious enough or occur frequently, they could affect the whole industry. For these reasons, the nuclear industry is making a heavy commitment to safety with special emphasis on the acquisition and operation of 'high reliability' systems and equipment. This involves implementing well-conceived reliability programs and the application of disciplined analysis methods and controls. In addition to the design of highly reliable equipment, the industry is applying a great amount of system redundancy and functional diversity within each facility to help ensure the safety of the overall operating plant.

The achievement of reliable systems requires careful planning followed by a well-executed program with engineering tasks that start when the design concept emerges and continue through development, plant construction and operation. Reliability must be treated as a basic design parameter on a par with performance and cost. It requires establishing adequate requirements, providing necessary resources (i.e. skills, tasks, etc.) for its realization and exercising authority to assure that the consequences in reliability are made part of each decision during the acquisition process. Reliability maturity for a new system is only reached through the application of systematic, well-planned and controlled, life cycle tasks performed by the system designer, the hardware manufacturer and the procuring operating utility (see Fig. 1.1). A brief summary of the specific tasks for each life cycle phase is given below:

1. Conceptual—involves reliability planning, trade-off studies and identification of areas of high technical risk
2. Specification—involves defining reliability requirements, preparing work statements and hardware specifications and evaluation of hardware proposals
3. Development—includes all elements of reliability engineering, analysis, and test, as well as utility management and monitoring
4. System integration—includes tasks to assure that system reliability is controlled during plant construction, installation and pre-operational testing
5. Operational—includes efforts to collect and analyze operational and failure data, to develop equipment improvement recommendations to assure reliability/safety integrity during maintenance, to maintain configuration control and to establish logistics and spare provisioning requirements.

These tasks must be an integral part of the overall acquisition-operational process during which the required level of reliability is first planned by the utility, specified by the system designer and then implemented into the equipment by the manufacturer. In structuring a reliability program, prime consideration must be given bo ensuring safety by selecting and specifying the necessary program tasks to be applied individually or in sequence with other tasks. These tasks, as they are planned and applied during an equipment life cycle, will help in meeting requirements (both quantitative and qualitative) for design integrity and

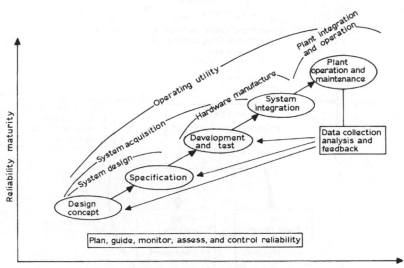

Fig. 1.1. System life cycle.

maintaining component availability. They also support the prevention of manufacturing and installation errors, contribute to the development of proper operating and maintenance procedures, and aid in controlling system configuration, in providing feedback from operating experience and in conducting personnel training. It should be noted that, although an effective reliability program includes tasks applicable to each phase of a system life cycle, current emphasis, because of the present state of the nuclear industry, is placed on the adequacy of the operational and maintenance procedures, on identifying and mitigating the aging mechanisms of safety related components, on recording data and analyzing failures, on implementing configuration management controls and on performing other tasks during plant operation.

1.1 OVERVIEW OF THE NUCLEAR INDUSTRY

The role and relationships of the major organizational elements involved in the acquisition, operation and regulation of systems used in US nuclear power plants is shown in Fig. 1.2. The figure shows that the US nuclear power industry consists of the utilities, architect–engineering (A–E) firms, manufacturers and suppliers (including nuclear steam supply system (NSSS) vendors), and the Department of Energy/Department of Defense

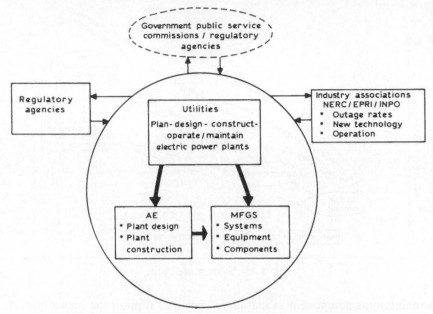

Fig. 1.2.   Nuclear power generating industry (conceptual).

(DOE/DoD) and trade associations (National Electric Research Council (NERC), Electric Power Research Institute (EPRI), Institute of Nuclear Power Operations (INPO) and Institute of Electrical and Electronic Engineers (IEEE)). Each in various ways supports the overall objective of the industry to supply safe and reliable electric power to the public at the lowest cost while complying with regulatory agencies, including the US Nuclear Regulatory Commission (NRC).

As of December 31, 1981, there were 76 operating nuclear power plants in the United States with an additional 90 under construction or on order. These nuclear power plants were, are, or will be, planned, designed, constructed, operated and maintained by electric utilities under license from the NRC. The 58 electric utilities with one or more nuclear power plants operable, under construction, or on order, as of December 31, 1983, are listed in Table 1.1.

The nuclear reactors for the power plants are supplied by NSSS vendors. There are currently five active NSSS vendors in the US—Westinghouse Electric Corp., Combustion Engineering, Babcock & Wilcox Co., General Electric Co. and General Atomic Co. The first three of these vendors supply pressurized water reactors (PWRs); General Electric supplies boiling water

TABLE 1.1
US Electric Utilities with Nuclear Power Plants Operable, Under Construction, or on Order, as of December 31, 1983

Alabama Power Co.[a]
Arizona Public Service
Arkansas Power & Light Co.[b]
Baltimore Gas & Electric Co.
Boston Edison Co.
Carolina Power & Light Co.
Cincinnati Gas & Electric Co.
The Cleveland Electric Illuminating Co.
Commonwealth Edison Co.
Connecticut Yankee Atomic Power Co.
Consolidated Edison Co.
Consumers Power Co.
Dairyland Power Cooperative
Detroit Edison Co.
Duke Power Co.
Duquesne Light Co.
Florida Power & Light Co.
Florida Power Corp.
Georgia Power Co.[a]
GPU Nuclear Inc.
Gulf States Utilities Co.
Houston Lighting & Power Co.
Illinois Power Co.
Indiana & Michigan Electric Co.[c]
Iowa Electric Light & Power Co.
Kansas Gas & Electric Co.
Long Island Lighting Co.
Louisiana Power & Light Co.[b]
Maine Yankee Atomic Power Co.
Mississippi Power & Light Co.[b]

Nebraska Public Power District
Niagara Mohawk Power Corp.
Northeast Utilities
Northern States Power Co.
Omaha Public Power District
Pacific Gas & Electric Co.
Pennsylvania Power & Light Co.
Philadelphia Electric Co.
Portland General Electric Co.
Power Authority of the State of New York
Public Service Co. of Colorado
Public Service Co. of New Hampshire
Public Service Electric & Gas Co.
Public Service Indiana
Rochester Gas & Electric Corp.
Sacramento Municipal Utility District
South Carolina Electric & Gas Co.
Southern California Edison
Tennessee Valley Authority
Texas Utilities Generating Co.
Toledo Edison Co.
Union Electric Co.
Vermont Yankee Nuclear Power Corp.
Virginia Electric & Power Co.
Washington Public Power Supply System
Wisconsin Electric Power Co.
Wisconsin Public Services Corp.
Yankee Atomic Electric Co.

[a] Subsidiary of The Southern Co.
[b] Subsidiary of Middle South Utilities Inc.
[c] Subsidiary of American Electric Power Company Inc.
(Source: World list of nuclear power plants, *Nuclear News*, February 1984).

reactors (BWRs); General Atomic supplies high temperature gas-cooled reactors (HTGRs).

In most cases, the utilities are assisted in the design and construction of the plants by architect–engineer (A–E) firms. For the plants operable, under construction, or on order as of December 31, 1981, there were 10 different A–E firms involved: Bechtel Power Corp., Black & Veatch Consulting Engineers, Burns & Roe Inc., Ebasco Services Inc., Fluor Power Services Inc., Gibbs & Hill Inc., Gilbert Associates Inc., Sargent & Lundy Engineers, Stone & Webster Engineering Corp. and United Engineers and Constructors Inc.

The system, equipment, and components for the plants come from any of many thousands of manufacturers, assemblers and suppliers. These suppliers may interact in their work with the utilities, the NSSS vendors, the A–E firms, or any combination of them.

The US nuclear industry is supported technically by several organizations. The DOE sponsors research and development, including the construction and operation of demonstration plants, in the nuclear power area. The electric power industry itself has formed the EPRI to carry out research and development in all areas of the industry; a portion in the nuclear power field. Following the accident at the Three Mile Island, Unit 2, Nuclear Power Plant, the US nuclear industry formed two additional supporting technical organizations: the Nuclear Safety Analysis Center (NSAC) at EPRI to improve nuclear plant safety and INPO to improve nuclear plant operations.

The *Electric Power Research Institute* (*EPRI*) began operation in 1973 under the voluntary sponsorship of the US electric utility industry—private, public, and cooperative—for the purpose of expanding electric energy research and development. Its objective is to advance capability in electric power generation, delivery, and use in the public interest, with special regard for efficiency, reliability, economy, and environmental considerations. NSAC, which is part of EPRI, was established in April 1979 after the TMI incident; its efforts are focused on (1) the analysis of current safety concerns, (2) evaluation of the work being done to address these concerns, and (3) identification and implementation of any additional work that is needed to help improve nuclear plant safety. NSAC also develops position papers on those issues that can be used by utilities in regulatory matters. NSAC concentrates on events of potentially serious consequences but relatively low probability which have a hardware and procedural orientation.

A specific area of NSAC interest is that of developing accelerated equipment aging techniques. A major goal in the evaluation of safety-related equipment is the prevention of failures due to common causes in redundant safety systems. Since equipment aging is a potential common-mode failure mechanism, it is essential to demonstrate, during the design and manufacture of safety-related electrical equipment, that the equipment can function under design-basis condition—not only in an 'as new' condition but also after the degrading effects of in-service aging have occurred. NSAC efforts include reviewing equipment aging theory and assessing qualitatively the vulnerability of equipment with respect to aging, particularly with respect to common-mode failures. This organization has also looked at the possibility of artificially aging equipment based on the Arrhenius relationship in order to demonstrate that aged equipment can function after an accident. NSAC activities are driven by *IEEE STD-323-1974* (Qualifying Class IE Equipment of Nuclear Power Generating Stations). This standard calls for accelerated aging in a time-correlated fashion such that a 'qualified life' may be demonstrated.

While EPRI is the R&D arm of the utilities, the *Institute of Nuclear Power Operations (INPO)* is their operational arm geared to establish industry-approved standards for the operation of nuclear power plants and to ensure that the utilities meet those standards. It is, in effect, a self-imposed policeman of the nuclear power industry; all of the utilities operating nuclear power plants in the US belong and provide financial support, to INPO. INPO became operational in December 1979. It was created by the utilities in rapid response to the TMI accident in affirmation of the nuclear industry's commitment to safety.

INPO's primary duties are evaluating the management and operation of US nuclear plants and recommending improvements. The goal is to evaluate all US nuclear plants at least once a year. After each evaluation, INPO drafts a preliminary report which it sends to the utility for its review and response. INPO staff members then meet with utility management to agree on actions to be taken in response to INPO's recommendations, and target dates are set for each item. These are included in the final report sent to the utility. The utility is asked to inform INPO when the targets are met. INPO continually informs the industry of 'best practices' that it has found as a result of its activities.

Other work performed by INPO includes studying human factors improvements and developing human reliability data for probabilistic risk assessment. They also operate a computer communications network called NOTEPAD. Utilities that join the network can talk back and forth, querying one another about equipment and practices and receiving information from INPO. Licensee event reports (LERs) and other unusual occurrences reported by utilities are reviewed by INPO as a first step in their 'Significant Event Evaluation and Information Network' (SEE-IN) program. The SEE-IN program was directed primarily at screening and analysis of LERs. As the SEE-IN program developed and matured, it became increasingly apparent that a comprehensive, effective experience feedback program must be based on screening and analysis of significant plant events and equipment, and component failures, and on the identification and reliability assessment of key equipment in critical accident sequences. Ready access to an effective component-reliability database is a vital part of such a program.

In light of the need for a component reliability database to support INPO activities. INPO, in January 1982, |assumed the management, technical direction, and funding of the 'Nuclear Plant Reliability Data System' (NPRDS). NPRDS collects and disseminates operating reliability statistics for safety-related components and systems in commercially operated US nuclear power plants. Information is collected on 29 major categories of components of mechanical and electromechanical designs. The information may be used for reliability and maintainability prediction and assessment, and for design improvement programs. Participants in NPRDS are provided with access to (1) complete engineering data on components and system, (2) unit and system operating hours, (3) statistics on reliability performance of equipment, and (4) complete description of component failure, including mode, type, cause, effect, and detection.

In addition to the NPRDS data function, the *National Electric Research Council (NERC)* is responsible for the operation of the 'Generating Availability Data System' (GADS), formerly the EEI Data System. GADS is a source of summary unit performance data on all types of electric power generating equipment, e.g. nuclear, fossil, hydro, combustion, and combined cycle units.

Information is reported to NERC/GADS by the participating utilities on a quarterly basis. Included in GADS are unit statistics, outage event types, outage causes, and unit performance information. Input data are subjected to validation checks through programs developed by NERC/GADS, and errors are corrected by the utilities prior to being entered into the master data files. Output data include forced outage rates, capacity factors, equivalent availability factors, and operating availability.

Furthermore, many utilities driven by the significant cost benefits to be realized from improved plant availability, are either designing, or already have, operational in-house systems for the collection and analysis of failure and maintenance data. Most of the systems either are, or will be, computerized with interactive terminals at each plant. They are based upon plant-specific data within each utility supplemented by data from other sources such as IEEE, LERs, NERC/GADS, NPRDS, etc.

In addition to the regulatory agencies associated with the electric power industry in general, the US nuclear power industry is also regulated by the NRC. Its mission is directed toward protecting the health and safety of the public. The NRC interacts primarily with the utilities, who are the licensees of the nuclear power plants, and it is the utilities' responsibility to assure that the NSSS vendors, A–E firms, and suppliers produce a nuclear power plant, systems, equipment and components acceptable to the NRC.

One of the authors has outlined for the NRC a safety-oriented reliability systems engineering program (RAP) for implementation (see Fig. 1.3); the program supports NRC's recently published safety goals and the one-step licensing process to assure plant safety and systems reliability and is tailored according to NASA, FAA and DoD proven experiences. This program, entitled 'Safety Assurance and Failure Elimination' (SAFE), is a comprehensive, multidisciplinary cost-effective effort aimed at the earliest possible elimination of obscure or hidden hardware or procedural deficiencies missed by other techniques such as the probabilistic risk assessment (PRA), quality assurance and surveillance testing. A key objective is to decrease the number of accident sequences and dominant system/equipment failure contributors as identified by current PRA efforts. Plans include adopting a preventive approach to safety and reliability assurance by stressing remedial actions at the design and operational stages. The process is structured in a life-cycle cost context to provide a firm basis for cost-effective decision making. The program is oriented toward the adaptation of existing reliability management and engineering practice from other high-technology fields.

The DoD and the National Aeronautics and Space Administration (NASA) have pioneered much of the work in the reliability and maintainability areas. These agencies have developed and published a number of specifications, regulations, handbooks, and standards dealing with the subject. This extensive and comprehensive series of documents provides the framework and methodology for quantitatively specifying, allocating, predicting, analyzing, demonstrating, and maintaining the required equipment/system R&M during each phase of the life cycle. They also include detailed design procedures for use by the designer to ensure that the required R&M requirements will be met. Appendix B includes a brief description of some of the key R&M documents used by DOD and

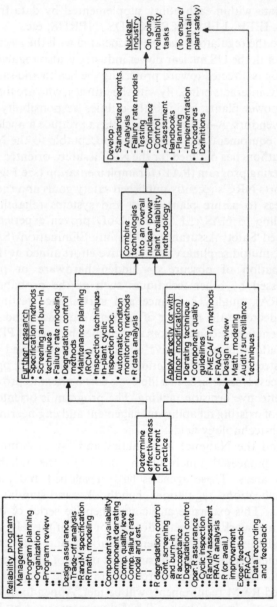

Fig. 1.3. NRC (proposed) SAFE program.

NASA. Many of these documents could be applied with slight modification in establishing and analyzing the reliability of nuclear power plants. The SAFE program incorporates many of these documents as well as others developed by the Federal Aviation Administration (FAA).

1.2 RELIABILITY ASSURANCE PROGRAM ELEMENTS

To ensure high operational reliability, sound engineering practice and program control elements, with defined corrective action criteria at key points, must be applied throughout a system/equipment life cycle. Cost must be evaluated in terms of the total life cycle and tradeoffs must be performed during acquisition to determine 'design to' requirements for the significant elements that comprise acquisition cost and operational maintenance (O&M) support cost.

Reliability must be treated as a fundamental performance parameter beginning with the design and specification phases of system development and continuing into and during the useful operating life of the system. Throughout, effective reliability engineering requires an allocation of adequate resources applied through appropriate levels of management authority. The connected flow of reliability engineering activities throughout the total system life cycle is diagrammed in Fig. 1.4 in the manner which also depicts the distinct role of the utilities, the A–E firms and the manufacturers. The utility, operating in a regulated environment has overall responsibility for the achievement of required levels of reliability. This role is carried out throughout the internal functions shown in the figure and across the interface between the utilities and A–E and manufacturing companies.

To assure the reliability of the operating systems and to maintain required plant safety levels, the NRC must monitor the industry's reliability programs through implementation of the proposed SAFE program. NRC system reliability activities must include the following seven functions:

1. Ensure design integrity: This involves reviewing reliability analysis reports, FMEA, FTA and common mode failures as well as participating in program reviews and witnessing demonstration and acceptance tests.
2. Maintain component availability: This involves reviewing component application data, logistics analysis and component failure analysis reports.

System Design Activities
(Conceptual Design)
- Perform PRA, A, R and LCC studies
- Perform R assessments of similar equipment
- Identify R levels and LCC
- Formulate R and LCC guidelines
- Structure R requirements for procurement specifications, design, test (R growth, demo, acceptance), program

Manufacturer's Activities
(Development/Production)
- Perform detailed R studies
- Evaluate R of design and establish R attributes—part quality, derating level, redundancy
- Perform R allocations, predictions and assessments
- Perform FMEA
- Identify critical and non-standard parts
- Determine logistics and spare provisioning requirements
- Prepare integrated test plan—R growth, R demo, and R acceptance
- Build test hardware
- Perform FRACA
- Track R growth
- Perform R demonstration tests
- Assess degradation factors
- Design and implement screening and burn-in program (parts, assembly, equipment)
- Perform factory acceptance tests

Utility Activities
- Develop reliability evaluation criteria
- Evaluate proposed mfg. R programs
- Perform tradeoff analysis
- Prepare R specifications
- Perform independent FMEA, FIA, CMF and CCF analyses
- Perform independent R predictions and assessments
- Conduct design and program reviews
- Establish part control program—approve non-standard and critical part control procedures
- Approve mfg. test plans
- Witness R tests
- Prepare FRACA reports
- Conduct source inspections
- Establish product improvement programs
- Perform failure analysis
- Establish data recording analysis and feedback (including tie-in with industry-wide databases)
- Initiate configuration management and change control program
- Conduct R training and indoctrination

Plant Integration/System Acceptance
- Plant acceptance
- R degradation control
- Cyclic inspection
- Perform R assessment

TRANSITION
TO OPERATION

┐
↓

┌───┐
Nuclear Power Plant Activities
● Perform periodic PRA and R&M (or availability) assessments
● Implement FRACA
● Prepare LERs
● Maintain configuration mgmt. and change control
● Establish critical item list (and control)
● Audit and survey R program
● Maintain data recording, analysis and feedback program
● Perform in-service cyclic inspection
└───┘

Fig. 1.4. Life cycle reliability activities.

3. Prevent manufacturing errors: This involves witnessing in-plant inspections, reviewing production degradation control programs including the application of controlled screening and burn-in and performing independent system reliability assessments.

4. Prevent installation errors: This involves witnessing inspection and verification tests, reviewing degradation control programs, and establishing workmanship standards.

5. Ensure proper operation and maintenance procedures: This involves reviewing normal and emergency procedures, maintenance procedures, cyclic inspections and evaluating maintenance/logistics plans.

6. Manage system configuration: This includes maintaining baseline system configuration controls including the approval of changes and the review of the critical item list.

7. Provide feedback from operating experience: This involves analyzing LERs, PRA/R&M assessments, reviewing failure analysis reports, evaluating corrective action and reviewing system improvement recommendations.

As shown in Fig. 1.4, reliability requirements are defined by the utilities at the beginning of the acquisition process during the conceptual design phase. This involves estimating the performance of the proposed equipment through comparison with field performance of existing or similar equipment, identifying any areas of risk or possible problem areas and identifying critical reliability/safety characteristics and criteria for subsequent development phases. Specific reliability engineering tasks performed during the early conceptual design phase include:

1. Estimating equipment complexity.

2. Estimating reliability performance of the proposed equipment through comparison to field performance of similar equipment.

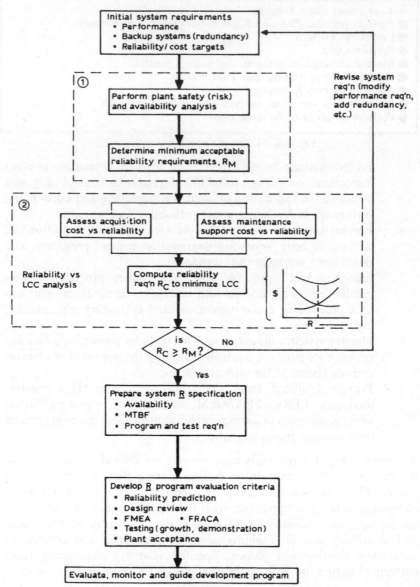

Fig. 1.5.　Reliability planning flowchart.

3. Estimating the cost associated with the predicted range of reliability values.
4. Performing reliability and cost tradeoffs.
5. Determining optimum reliability design points.
6. Identifying areas of risk and/or potential problem areas and initiating programs for their mitigation.
7. Identifying critical reliability/safety characteristics and criteria for overall system test and program activities.
8. Structuring program schedules recognizing the need for reliability growth and development.

More detailed tradeoff studies are performed as the design progresses to determine the most cost-effective reliability level. Programs are also initiated to alleviate potential problems in those areas identified as critical to performance, safety and cost. Requirements for inclusion into the hardware procurement specification are then formulated.

It should be noted that the procurement specification is fundamental to an effective acquisition process. The procurement specification provides a concise description of the equipment requirements that are necessary to meet the functional needs and defines the applicable reliability, safety and test requirements that govern the hardware acquisition. It establishes a basis for the acquisition process to proceed to the development phase and it generally includes a schedule that recognizes the need for reliability growth.

A reliability planning flowchart, which highlights some of the major steps in the hardware acquisition process is given in Fig. 1.5. The figure shows, based on the initial system requirements, that two separate analyses are performed to determine reliability requirements. The specific purposes of these efforts are: 1. to meet safety/plant availability requirements and 2. to minimize cost. The first analysis addresses plant safety (from PRA) and availability requirements and results in determining the minimum reliability level, R_M, that is necessary to meet a plant's overall availability/safety needs. The second analysis considers the type of system needed; its state-of-the-art (SOA), complexity and other performance factors to assess acquisition cost and maintenance support cost. This second analysis results in the determination of the reliability requirement that is optimum from a total life cycle cost (LCC) standpoint.

Initial computations are based on meeting minimum performance requirements. The safety/availability reliability requirement is then compared with the LCC reliability requirement and a tradeoff environment is produced that considers the reliability of the system in the context of plant performance and the associated LCC of the system. When a tradeoff evaluation is necessary, the usual iterative procedure may require some revision of the initial system requirements, e.g. to incorporate back-up equipment, functional redundancy or perhaps to modify the performance requirements. The reliability specification is formulated, as shown in Fig. 1.5, once a balance has been reached. This includes defining program elements to be performed by the manufacturer that are consistent with the specified reliability requirements. Reliability program elements cover: 1. reliability allocation, prediction, and assessment; 2. failure mode, effects, and criticality analysis; 3. maintenance concept; 4. maintainability prediction; 5. component control and standardization; 6. supplier audit and surveillance; 7. design review; 8. reliability tests (demonstration/acceptance; 9. failure reporting, analysis, corrective action and data recording and analysis. The scope, extent and depth of the activities are governed by the criticality of the equipment, the equipment design configuration, the state-of-the-art, the maintenance concept and the cost limitation. The relationship of the specific reliability level to the state-of-the-art and the required reliability improvement attributes provide a basis for determining the scope and rigor of the reliability program and test provisions. Chapter 3 further describes the reliability specification process and presents information that can be applied in determining cost effective requirements for incorporation into specific system specifications.

It must be emphasized that the establishment of cost-effective requirements and incorporating them into the basic equipment procurement specifications is key to obtaining reliable equipment. This means that reliability is treated as a design parameter and that quantitative requirements such as mean time between failures (MTBF) are optimized with respect to total life cycle cost while meeting plant safety requirements. When redundancy is used to meet an overall system reliability requirement, a minimum acceptable MTBF is also specified, and test programs are generally required to provide MTBF verification.

Although the specific requirements in each new hardware specification are, of course, peculiar to that item, the attributes and controls indicated in Table 1.2 represent the minimum set. The specification of any or all of these techniques in the design will increase the initial cost of the equipment but a decrease in support costs during the life of the equipment should more than offset the increased first cost. The minimum MTBF requirement to be specified must be supported by preliminary studies comparing total life cycle costs expected for various levels of MTBF.

In order to assure that the specified level of reliability is attained, there

TABLE 1.2
Reliability and Maintainability Improvement Attributes and Controls

1. Select parts, components and equipment of proven reliability and durability.
2. Derate parts and components.
3. Use carefully designed-in redundancy (or functional diversity) where feasible and cost-effective.
4. Design to incorporate easily accessible and interchangeable components, modules, assemblies and equipment.
5. Provide automatic detection, location and diagnostic features to the maximum extent possible.
6. Apply well planned and documented reliability testing, including reliability growth, demonstration and acceptance.
7. Apply effective reliability and maintainability controls, disciplines and provisions during equipment/component development.
8. Specify adequate and consistent quality controls to ensure that the inherent reliability is maintained during construction.
9. Define areas for preventive maintenance based on failure trend data.
10. Perform ongoing monitoring of supplier activities to ensure adherence to reliability and quality requirements.
11. Analyze failures with rapid feedback for correction.
12. Redesign to simplify equipment and eliminate areas of unreliability.

needs to be close interaction, during system development, between the operating utility, the system designer and the manufacturer. The hardware development and test efforts are monitored and evaluated in terms of evaluation criteria and guidelines developed relative to the specified reliability requirements. Chapter 4 presents some general evaluation guidelines and describes how to effectively evaluate the adequacy of system development programs.

After the system has been specified and a hardware manufacturer is selected, the system is transitioned to development as shown in Fig. 1.4. Reliability efforts, in general, change from reliability planning by the utility and system designer to reliability implementation by the manufacturer. Reliability engineering tasks are performed by the manufacturer on a continual basis throughout development and testing. Initial efforts involve performing broad system analyses and life cycle cost tradeoffs with emphasis on reliability. As the system proceeds through design, more detailed reliability analyses are performed including reliability allocations, predictions, failure modes, effects and criticality analysis, that share the design configuration, determine the extent of derating and dictate the quality and life characteristics of constituent component parts.

During operation, the focus shifts to reliability data recording, assessment, failure analysis and feedback, and configuration management. A data system supported by failure analysis and corrective action is critical to an effective reliability program. It provides a means to assess actual operating reliability, to identify problem areas and to set corrective action priorities. Also, a well-documented configuration management system to provide complete traceability of equipment configuration as well as control of hardware and procedural changes must be rigorously maintained during plant operation.

The overall objectives of a reliability engineering program are to: 1. support the design and development process by establishing an inherent reliability consistent with the specified requirements and 2. ensure that the demonstrated (or specified) reliability level is not appreciably degraded during system integration and is maintained throughout operation. Accomplishment of these objectives requires that the operating utility/licensee plan and implement a comprehensive and highly detailed life-cycle reliability program that meets regulatory requirements. The program as discussed above must consist of effective, systematic and timely management activities, engineering tasks and controlled tests. Some of the essential elements that must be thoroughly planned and completely integrated into the program are:

(a)   Definition and implementation of effective management controls that directly enable reliability personnel to influence design, provide timely outputs consistent with major design and program decision points, and in general, provide the means to develop systems that meet specified requirements most effectively.

(b)   Performance of detailed reliability analyses and cost tradeoff studies during the early design phase.

(c)   Application of systematic and highly disciplined engineering tasks continually during the design phase. The purpose is early identification and correction of problems which will force the design to be iterated as often as necessary, prior to the build-up of system hardware.

(d)   Early procurement, build-up, reliability growth and qualification testing of critical components. Emphasis should be given to failure analysis and corrective action as well as the application of a test cycle that reflects operating plant environment, including mechanical stresses and climatic extremes.

(e)   Preparation of maintenance plans that identify the reliability of critical components requiring maintenance and classifies them in accordance to the type of maintenance (i.e. time replacement or on the basis of condition).

(f) Performance of plant acceptance and in-service testing and surveillance.

(b) Implementation of a reliability assurance program that provides controls and procedures which allow a smooth transition from development to system integration, operational checkout and acceptance and plant operation and maintenance without degrading reliability.

Each task contributes not only to the total reliability program, but also provides timely inputs to other reliability tasks in relation to hardware and acquisition milestones. For example, reliability management begins with the definition of requirements and extends through preparation of detailed reliability plans during development and implementation. Reliability analysis performed initially by the system designer (concurrent with early PRA safety analyses), to establish the specified level of reliability, are also performed by the manufacturer during development to further define reliability, to establish the basis for meaningful tests and to assess achieved levels of reliability during plant operation. Component control includes efforts to select, specify and identify critical parts, and monitor component qualification throughout acquisition and plant operation. Test planning includes specifying development tests that will ensure that the required reliability levels are achieved. Also, the effectiveness of the reliability program can be monitored by:

—Comparing reliability analysis results against specified requirements.
—Evaluating design growth through monthly reliability assessments.
—Evaluating reliability test (and failure analysis) results.
—Assessing the effectiveness of corrective actions steps.

Chapter 5 discusses the design process from a reliability standpoint and Chapter 6 describes the basic engineering procedures for implementing a reliability program during system development.

In addition, the availability of electric power to meet load demands is a prime consideration in evaluating power generating systems. Availability engineering concepts and techniques are applied to assess the performance of major components, to aid in the modification of operating systems and to help design new systems. Past experience, expressed as outage frequency and duration is used to identify availability-related deficiencies in hardware, software and personnel actions as well as to assess system/plant availability. Corrective actions that can achieve an availability goal while considering potential savings and costs are identified through availability analyses. Controls that must be imposed to prevent degradation during installation, operation and maintenance are also identified through availability analysis. Availability engineering involves performing systematic and highly disciplined efforts to set and achieve a quantitative availability goal at minimum cost. Availability concepts are further discussed in Chapter 2.

Finally, maintaining accurate and up-to-date records through a formal data recording and analysis system is essential in assessing operational reliability and availability performance. Comparative evaluations between the reliability specified, estimated and measured during acquisition, and the actual operating reliability provides criteria for improvement, if necessary. An effective data system requires that all failures occurring during plant operation are recorded and analyzed and that the results are fed back to management, engineering and manufacturing activities for corrective action. Complete reporting provides data on such things as accumulated operating time, time to failure, test conditions, on–off cycling, adjustments, replacements and repairs related to each equipment and component. Failures are analyzed to determine their basic causes, i.e. part/material circuit design, workmanship, etc., and to provide a basis for effective corrective action involving hardware replacement, reliability assurance controls and operational/maintenance procedural changes.

Chapter 2

Reliability Analysis

This chapter discusses analytical methods that are fundamental to the reliability engineering discipline and defines those assumptions which are basic to reliability theory and its application to hardware design and engineering. This discussion is not intended to provide a complete theoretical foundation for reliability. Rather, it is intended to provide an overall basis for those design, engineering and test activities necessary to the development of highly reliable and maintainable systems and equipment. As such, this chapter serves as a general preamble to the topics covered in subsequent chapters. Appendix B (Bibliography) identifies some of the sources that present fundamental concepts and considerations, including statistical methods, probability theory and system engineering principles.

An effective system reliability engineering program requires careful planning followed by well-executed engineering tasks that start with design and continue through development, manufacturing, plant construction and operation. A reliability program begins during design with the selection of proven, high quality, well-derated, long life parts. Early emphasis is placed on the use of adequate design margins and ample means for heat dissipation and shielding. Attention is focused on good construction practices and on the application of development tests to identify potential problems. The process continues during development with the use of well-established, systematic reliability engineering techniques, supported by tests to stimulate reliability growth and to demonstrate the level that has been achieved.

During manufacturing, the focus shifts to control of the processes through the application of well-planned and carefully executed quality inspections and the reliability screening of incoming components, fabricated boards and modules and finished assemblies and subsystems.

The screening tests, which are central to the manufacturing inspection and acceptance process, consist of a family of techniques in which electrical, thermal and mechanical stresses are applied to accelerate the occurrence of potential failures. By this means, latent failure-producing defects, which are not usually caught during normal quality inspection and testing, are removed from the production stream. Burn-in is a specific subclass of screen which employs stress cycling for a specified period of time.

Frequent inspection should be conducted over long periods of plant dormancy to assess the functional status, or readiness, or hardware items. When followed up by a properly applied preventive maintenance program, plant deterioration can be controlled and required levels of equipment and component reliability can be maintained.

Design establishes the inherent reliability potential of a system or equipment item. The transition from the paper design to actual hardware and ultimately to operation frequently results in an actual reliability that is far below the inherent level. The degree of degradation is directly related to the inspectability and maintainability features designed and built into the system as well as to the effectiveness of the tests and controls that are applied during production, and subsequent periods, to eliminate potential failures, manufacturing flaws and deterioration factors. Lack of attention to these matters can result in an actual system reliability as low as 10 % of its potential.

In addition, in-service component aging, which can be caused by a variety of environmentally induced mechanisms, can severely degrade reliability and, if not controlled, can significantly affect power plant safety. Aging effects must be addressed during design, and protection provided to assure adequate system functional capability (to meet safety requirements) over the entire plant life (generally of the order of 40 years). A major goal in the evaluation of safety-related equipment is the prevention of failures in redundant safety systems due to common causes. Since equipment aging is a potential common-mode failure mechanism, it is essential to demonstrate that the design and manufacturing of safety-related systems are such that the systems can function under a design basis condition—not only in an 'as-new' condition, but also after the degrading effects of in-service aging have occurred.

The impact of production, shipment, installation and operation, and maintenance degradation factors on the reliability of a typical system or equipment item and the life cycle growth that can be achieved is conceptually illustrated in Fig. 2.1. The figure depicts the development of a hardware item as it progresses through its life cycle. The figure shows that an upper limit of reliability (e.g. MTBF) is established by design. As the item is released to manufacturing, its reliability will be degraded and as production progresses, with resultant process and inspection improvements and manufacturing learning factors, reliability will grow. The figure further shows that when the hardware is integrated into the plant, whether it is dormant, intermittently on-line or fully operational, its reliability will again be degraded. The extent of degradation is dependent on the severity of the application environment of the particular power plant, the strength–aging characteristics of the components, the effectiveness of the inspection and control program, and on the operational familiarity and maintenance experience of the power plant personnel.

The reliability engineering program must include efforts to maximize the inherent design reliability potential by forcing the design to be iterated and to minimize degradation by eliminating potential failures and workmanship flaws prior to operation. Such measures demand that all reliability engineering activities be effectively managed and executed during the acquisition–operation cycle of the system. Specifically, a reliability engineering program is designed to:

—Assure that the inherent reliability embodied in the design can be retained during manufacturing and storage/dormant periods.
—Determine the need for additional screen and surveillance tests or improved inspection techniques.
—Assure a smooth transition from design to production to operation.
—Assess, grow and control actual reliability during development, production and operation including non-operating periods.
—Assure safety system operation after the degrading effects of in-service component aging.

This chapter focuses on reliability analysis methods. Sections 2.1 and 2.2 describe some of the fundamental reliability and maintainability concepts applicable to reliability analysis. Section 2.3 discusses availability analysis concepts. Section 2.4 discusses degradation and control. Section 2.5 addresses some of the methods and techniques that can be used to analyse software reliability and to minimize errors that can lead to system failure. Section 2.6 presents a description of currently available techniques applicable to the assessment of system reliability. Procedures for their implementation are given in Chapter 6.

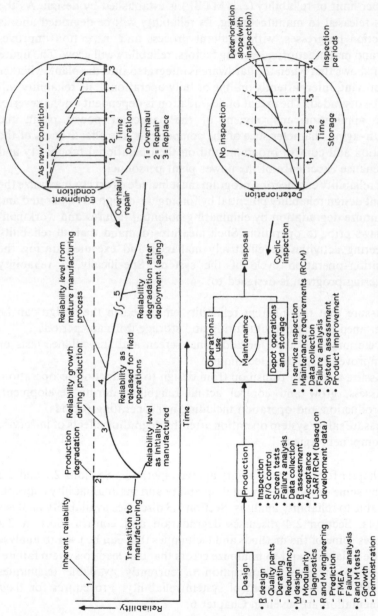

Fig. 2.1. Life cycle reliability model (conceptual).

2.1 BASIC RELIABILITY CONCEPTS

Reliability is defined as the probability that an item will perform satisfactorily for a specified period of time under a stated set of use conditions. From a functional point of view, in order for a hardware item to be reliable, it must do more than meet an initial factory performance or quality specification—it must operate satisfactorily for an acceptable period of time in the actual application for which it is intended.

Since reliability is not a deterministic parameter, it must be expressed as a probability, i.e. in terms of the likelihood that the system is operational. Reliability is a function of three conditions: satisfactory performance, time and conditions of use. Satisfactory performance is that set of criteria that is necessary to accomplish the intended function—any observed performance outside the stipulated criteria is defined as failure. Time is that operating period over which the system is expected to operate to accomplish the desired function. Use conditions are the total environment (atmospheric, physical, thermal and electrical), including transient, cyclical and steady state, under which the system is expected to function.

Determining and ensuring reliability, therefore, involves the understanding of a broad range of factors which relate to these conditions and the use of a number of widely accepted indices. One key index is the system hazard rate, which can vary as a function of age. Hazard rate is a measurement of the number of system failures occurring per unit time.

The manner in which systems and their parts fail, together with the underlying cause of failure, are important in controlling reliability. In general, the total distribution of failure over the lifespan of a large population of a hardware item can be separated into quality, reliability (or stress related), and wearout failure distributions. These failure distributions combine to form the infant mortality, useful life and wearout life periods shown in Fig. 2.2. They are characterized as follows:

Period I: Infant mortality period—The initial high hazardous rate during this period is typically due to latent design defects, quality/manufacturing deficiencies and defects introduced during plant construction. From a design point of view, the approach to reducing the hazard rate and minimizing the duration of this period or eliminating it entirely (prior to plant operation), are the use of reliability growth testing during development followed by controlled screening and burn-in tests, coupled with process and inspection controls during fabrication. In development, reliability growth testing primarily acts to remove latent design

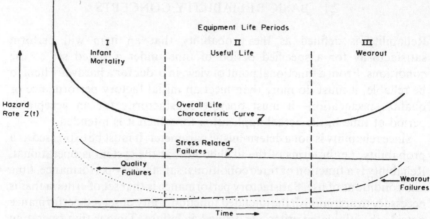

Fig. 2.2.   Components of failure.

shortcomings, while in production, the use of screening tests basically removes latent quality and fabrication defects. If necessary, follow-on equipment improvement efforts on operational equipment will remove any latent deficiencies which escaped detection during manufacturing and pre-operational testing.

Period II: Useful life period—This period is characterized by random failures and a relatively constant hazard rate. (A constant hazard rate is defined as the failure rate.) Failures during this period are primarily caused by random occurrence of stresses that exceed strength levels, but can include failures due to improper maintenance as well as latent design or fabrication deficiencies which escaped detection during manufacturing. Figure 2.3 shows the relationship between stress and

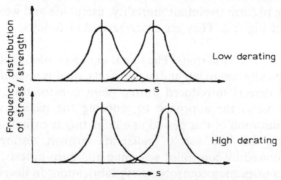

Fig. 2.3.   Stress/strength frequency distributions.

strength. This is a statistical relationship which assumes a normal frequency distribution for the scatter of the strengths of the item population and the scatter of the stresses applied to these items. The shaded overlapping area between the two curves indicates the extent to which a simultaneous occurrence of high stress and low strength can occur and result in random failures. The probability of random failures can be reduced by increasing the separation between the two frequency distributions. This can be accomplished by: 1. selecting greater strength parts (derating) and/or 2. reducing the scatter of strength through tighter part manufacturing processes and inspection controls.

Design/reliability prediction iterations, followed by failure mode, effects and criticality analysis (FMECA), are applied during development to determine and mitigate the impact of stress-related failures on inherent hardware reliability. In essence, the reliability prediction methodology quantitatively shows the ability of the design (in terms of a reliability numeric) to withstand the effects of electrical stress (power, voltage, current) and environmental stress (temperature, humidity, vibration, etc.). In production, screen stress testing can be used to force out low strength items (insufficient design safety margins) in addition to detecting unexpected failure modes. With regard to unexpected failure modes, it should be noted that the reason architectural structures, for example, generally exhibit high reliability is that their failure modes are few and are well-understood in terms of the physics of failure. In contrast, complex electronic equipment, especially those systems employing new technologies, can have many modes of failure, the physics of which are not yet well understood. Thus, screen-testing can be applied to reveal inadequate safety margins, as well as new and unexpected failure modes.

Period III: Wearout period—This is the end of life or the wearout phase, during which the failure rate increases significantly. The increase is due to a gradual physical or chemical change of the system components with time resulting in a decrease in strength. The point in time at which the failure rate becomes unacceptably high essentially determines when the equipment is replaced. The approach to preventing or delaying the onset of wearout is the use of scheduled replacement of limited life components. In addition, aging mechanisms which could result in the degradation of safety functions of critical components must be identified and specific criteria applied to detect and mitigate these aging mechanisms. It should be noted that premature wearout or aging can be caused by: abuse during earlier operating periods; undiscovered latent deficiencies; insufficient safety margins; and unaccounted for environmental influences.

The overall approach to reliability engineering is to minimize initial failures (Period I) by emphasizing screen tests and inspections at both the part and assembly level, to minimize wearout and aging effects (Period III) through inspection and replacement of short life components and to concentrate design and reliability engineering action on the normal or useful life period (Period II) characterized by stress-related failures.

The normal operating period of the life characteristics curve should constitute the greater portion of the equipment life cycle. During this period failure rate is considered to be constant. A population of hardware items which have survived the initial burn-in period can be considered to have a constant failure rate as described by the exponential failure distribution.

The exponential reliability expression is given below. When appropriate values of λ and t are inserted into the expression, the probability of success (i.e. reliability) is obtained for that time period,

$$R(t) = \exp(-\lambda t)$$

where: $R(t)$ is the probability that the item will operate without failures for the time period, t (usually expressed in hours), under the stated operating conditions; and λ is the item failure rate (usually expressed in failures per hour) and is a constant for any given set of stress, temperature, and quality level conditions (it is determined for parts and components from large scale data collection and/or test programs).

Used as a theoretical prediction mechanism, this expression permits a determination of the probability of survival for any single system or equipment for which the failure rate, λ, has been determined either by testing, statistical analysis of historical performance data or from other estimation techniques.

The probability of failure in time due to the unreliability, $Q(t)$, is computed from the reliability function as shown below.

$$Q(t) = 1 - R(t)$$
$$Q(t) = 1 - \exp(-\lambda t)$$

Both $R(t)$ and $Q(t)$ apply to the probability that a single item or group of items will survive (or fail) at some point in time. The value of the function ranges between zero and one and expresses either probability of survival or percentage surviving after time t. This is graphically illustrated in Fig. 2.4.

The reciprocal of the failure rate, λ, during the useful life period, is defined as the mean time between failures (MTBF). The MTBF is primarily a figure of merit by which one hardware item can be compared to another.

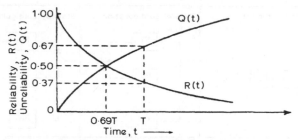

Fig. 2.4. Exponential failure rate.

Expressed mathematically, $MTBF = 1/\lambda$. It is the mean time between system failures, where all failures that result in an unscheduled maintenance action are included. MTBF, as a reliability parameter, serves as a popular index of failures in the useful life period. Though it does not indicate which subsystem will fail or when a system will fail, it can indicate the expected number of unscheduled maintenance actions during a given time period.

System reliability is dependent on the individual reliabilities of the constituent elements. The probability relationships which show how the individual reliabilities or failure rates are combined upward to higher assembly levels are described in the following paragraphs. These relationships apply regardless of whether the entry level for performing the integration is at the part, circuit, module or equipment level. Basic to the use of these relationships is the assumption that all elements fail in accordance with the exponential distribution and that for maintained systems, a repair upon failure policy is in effect.

To evaluate system reliability, formulae and techniques have been derived relating the reliability of parts and components to the overall system. In general, parts and components can be connected in one of two ways, either in series or in parallel. If the parts are in series, then all of the parts must operate successfully if the system is to operate. On the other hand, if redundant parts are in parallel only one of them must operate for the system to operate.

These concepts can be applied at any level of assembly, and even items of equipment can be arranged in series, or in parallel configurations (to allow higher system reliability). Reliability of a serial configuration is a product of the separate configurations as shown in Fig. 2.5. Reliability block diagrams that depict these configurations are also shown in this figure. The concepts of serial and parallel reliability allow requirements to be specified that meet the various operational requirements in the operating situation.

Reliability formulae as given in Fig. 2.5 do not account for repair. When

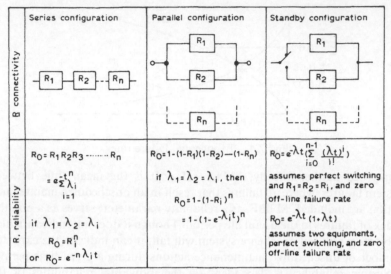

Fig. 2.5.   Reliability configurations.

considering repair the concept of system availability is used. System availability is a property of an equipment or system that provides a combined measure of the reliable operation of the system and its ability to be efficiently maintained. Availability is a probability, like reliability, of the system being in service when required. It has a similar meaning for repairable systems as does reliability for non-repairable system. The difference, however, is that reliability only accounts for the single event, failure and availability accounts for both failure and repair events. Availability analysis methods will be discussed later in the chapter.

The serial and parallel reliability concepts establish the mathematical framework for analyzing system reliability. Figure 2.6 further illustrates their application. The figure depicts a simplified equipment composed of three operating elements. Elements 1 and 2 are identical and represent functional redundancy operating in series with element 3. Reliability block diagrams can be defined corresponding to non-redundant serial, functional and unscheduled maintenance reliability.

The figure indicates that the use of redundancy provides a significant increase in functional reliability above that of a series or non-redundant configuration; however, it imposes a penalty by adding an additional serial element in the unscheduled maintenance chain.

The MTBF of a system is the reciprocal of its total failure rate. Included in this definition are all failures which constitute an unscheduled

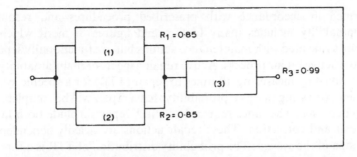

Reliability requirement	Reliability block diagram	Calculated values
1. Serial (non-redundant) reliability		$R = R_1 R_3 = 0.84$
2. Reliability		$R = (2R_1 - R_1^2)R_3 = 0.95$
3. Unscheduled maintenance reliability		$R = R_1 R_2 R_3 = 0.72$

Fig. 2.6. Reliability criticalities.

maintenance action. Therefore, regardless of the system configuration that is employed to meet availability requirements, for MTBF purposes, *all* elements are considered in series as shown in Fig. 2.6 (3). MTBF is expressed as:

$$\text{MTBF} = \frac{1}{\sum\limits_{i=1}^{n} \lambda_i}$$

2.2 BASIC MAINTAINABILITY CONCEPTS

Maintainability is a characteristic of a system or equipment which describes the ease and speed with which maintenance actions can be performed in accordance with prescribed procedures and resources. Maintainability includes many factors and figures of merit which are logically associated with maintenance actions but are functionally separate from the active repair process. Active repair aspects of maintainability can be quantified by measuring all times to repair (TTR) for a specific system, and then expressing this as a probability that a repair will be complete by a designated time. The times required to effect repairs include both failure diagnosis and correction. These repair actions are heavily dependent on the design of the system. The probability distribution of TTR is a function of the system design and can be used to determine and specify its maintainability level. Designing for ease of maintenance involves detailed consideration of such features as automatic failure detection, diagnostics and the incorporation of easily accessible and interchangeable modules and subassemblies.

Figure 2.7 depicts the relationship of maintenance downtime elements. As indicated by the figure, the total maintenance downtime is divided into several groups, the principal ones being active maintenance downtime and corrective maintenance downtime, of which active repair time is a principal quantifiable element.

Corrective maintenance is the action performed, as the result of a failure, to restore an item to a specified condition. The element of corrective maintenance which is important in terms of design is the active repair time.

The quantification of ease and speed of repair takes place at this level and is expressed as TTR. The frequency distribution of TTR has an important parameter, MTTR, which is mean time to repair, and is the single measure most often used to quantitatively describe the maintainability of a system. MTTR is determined by the use of failure rate data obtained from reliability studies as a statistical weight factor and maintenance time factors derived from a review of system or component maintenance design characteristics. Conceptually, the repair of hardware items after the occurrence of a failure necessitates the initiation of corrective maintenance tasks which ultimately result in the interchange of a replaceable part or assembly. In order to achieve a complete 'repair', various activities before, during and after the actual interchange are necessary. The calculation of MTTR is based only on the active repair time associated with the four time elements of: fault isolation, fault correction, calibration and checkout. Those time elements which are related to preparation and delay, while quantifiable in terms of the total maintenance scenario, do not permit insight into the maintainability design aspects of the operating system, and are therefore less valuable in describing the maintainability of the system.

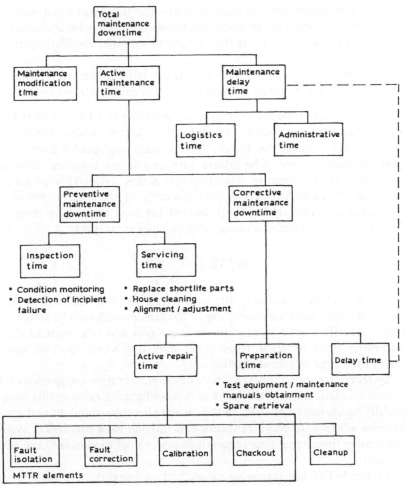

Fig. 2.7. Maintenance time elements.

The specifications associated with the four active repair time elements are:

1. Fault isolation time—the time associated with activating diagnostic equipment and interpreting outputs to determine fault isolation.

2. Fault correction or removal and replacement time—the time associated with gaining access to the replaceable item, up to the point of interchange, physically removing the failed item, replacing it with a new item, and performing the steps associated with disassembly in reverse order.

3. Adjustment and calibration time—the time associated with the manipulation of operating and maintenance controls and mechanical parts so as to bring the equipment within its specified operating ranges.

4. Checkout time—the time associated with the verification that the repair has restored the equipment's normal performance.

The composite time for all the above activities is TTR or repair time, R_p. In order to provide weight factors for the expected number of corrective maintenance actions, the failure rate of each replaceable component/part/assembly is used. The failure rate and repair time are combined to arrive at a corrective maintenance action rate. This process is repeated for each replaceable part/assembly in the system. From the maintenance action rates $(R_p\lambda_p)$ derived for each replaceable item, the MTTR can be determined using the following expression.

$$\text{MTTR} = \frac{\sum [R_p\lambda_p]}{\sum \lambda_p}$$

In order to further clarify the distinction between MTTR and other maintainability figures of merit, Fig. 2.8 is presented. As can be seen in this figure, MTTR is associated with on-line repair and is a maintainability parameter quite different from other numerics which describe system restoration time or offline repair.

System restoration time is that time associated with the reinitiation of the system's functional capabilities. For non-redundant systems this time is usually equivalent to MTTR. In the case of standby redundant systems or systems where a different hardware can provide back-up service, system restoration time is that time required to switch operation over to the back-up unit.

System MTTR is that time associated with on-line corrective maintenance. It is the time required to localize and isolate a fault to a replaceable module or subassembly, disassemble the equipment to the extent necessary to gain access to the failed item, replace the defective item, and reassemble, align, and checkout the repaired system.

Off-line repair involves the time to localize and isolate a fault on a replaceable item to the part level (i.e. the smallest element to be repaired or replaced), disassemble the replaceable item to gain access to the failed part, replace (or repair) the defective part and reassemble, align and checkout the fully repaired item.

Preventive maintenance consists of actions performed to retain a system

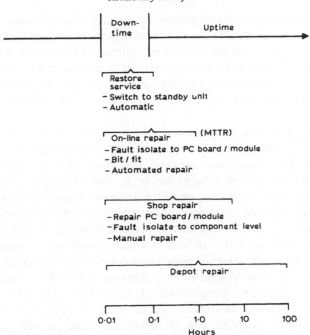

Fig. 2.8. Maintainability figures of merit.

in a specified operating condition. Systematic inspection of condition and performance are coupled with services designed to prevent or substantially reduce incipient failures. New systems can be designed with the capability to remotely monitor performance, predict imminent failures, and to make compensating adjustments or corrections, and thus greatly simplify the preventive maintenance task. The inspection and service time elements associated with preventive maintenance actions are shown in Fig. 2.7. It is possible, as in the case of corrective maintenance, to quantify preventive maintenance actions. This can be accomplished by measuring the time required to perform these tasks using a frequency distribution parameter, mean-preventive-maintenance-time, as the most descriptive measure of preventive maintenance efficiency. As indicated in Fig. 2.7, preventive maintenance includes those time elements expended to replace short life parts, to perform clean-up and housekeeping tasks, and to perform routine inspection and alignment. Preventive maintenance time, as with corrective maintenance, includes only the actual active time during a preventive maintenance action. It also includes preventive maintenance time which is expended while equipment is in operation.

The implementation of well-planned preventative maintenance results in the detection of potential faults and shifts in performance specifications for correction prior to actual equipment failure. Thus, many delay time elements which affect the total corrective maintenance repair time, such as travel time, part obtainment and other logistics and administrative factors, are avoided or minimized thereby increasing the ultimate availability of the equipment.

Periodic preventive maintenance also increases the familiarity of the maintenance technician with the functional and service aspects of the equipment. This experience provides for more efficient restoration of service when equipment failures do occur and corrective maintenance becomes necessary. However, preventative maintenance introduces a risk that the service action will precipitate or accelerate a failure at some other point in the system. This risk is also present, however, during corrective maintenance actions and it should be small enough not to outweigh the advantages of a well designed preventive maintenance program.

A relatively new program for deriving preventive maintenance requirements is reliability centred maintenance (RCM). RCM is based on the premise that more efficient and cost-effective life-time maintenance and logistic support programs can be developed using a well disciplined decision logic process which focuses on the consequences of failure. The resultant maintenance program provides the desired or specified levels of safety and reliability at the lowest possible overall cost. RCM concepts are described in Section 6.2.

2.3   AVAILABILITY ANALYSIS AND IMPROVEMENT

Availability is a property of an equipment or system which provides a single combined measure of its reliable operation and its ability to be efficiently maintained. While it is possible to confine system availability to the hardware aspects of the system, it is equally possible to express system availability in the context of the overall system configuration and its support infrastructure. This broader approach includes the effects of such planning and organizational strategies as: spare provisioning, communications, operator/maintenance training and administrative times required to effect repairs as well as the ability of the power plants to provide electric power to meet load demands.

Availability engineering techniques are applied to assess the overall performance of major components, to aid in the modification of operating systems, and to help design new systems. Past experience, expressed as outage frequency and duration, is used to identify availability-related deficiencies in hardware, software, and personnel practice as well as to assess system availability directly. Corrective actions which can lead to achieving an availability goal, along with the associated cost consequences, are identified through availability analyses. Controls that must be imposed to prevent degradation during installation, operation and maintenance are also identified through availability analyses. Availability engineering involves performing systematic and highly disciplined tasks to set and achieve a quantitative availability goal at minimum cost.

The availability of electric power to meet load demands is continually assessed and analyzed by most utilities. The analysis is based on system outage rates (forced and planned) compiled by the utilities and supplemented with industry-wide data from NERC/NPRDS. These assessments provide an effective and viable basis to correct and improve the plant/system.

Usually expressed as a percentage, availability is a figure that reflects the ability of an item to meet its intended duty cycle. It is the percentage of time that an item is neither forced nor scheduled out of service. Formulae used to analyze availability include:

Operating availability:

$$\text{Operating availability} = \frac{\text{Available hours}}{\text{Period hours}} \times 100\%$$

where Available hours = the time in hours during which a unit of major equipment is available for service whether or not it is actually in service (this includes both service hours with breakers to the system bus closed and reserve shutdown hours during which the unit is shut down but available for service); Period hours = the clock hours in the period under consideration.

Capacity factor:

$$\text{Capacity factor} = \frac{\text{Total electric generation in MWh}}{\text{Period hours} \times \text{Maximum dependable capacity in MW}} \times 100\%$$

where Maximum dependable capacity = Dependable capacity winter or summer, whichever is smaller in MW electricity.

Forced outage rate:

$$\text{Forced outage rate} = \frac{\text{Forced outage hours}}{\text{Forced outage hours} + \text{Service hours}} \times 100\%$$

where Forced outage hours = time in hours during which a unit or major equipment was unavailable due to a forced outage; Service hours = total hours the unit was actually operated with breakers closed to the station bus; Forced outage = the occurrence of a component failure or other condition which requires that the unit be removed from service immediately or up to and including the very next weekend.

Analyses are performed by the utilities to estimate values for system/component availability and capacity factor. The operating and outage experience found in the utilities' internal databanks and/or in the NERC's availability data reporting system are used as input. These analyses account for frequency, duration and effect on output power of the major outages and the various modes of operation. An availability model is developed as part of the analysis. It allows the various outage rates and other parameters associated with each of the subsystems and components that constitute a power generating unit to be combined.

Many utilities also use a loss of load probability (LOLP) criterion in analyzing power plant availability. LOLP represents the expected number of days for a given time period during which system load may exceed available generating capacity within the system. LOLP accounts for installed capability relative to peak annual load and is especially useful for power plant planning and component design tradeoff studies because it is sensitive to the sizes and outage rates of the individual generating units, the annual load profile, maintenance scheduling and load forecast derivations. A one-day loss of load in 10 years is the generally accepted industry standard.

The availability of a power generating system is a direct result of both planned and forced outages. The result is that the total power required at any particular time may not be available and must be produced by other, perhaps less economical, units in the plant or purchased at higher cost from neighboring systems.

Planned outages include nuclear refueling, normal maintenance, inspections and equipment overhauls at specified time intervals. Typically, planned maintenance (PM) procedures and schedules are recommended by the manufacturer and followed at the discretion of the utility, depending on their own experience and judgement. As much as possible, PM is usually done when the plant is down for nuclear refueling. These planned outages generally follow a pattern of increasing complexity, depending on the operating times accumulated by the unit.

Forced outages occur when equipment malfunctions interrupt equipment usage during normal operating periods. Forced outages include such things as emergency tripouts, unscheduled shutdowns and failure to start. Each piece of equipment or component in an electric generating unit has an associated historical forced outage rate (FOR) that contributes to total unavailability. Most large utilities implement a computerized data collection and recovery system to provide historical FORs that reflect their specific operating experience and environmental characteristics. In addition, NERC collects and publishes industry-wide generic FOR data. As previously indicated, historical FORs and duration times are essential in computing and evaluating electric power generating availability. FOR data provides the utilities with a basis for assigning priorities, conducting life cycle cost–benefit studies, and implementing product/component programs. Although FORs for power generating systems and components vary widely with plant location and operational factors, the NERC data show that across the industry, the turbine-generator and boiler account for about 90 % of the forced outage hours and all other components constitute the remaining 10 %. This explains the priority given by the utilities, manufacturers and EPRI to improve the reliability of the turbine-generator and boiler.

Electric power generating units and components generally exhibit an increasing failure (or hazard) rate as they age in contrast to electronic components, which exhibit constant failure rates. Thus, with power generating units and components, it is beneficial to inspect (using NDT techniques) for indications of impending failure as part of the overall need for preventive maintenance by replacing components whose performance and physical attributes are beyond predetermined acceptance levels (derived from failure rate studies) and would fail during use. A tradeoff exists between preventive and corrective maintenance and cost–benefit analyses can be and frequently are performed to determine the optimum maintenance program.

Availability requirements or goals are met by reducing the frequency and duration of outages. R&M engineering methods are used to predict the frequency of failure (reliability) and the duration of the failures (maintainability) and to identify areas where design improvements can be made most cost-effectively. Many analysis methods are available; they differ in the level of part and component attribute data required for their application. Figure 2.9 depicts some of these procedures and illustrates when they are applied during the development of a power plant generating unit to meet (or improve) availability goals. The figure shows that as the development program progresses, reliability allocation, prediction and availability analysis, as part of an overall iterative process, are continually

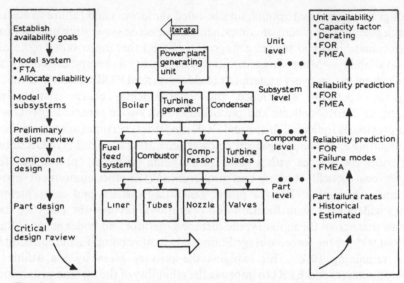

Fig. 2.9.   Reliability/availability methodology (FOR: forced outage rate).

updated to reflect the more detailed level of design. Some of the data necessary to support the application of the particular technique are indicated.

Availability analyses are also performed during system design and development prior to actual plant operation. These analyses use generic historical data at the component level derived from the utilities' internal data bank and/or NERC/NPRDS data collection systems. They are based on reliability estimates (at start-up as well as when running) and identify problems during development when corrective changes can be implemented most cost-effectively. These predictive efforts provide inputs for budgeting program activities, developing LCC estimates and performing cost–benefit studies. Standard reliability prediction techniques and other reliability analyses, including FMECA, FTA and design review, are performed during development to ensure operational reliability. The results are also used to support availability analyses. A flow chart of a typical reliability/availability analysis procedure is shown in Fig. 2.10.

The total cost and availability of a system throughout its life cycle are related to the level and combinations of designed-in R&M performance measures. The determination of the reliability level which is sufficient to meet or exceed the availability objective for the system while minimizing the life cycle cost (LCC) is possible through R&M and LCC analysis.

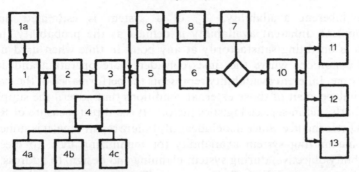

1. Determine Availability Requirements (Goals) — Base on economics and consideration of actual availability achieved under similar conditions.

1a. Monitor External Forces — Regulatory requirements may dictate plant availability goals.

2. Prepare Allocation Model — Construct model consistent with data available and design stage.

3. Allocate Goals — Allocate system goal to sub-elements.

4. Process Data — Data used as predictive input in subsequent analyses.

4a. Industry Data — Compiled from various sources such as NERC/EEI, NPRDS, and Institute of Electrical and Electronic Engineers.

4b. Vendor Data — Used preferentially when they can be verified.

4c. Plant Data — Information on Product performance in the O&M environment.

5. System Failure Analysis (FMEA/FTA) — Identify the effects of possible failures, faults, or mistakes.

6. System Reliability Assessment — Combine the allocation model and system failure analysis to provide a means for predicting R.

7. Identify Problem Areas (critical items list) — Base on failure analyses and numerical predictions.

8. Conduct Critical Item Reviews — Review periodically.

9. Develop Corrective Action — Develop Corrective action plan based on information derived during the critical item reviews.

10. Implement Corrective Action — Increase availability by changing one or more of the following:
    • Hardware
      - Component/system service equipment reliability
      - Maintainability/accessibility
      - Spares/redundancy
    • Software
      - Installation/start-up/test procedures
      - Maintenance schedules
      - Repair and overhaul procedures
      - Inspection intervals and instructions
    • Personnel
      - Skill levels and training
      - Motivation

11. Reliable System Design — Changes in design configuration components must be fully documented.

12. Reliability Requirements in Procurement Specifications — Write to ensure that procured equipment can meet its availability goal in terms of its R&M ability characteristics. Wherever possible, quantitative requirements should be specified.

13. Improved O&M Procedures — Operator error can defeat the best intentions in engineering design.

Fig. 2.10. Typical reliability/availability analysis procedure.

The inherent availability, A_i, of a system is estimated during development. Inherent availability is defined as the probability that a system is operating satisfactorily at any point in time when used under stated conditions where the times considered are operating and active repair time. Inherent availability represents the performance of the system itself, independent of those external conditions that govern the supply of spare parts and in-service logistics factors. It is a direct measure of R&M design performance. Since it includes only failure repair time, it is sufficient, when maximizing system availability (or minimizing LCC to meet an availability objective) during system planning and design, to address only the inherent availability of the system. This is accomplished through application of R&M design attributes and by incorporating backup design configurations, and is not contingent upon variables outside the system itself. The inherent availability, A_i, can be stated as:

$$A_i = \frac{MTBF}{MTBF + MTTR}$$

where MTBF = mean time between failure and MTTR = mean time to repair. This formula represents an average availability, which is based on the continuous duty cycle of a system having a constant failure rate and repair rate. As the formula indicates, when downtime is zero (i.e. when MTBF approaches infinity or when MTTR nears zero) then availability, A, is one. Thus, near zero downtime becomes an engineering objective and MTBF and MTTR are the parameters which directly affect the percentage of time that a system is available for use. While any discussions of R&M theory may tend to treat reliability and maintainability as separate topics for reasons of clarity, the combined effect of both must ultimately be considered in the planning stages of any system. A number of combinations may be possible for achieving the same desired level of system availability. The specific tradeoff which is selected will also be influenced by environment, cost, personnel training or other external constraints which are also important to the utilization of the equipment.

Each specific design configuration and its associated MTBF/MTTR values define an inherent availability level. The optimum design configuration is that which meets the availability requirements at lowest cost (see Fig. 2.11). Both MTBF and MTTR are influenced by the design strategies defined in the early design phase of hardware development. Tradeoff analyses are performed to determine the MTTR/MTBF mix that will achieve maximum availability while minimizing the total LCC of the equipment. An item can be designed and built to have high MTBF with respect to MTTR, or ease of maintenance can be designed into the item,

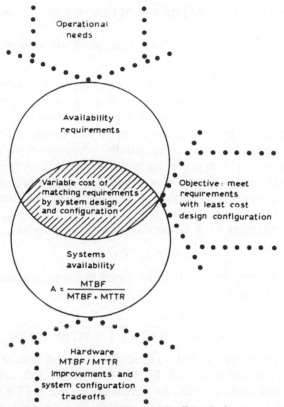

Fig. 2.11. Availability tradeoff analysis.

which would result in short maintenance time elements and a low MTTR with respect to MTBF. Frequently, the most practical way to achieve high availability is to supplement the design for reliability with a design for efficient and rapid repair and a high degree of maintainability. Determining the most cost effective MTBF/MTTR mix is critical in developing hardware that meets availability requirements at minimum life cycle cost.

The complete tradeoff process is an iterative process, which may begin with existing equipment/design concepts and extend to the consideration of the application of state-of-the-art technology. The method is used to converge on a totally optimized system that considers:

1. Required system availability as defined through safety (risk) analysis.
2. Inherent system availability as a function of equipment design.

3. Optimum mix of MTBF and MTTR that will minimize life cycle cost.
4. Use of redundant backup equipment to achieve availability requirements.

The operating modes of a system or equipment can be viewed as consisting of two mutually exclusive states, operating and non-operating. In each state additional distinctions can be made. If the unit is operating it may be in an active or standby mode. In the non-operating state, the unit may be in storage, non-operating standby or down for maintenance or repair. If the unit is in an operating standby mode the further implication may be that the unit is operating as one element of a redundantly designed system. All of these operating modes must be considered when calculating the availability of the system.

System availability for a simplex system (non-redundant) is

$$A_S = 1 - P_f(A)$$

where $P_f(A)$ is the probability that the system (designated as unit A) will fail and can be estimated by the formula:

$$P_f(A) = \frac{MTTR}{MTBF + MTTR} = \frac{1}{1 + MTBF/MTTR}$$

Making the probability of failure as small as possible consists of making the R&M numerical ratio, MTBF/MTTR, larger. This ratio becomes numerically larger when the MTBF is increased, or the MTTR is decreased.

The availability of a redundant system is increased because the probability of simultaneous failure of the backup unit is statistically remote. High availability of systems with marginally reliable equipment can thus be attained using backup units. If we regard the behavior of each unit operating in an active backup configuration as independent in terms of its failure characteristics, then the availability of the active redundant system is one minus the probability of simultaneous failure of both units. This is stated as:

$$A_D = 1 - P_f(A)P_f(B)$$

where P_f is the individual probability of failure for unit A and a second unit B. The term $P_f(A)P_f(B)$ is the joint probability that both units will fail simultaneously. If, by similarity, $P_f(A) = P_f(B) = P_f$, then:

$$A = 1 - P_f^2$$

Determining the quantitative impact of the application of backup units

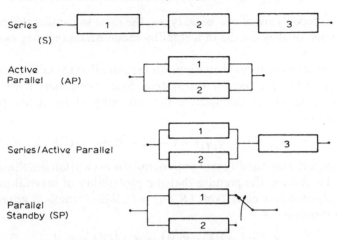

Fig. 2.12. Reliability configurations.

involves constructing detailed availability models that depict the system's R&M configuration. This involves, for example, considering three basic reliability configurations: series, parallel and series–parallel combinations, as shown in Fig. 2.12.

The boxes in Fig. 2.12 represent the equipment that form a basic system. Each box has some known or estimated failure rate, λ (or MTBF where MTBF $= 1/\lambda$), which permits the calculation of the reliability of the particular configuration. A more complex model can be assembled by groupings of these basic configurations. The system model, once fully defined, is used for the computation of system reliability and as input to the overall system availability model.

Series model reliability calculations are possible when the failure rate is known for each unit in the series. Reliability must be considered with respect to the same time interval for the entire system as well as each unit. The reliability of each unit in the system can be expressed as:

$$R_1 = \exp(-\lambda_1 t_0)$$
$$R_2 = \exp(-\lambda_2 t_0)$$
$$R_3 = \exp(-\lambda_3 t_0)$$

where R_1, R_2 and R_3 are the respective failure rates for each unit and t is operating time. The failure of any unit constitutes the failure of the system, the total reliability of which can be computed as:

$$R_S = R_1 R_2 R_3 = \exp(-(\lambda_1 + \lambda_2 + \lambda_3)t_0)$$

Parallel models are of two basic types; the units in parallel can be in an active mode, or they can be in a standby mode which requires switching between units.

Active mode reliability calculations also require that the failure rates and time interval be specified for the system. Since both units must fail to constitute a failure of the system, the reliability of the active parallel configuration is:

$$R_{AP} = 1 - (q_1 q_2) = 1 - [(1 - R_1)(1 - R_2)]$$

where $q_1 q_2$ is the probability that both parallel units will fail simultaneously and $q = 1 - \bar{R}$ from the premise that the probability of survival plus the probability of failure equals one ($R + Q = 1$). This formula is generalized to n parallel units as

$$R_{AP} = 1 - [(1 - R_1)(1 - R_2) \cdots (1 - R_n)]$$

Standby mode reliability calculations are also possible when the failure rate for each equipment in the configuration along with the time interval to be specified is known for the system. In order to simplify the mathematics involved in computing standby parallel reliability, it will be assumed that the off-line unit cannot fail until it is switched on and that the switching element cannot fail (their reliabilities are equal to one ($1 \cdot 0$)). Thus, a system failure can only occur if both units of the configuration have failed. Assuming both units have the same on-line failure rate, λ, the reliability of the standby parallel configuration* can be expressed as:

$$R_{SP} = R_1 (1 - \ln R_1)$$

Generalizing to n units in standby parallel redundancy, the formula becomes:

$$R_{SP} = R_1 \sum_{i=0}^{n-1} \frac{(-\ln R_1)^i}{i!}$$

A comparison of the active and standby parallel redundancy configurations yields the following conclusions:

(a)   If the assumptions made for the standby parallel configuration are present, the system MTBF for the standby configuration is greater than that of the active parallel configuration.

* The formulae for standby parallel redundancy can be derived from the Poisson failure distribution function.

(b) The switching element required in the standby configuration may, in reality, have a high failure rate associated with it and thus introduce a high level of unreliability into the system.

(c) The failure rate of the unactive unit in the standby configuration will always have associated with it a failure rate greater than zero and may in some instances have a failure rate equal to the active unit.

(d) Due to points (b) and (c) above, the reliability of a standby configuration may in some instances be less than that of an active parallel configuration.

(e) Protective circuitry may be required in some active parallel configurations to eliminate the possibility of total system outage due to the failure of one unit.

(f) Equipment inefficiency may result in the active parallel configuration. Each unit may be operating at only half load when both units are operational but must be able to operate at full load upon the failure of either unit.

Series/active parallel model reliability calculations are a straightforward combination of the series and parallel formulae. Thus, for these elements shown in Figure 2.12 the formula becomes:

$$R = R_P R_S$$
$$R = \{1 - [(1 - R_1)(1 - R_2)]\}R_3$$

and in the general case:

$$R = \{1 - [(1 - R_1)(1 - R_2) \cdots (1 - R_n)]\}R_{n+1}R_{n+2}R_{n+k}$$

for n parallel elements and k series elements.

It should be emphasized that the series–parallel reliability formula, as given above, does not account for repair. When considering repair, system availability should be computed as has been previously discussed in this chapter.

Availability can be improved by increasing the MTBF or by decreasing the MTTR of the equipment. MTBF and MTTR improvement can be considered in combination with, or as an adjunct to, backup system design configurations. Note that MTBF, as considered here, takes into account all failures requiring corrective maintenance actions. Redundancy is considered, then, only in terms of an availability improvement attribute. MTBF is primarily a figure of merit by which one hardware item can be compared to another.

Various MTBF improvement techniques presently exist and some of the more widely used techniques include:

(a)   Better parts and materials—Suppliers apply special processes, tests and controls to manufacture higher quality parts and materials.

(b)   Derating—Through the operation of parts at less severe stress levels than those for which they are rated, the part reliability and, ultimately, overall hardware reliability can be improved.

(c)   Production reliability screening or burn-in tests—Used to aid in the elimination of such failure modes as workmanship errors, premature failures and quality defects in order to better assure the delivery of equipment with higher reliability levels. Though part screening may be performed, the performance of production screening helps eliminate the defects induced during the assembly of systems (as well as to eliminate any defective parts which pass initial screens).
Environmental control (e.g. cooling)—The operation of a system in a temperature- and atmospherically-controlled environment helps in reducing failure due to increased part stresses (caused by such things as temperature rise in system operation, sudden room temperature changes, etc.).

(e)   Environmental hardening (e.g. shock, vibration)—The ability of an equipment to withstand the stresses imposed upon it during normal operation, as well as during testing, due to shock and vibration will be determined by intrinsic and specifically designed environmental hardening features. Such environmental conditions can cause damage to parts when stress levels exceed the allowable working stress levels. They should be anticipated during design, development and production.

Various maintainability improvement techniques presently exist which have a direct impact upon MTTR. Some of the more widely used techniques include:

(a)   Modularity—Refers to the separation of components of an equipment into physically and functionally discrete groups to facilitate their removal or replacement upon failure. Thus, the greater the degree of modularity, the greater the reduction in MTTR that will be experienced due to improved ease of repair in fault detection and isolation.

(b) Diagnostics—Refers to the built-in test (BIT) and fault isolation test (FIT) features for the detection of subassemblies, boards, parts, etc., performing below minimum acceptable levels.

(c) Other maintainability improvement techniques which do not directly impact system MTTR include:

  (i) Performance monitoring—to aid in the identification of degraded equipment performance which may be an indicator of impending system failure (thus preparing maintenance personnel for the type of repair task to be encountered) or may identify an existing fault.

  (ii) Personnel training—the level and expertise of maintenance personnel trained for the proper performance of maintenance tasks can directly influence the duration of the various maintenance task elements.

  (iii) Levels of support (on-line/base/depot)—the maintenance philosophy prescribed for repair of a failed hardware element, whether it requires on-line, or shop repair, is a direct influence on system repair and may impact system downtime due to spares availability considerations.

  (iv) Spares support—the levels of spares support for failed components can directly influence system downtime. Cost tradeoff analysis may be required when determining the amount of safety stock to have on-site (and thus safety stock cost) versus cost of an equipment down period due to unavailability of spares.

In addition to the above MTBF and MTTR improvement techniques, engineering practices and controls applied during system acquisition to improve availability are:

(a) Warranty provisions—To increase the R&M of field equipment by providing the manufacturer with incentive during acquisition.

(b) R&M testing—R&M tests serve to: (i) demonstrate R&M compliance; (ii) help improve R&M growth and (iii) aid in the conditioning of hardware for field use.

(c) R&M control—R&M levels are controlled by the application of reliability engineering tasks, tests and surveillance activities at key points during the system acquisition cycle.

(d) R&M growth—An R&M level growth program is undertaken in order to help assure that inherent R&M potentials are approached.

Included are R&M growth tests along with a well-executed failure reporting analysis and corrective action methodology.

(e)   Critical parts control—Those parts which have been identified as having a significant impact on system operation and reliability are classified as critical. Documented control of such parts throughout the system life cycle is essential in tracking R&M and meeting availability requirements.

(f)   Failure modes, effects and criticality analysis (FMECA)— Represents a systematic approach to identifying and tabulating failure modes and then determining their effects at higher levels of assembly. The FMECA provides a disciplined approach for a part-by-part analysis of failure consequences.

(g)   Vendor control program—Significant improvements in R&M

Baseline system

Improved reliability

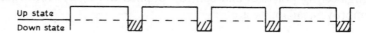

Improved maintainability

Improved R & M

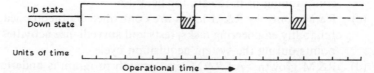

Units of time

Operational time ⟶

Fig. 2.13. Availability improvement through R&M improvement.

levels may be brought about through a comprehensive vendor control program. An effective program involves continued surveys of contractor facilities and program activities during development and protection from an R&M and quality assurance standpoint.

The effect of R&M improvement is displayed graphically in Fig. 2.13 where three different approaches to availability improvement are illustrated. The first approach, improved reliability, demonstrates how lengthened periods of uptime (without any change in the total length of the period downtime) can result in improved system availability as measured against the baseline for that system. Improved availability may similarly be achieved by improving the maintainability of the system producing a corresponding reduction in the length of downtime periods. These effects can also be combined to produce an even greater improvement in system availability. By increasing both the reliability and maintainability of the system, the length of the total uptime period is increased directly as a result of improved reliability and indirectly due to improved maintainability. In application, the availability objective may also be a function of the specific application, the system, environment, cost and other conditions which dictate or inhibit the degree to which these availability improvement approaches are applied.

2.4 RELIABILITY DEGRADATION AND CONTROL

The need for a program to control reliability degradation becomes evident when one considers that:

1. manufacturing and maintenance processes introduce defects into hardware that are not ordinarily accounted for by reliability design engineering efforts; and
2. inspection and test procedures are imperfect and allow defects to escape which later result in operational failures.

Therefore, if the reliability that is designed and developed into a system is to be attained, efforts must also be applied during production and subsequent life cycle stages. To realistically assess and fully control reliability, life cycle degradation factors must be quantitatively measured and evaluated. This is especially important for new hardware items where learning is not yet complete and a high number of process induced defects can be expected which would, unless controlled, result in a high initial failure rate. It is also important for systems that have been in operation for a long time and where component aging mechanisms exist. If these mechanisms are not detected and component replacements made, degradation of safety function could result.

The extent of degradation during production is dependent on the effectiveness of the inspection and quality/reliability engineering control program. Program effectiveness must be measured and evaluated to determine the need for process improvement or corrective changes. The effective accomplishment of the production study and, more importantly, the adequacy of subsequent corrective measures will dictate the rate at which reliability degrades or grows during production. Specifically, reliability degradation is minimized during manufacturing and reliability grows as a result of improvements or corrective changes that:

(i)   Reduce process induced defects through:
—accelerated manufacturing learning
—incorporation of improved processes.
(ii) Increase inspection efficiency through:
—accelerated inspector learning
—better inspection procedures
—incorporation of controlled screening and burn-in tests.

As process development and test and inspection efforts evolve, problem areas are resolved. As corrective actions are instituted, the outgoing reliability approaches the inherent (design-based) value.

Assessing and controlling reliability degradation involves quantifying process-induced defects and determining the effectiveness of the inspections and tests designed to remove them, i.e. estimating the number of defects induced during assembly and subtracting the number estimated to be removed by the quality/reliability inspections and test. This includes estimating defects attributable to purchased components and materials as well as those due to faulty workmanship during assembly.

Process-induced defects can be brought about by inadequate production learning or motivation and from fatigue. Quality control inspections and measures are performed to weed out those defects that are generally apparent and detectable through standard inspection procedures. No inspection process, however, can remove all defects. A certain number of these quality defects will escape the production process, and the items will be accepted and released for storage or deployment.

In certain situations, these quality defects can be overshadowed by an unknown number of latent defects. Latent defects, which ordinarily pass factory quality inspection, are due to flaws, either inherent to the parts or induced during fabrication, that weaken the fabricated hardware such that it will fail later under the proper condition of stress during dormant periods or when deployed. Reliability screen tests are designed to apply a stress at a given magnitude over a specified duration to accelerate their occurrence and subsequent removal during manufacturing. As in the case of conventional quality inspections, screen tests designed to remove latent defects are not 100 % effective.

It should be noted that standard handbook reliability prediction and analysis methods are based primarily on system design characteristics and data emphasizing the attribute characteristics of the constituent parts. They use the exponential failure model to provide an estimate of the reliability potential of a system during its useful life period. These estimates do not specifically account for initial process induced defects or for the quality inspections and screen tests designed and applied to remove them. Consequently they do not yield early system life performance, particularly for newly designed or initially manufactured hardware.

In order to assess the reliability of a system or an associated component or subsystem during its early life period, it is necessary to evaluate the components of failure that comprise the item's overall life characteristics curve. In general, the total distribution of failure over the lifespan of a large population of a hardware item can be separated into quality, reliability (or stress-related) and wearout failure distributions. These failure distributions combine to form the infant mortality, useful life and wearout periods shown in Fig. 2.14.

The general approach to reliability is to address only the useful life period, where the sum of the distributions of failure result in a constant failure rate that can be described by the exponential failure distribution and standard handbook prediction techniques can be applied. Design action is focused on reducing stress-related failures and generally includes efforts to select high-quality, long-life parts that are adequately derated. Generally, no special effort is directed toward eliminating early quality and reliability defects.

For many new systems, a design-based approach in itself is not adequate to assure reliability. Examination of Fig. 2.14 shows that the infant mortality period (T_0 to T_S) is dominated by a high but rapidly decreasing quality-related failure distribution, a relatively high and decreasing latent stress-related failure distribution. Experience has shown that the infant mortality period can vary from a few hours to well over 1000 h, with the majority of early failures occurring in the period (T_0 to T_A) defined by the

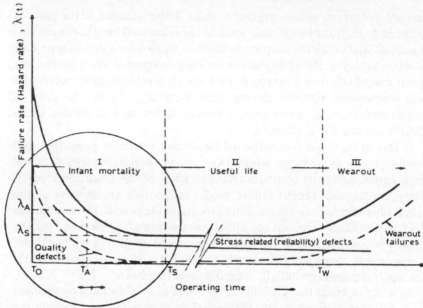

Fig. 2.14. Failure distribution (new hardware system).

break in the failure rate (mortality) curve. The duration of this critical phase in reliability growth is dependent on the complexity and maturity of the hardware and, if not controlled, would dominate the overall mortality behavior leaving the product without a significant high reliability period of useful life. Positive measures must be taken, beginning with system design, to reduce the amount of early failures and to shorten the time it takes to achieve a stabilized low level of mortality (failure rate). This includes evaluating the impact of intrinsic part defects and manufacturing process-induced defects, as well as the efficiency of conventional inspections and the strength of reliability screening tests.

The types of defects to which a system is susceptible during the infant mortality period are determined largely by its processing, while the presence of these defects in the finished item is a function of the tests and controls that are applied during production. These process-induced defects include such items as poor welds or seals, board defects (e.g. delamination, fracture, insulation cracking), part/board bond separation, solder problems (cracking, open, etc.), broken or nicked leads, part defects (intrinsic, overstressed), tolerance drifts, incorrect positioning of parts, dirt or contamination on surfaces or in materials and chemical impurities in metal or insulation or protective coatings.

Figure 2.1 showed the reliability impact of process defects. The figure indicated that reliability is first established by the basic system design configuration, its complexity and technology, the quality level of its constituent parts, the applied electrical/thermal derating margins and various other design application factors. The reliability of initially manufactured items will then be degraded from this inherent level as a result of defects induced by the manufacturing process and because conventional quality inspections are not designed to detect and remove from the production population early stress-related reliability defects. Subsequent reliability improvement and growth can only be achieved through careful inspection, systematic application of controlled screening and burn-in tests, extensive failure analysis and prompt and effective corrective action. The rigor with which the test, failure analysis and corrective actions are performed will determine the slope and degree of reliability improvement.

The identification of significant defects, the determination of reject rates, and the analysis of aging deterioration influenced by the plant environment and dormancy, are essential for minimizing performance degradation and ensuring system reliability and plant safety.

The assessment and control efforts must also include identifying components and equipment (in terms of their major critical characteristics) which exhibit deterioration or aging when either dormant, intermittently on-line or fully operational and preparing procedures for cyclic inspection to detect failure mechanisms at work in them and ultimately to maintain reliability. The inspection procedures are to identify components that do not meet acceptable levels of performance for significant parameters of unreliability (as defined by failure mode and deterioration analysis) and which are to be repaired or replaced. The results of these efforts are used in long term failure rate predictions, design tradeoffs, the definition of allowable test exposures, the necessity to retest after dormant periods as well as in repair and replacement plans.

Equipment items and component parts age and deteriorate over long dormant periods as well as during operating periods due to a number of failure mechanisms. Dormant failure mechanisms are of the same basic kind as those found in the operating modes, although they usually develop at a slower rate. Additionally, many system hardware failures which occur during service life are traceable to built-in (latent) manufacturing defects that accelerate specific aging mechanisms. These defects will, in all probability, pass quality inspection during manufacturing but finally become evident during plant operation.

Examples of aging failure mechanisms are the electrical contacts of relays, switches and contactors. These contacts are susceptible to the formation of oxide or contaminated films or the adherence of particulate matter that becomes attracted to the surface. During active use, the mechanical sliding or wiping action of the contacts is effective in rupturing the films or dislodging the foreign particles in a manner which produces a generally stable, low resistance contact closure. However, after long periods of non-operation the contaminant films and/or the diversity of foreign particles may have increased to such an extent that the mechanical wiping forces are insufficient for producing a low resistance contact.

Another failure mode which may become significant is the deterioration of lubricants used on the bearing surfaces of relays and solenoids. Lubricants can oxidize and form contamination products. Lubricants can also attract foreign particles, particularly when exposed to dust and this can lead to lubrication failures and excessive wear.

Over a period of time, many plastics (such as those used in the fabrication of electronic components, i.e. integrated circuits, capacitors, resistors, transistors, etc.) lose their plasticizers or other constituents, which may evaporate from the plastic, causing the material to become brittle and possibly to shrink. This can cause seals to leak, insulation to breakdown under electrical/mechanical stress and can precipitate other changes conducive to failure. Additionally, plastics may continue to polymerize after manufacture. That is, the structure of the molecules may change, without necessarily any change in chemical composition. This will result in changes in physical properties.

Many materials slowly oxidize, combine with sulfur or other chemicals or break down chemically over a period of time. These changes may affect electrical resistivity, strength, etc. In addition many of these materials, when exposed to condensed moisture, or high humidity, may through a leaching process lose essential ingredients, such as fire retardant additives, thereby causing a hazard to slowly develop.

Component parts and assemblies which are sensitive to contaminants are sealed during manufacture. These seals will often leak, partly as a result of flexing due to changing temperature and atmospheric pressure, allowing air, moisture or other contaminants to reach the active portions of the component. This leakage can be so subtle that the effects may not be discernible for years, but ultimately significant changes can occur.

Also rough handling during shipment, maintenance operations and aging and deterioration mechanisms can, if uncontrolled, lead to a variety of component and equipment failure modes.

A summary of some of the failure modes encountered with electronic

TABLE 2.1
Failure Modes Encountered with Electronic Components

Component	Failure mode
Connectors	Corrosion causes poor electrical contact and seizure of mating members. Moisture causes shorting of the ends.
Switches	Metal parts corrode and plastic bodies and wafers warp owing to moisture absorption.
Plugs, jacks, sockets, etc.	Corrosion and dirt produce high resistance contacts. Plastic insulation absorbs moisture.
Transformers, inductors	Windings corrode causing short- or open-circuiting. Moisture causes change in inductance and loss in Q. Moisture swell phenolic forms. Wax coverings soften at high temperatures.
Semi-conductors, diodes, transistors, microcircuits	Plastic encapsulated devices offer poor hermetic seal resulting in shorts, or opens caused by chemical corrosion or moisture.
Resistors	The values of composition-type fixed resistors drift and these resistors are not suitable at temperatures above 85°C. Enameled and cement-coated resistors have small pinholes which bleed moisture, accounting for eventual breakdown. Precision wirewound fixed resistors fail rapidly when exposed to high humidities and to temperatures of about 125°C.
Capacitors	Moisture permeates solid dielectrics and increases losses which may lead to breakdown. Moisture on plates of an air capacitor changes the capacitance.
Motors	Swelling and rupture of plastic parts and corrosion of metal parts. Moisture absorption and fungus growth on coils. Sealed bearings are subject to failure.
Relays and solenoids	Corrosion of metal parts causes malfunctioning. Dust and sand damage the contacts. Fungi grow on coils.
Batteries	Dry batteries have limited shelf life. They become unusable at low temperatures and deteriorate rapidly at temperatures above 35°C. The output of storage batteries drops as low as 10% at very low temperatures.

components as a result of long time deterioration or aging is given in Table 2.1. Protective measures must be applied to isolate hardware systems and their component items from deteriorative influences in order to eliminate or reduce failure modes such as those listed in the table.

Scheduled cyclic inspection and functional checkout are the key to assuring the actual reliability of hardware (during long-term non-operating and operating periods). These inspections will provide data on system reliability, detect deterioration and other deficiencies caused by aging and improper process methods, and provide a basis for hardware reconditioning or replacement. The type, extent and frequency of inspection is dependent on the deterioration susceptibility of the hardware.

An effective inspection program requires performance of a thorough and detailed visual inspection to identify defects and deterioration mechanisms due to environmental stresses and aging factors followed by functional checkout to verify operability. Emphasis is placed on inspecting for evidence of deterioration, contamination and corrosion as well as to look for loose or frozen parts, damaged parts, leakage, excessive moisture (fungus, mildew, rot) and damaged or deteriorated preservation and packaging. The functional test is defined such that it can be applied simply and quickly on sensitive hardware items to assure that they perform satisfactorily and are ready for use. The functional test is not intended to represent a complete and detailed inspection of the hardware to determine compliance to specified requirements but is designed to verify operability by fully utilizing end item functions to indicate operational readiness. The tests can range from a relatively simple checkout of a critical component subsystem or assembly to perhaps a full test of a complete system.

The inspection procedure recognizes the fact that hardware deteriorates due to age, environment and environmental conditions and when deterioration begins to take effect the reliability level will decline below that which was initially manufactured (as shown in Fig. 2.1). Although the effectiveness and adequacy of the reconditioning operations and controls will minimize the decline, the resultant reliability level of the reconditioned hardware would be somewhat lower than the 'as-new' condition. The inspection requirements should reflect:

1. Minimum acceptable performance deterioration parameters which are lower than the initially specified values.
2. Criteria that indicate the quality limits beyond which repair is not economically warranted.
3. Acceptance criteria for reconditioning cycle(s) at predetermined storage/dormant (as well as operational) milestones.

2.5 SOFTWARE RELIABILITY ANALYSIS

Although the control systems used in nuclear power plants may be operationally perceived as having a 'hardware' character, they are increasingly dominated by their 'software' content. Software engineering is a rapidly developing field in which well understood and universally accepted design concepts and standards of practice have not yet stabilized. This section addresses the reliability aspects of the software/firmware used in electronic control systems. It discusses some of the analysis methods and techniques that are used to minimize the many kinds of mistakes and errors that can lead to system reliability problems. It views software development as part of the overall system development process recognizing that complex computing systems are comprised of interacting hardware and software elements which are subject to a wide spectrum of failure modes and errors. This unified approach provides a sound basis for effecting a smooth transition and for assuring that operational experience is reflected in the evaluation of software reliability. Software reliability, like hardware reliability, is considered as an integrated system engineering process that includes tasks and activities that start during the conceptualization of every system and continues through system analysis, software design, coding, test and system operation. Emphasis is placed on initial specification, design, and program development where most of the software flaws and errors are introduced. Chapter 6 presents reliability engineering and control procedures applicable to both hardware and software.

Modern control systems are based on microcomputers (a computer system using microprocessors). Microcomputers, as compared to mini-computers, are relatively easy to design, are flexible, small, low cost, simple and inherently reliable. A microprocessor is a single integrated circuit that can be programmed with stored instructions to perform a wide variety of functions, consisting of, at least, a controller, some registers and an arithmetic and logic unit. It is usually contained in an integrated circuit (IC) package with 18-64 leads. Its functions include:

—Arithmetic calculations.
—Data manipulations.
—Equipment monitoring.
—Equipment testing.
— Process control.
—Data transfer and interdevice communication.
—Timing.
—Measurement.

TABLE 2.2
Microprocessor Types

Supplier	Microprocessor	Number of bits	Package size (pins)	Number of instructions	Execution time (ms)	Approximate cost (at time of publication)
Intel	8080A	8	40	78	1·5–3·75	$4·25 (100 Qty)
Intel	8085	8	40	80	0·8–5·2	$8–10 (100 Qty)
Motorola	M6800	8	40	72	1–2·5	$8·95, 1 MHz (25–95 Qty)
Motorola	M6802			72	2–5	$11·95, 1 MHz (25–99 Qty)
Motorola	M6809	8	40	59	2–5	$28·00 (25–99 Qty)
Zilog	Z80			150+	1–5·75	$8·90–$34·50 (pending freq. 100 Qty)
Intel	8086	16	40	97	0·4–37·8	$86·50 (100 Qty) 5 MHz
Motorola	MC68000	16	64	61	0·5	$249·00 (Qty 1)
Texas Instruments	TMS9980/ TMS9981	16		69	3·2–49·6	$19·10 (100 Qty)
Texas Instruments	TMS9900	16	64	69	2–31	$31·25 (ceramic— 100 Qty)
Zilog	Z8000	16	40	110+	0·75–90	$100 (100 Qty)

The logical structure or architecture of a computer as seen by a programmer working at the assembly level must be defined and an instruction set prepared and included as part of the operating system. The instruction set architecture (ISA) includes all information required by a programmer and provides a basis for implementation by independent vendors each using different techniques and speeds.

Microprocessors and peripheral devices are supplied by many manufacturers. Typical devices are listed in Table 2.2. This list is not exhaustive but is representative of the classes of available microprocessors. The quality level of these hardware devices has a direct effect on the reliability of the computer system. They can be procured to specified levels of quality as defined by *US MIL-M-38510* and *US MIL-STD-883* which are keyed to reliability failure rate prediction models given in *US MIL-HDBK-217*.

Software includes the programs which govern the processing of system data to command the computer to perform computational or control functions as well as self- or system-diagnostic functions. The programs for microcomputers are normally stored in ROM (read only memory) or EPROM (erasable programmable ROM) rather than in RAM (random access memory). ROM and EPROM store the program permanently, ROM is programmed during manufacturing and EPROM can be programmed in the field by the user and can be erased and reprogrammed with different input information. The basic program instructions for ROM and EPROM are generally referred to as firmware. RAM is a volatile form of memory in that the memory content is lost when power is removed.

Regardless of the memory type the task of developing the program is the same, i.e. the translation of instructions ultimately into the binary machine language. Developing the program, in general, involves the following steps:

Step 1: Establish program requirements.
Step 2: Develop program flowchart.
Step 3: Code program.
Step 4: Enter source program into development system.
Step 5: Load assembler or compiler into development system memory.
Step 6: Correct syntax errors in object program.
Step 7: Load object program into development system memory.
Step 8: Execute program with monitor/debug.
Step 9: Transfer the program to microcomputer system's memory.

Each step of the program development process is inherently error prone (see Table 2.3). Errors can be introduced from misinterpretation, mistakes

TABLE 2.3
Program Errors

Requirements
   Incorrect requirements:
   —Model does not fit physical situation well enough
   —Document references incorrect
   Untestable requirements:
   —Decision criteria, accuracy criteria, processing rates, error recovery requirements missing, incomplete or inadequately stated
   Inconsistent or incompatible requirements:
   —Two locations in specification give conflicting information
   —Conventions (e.g. coordinate systems) not consistent with other documentation

Design
   Deficient design representation:
   —Flowcharts being superseded by program design language, hipo charts, Nassi–Schneiderman charts
   Unstructuredness:
   —Module dependence, data base structure not designed for efficient access, nonstandard control structures
   Incomplete reflection of requirements:
   —e.g., error conditions not detected/circumvented, communicated as required
   —Partially inaccurate approximations

Code
   Missing or incorrect logic
   Misinterpretation of language constructs
   Erroneous or unjustified assumption
   Data structure defects
   Typographical errors
   Unchecked singularities

Misinterpretation of language constructs
   A programmer uses certain language constructs in a way he believes is correct but the compiler or assembler interprets them differently

Singularities and critical values
   An expression in which a mathematical formula has a singular point, such as a division by zero or an inadmissible region, such as the logarithm of a negative number, for which the programmer forgot to test and provide a response
   Every expression which includes a quotient has the *potential* for overflow/ underflow

Typographical errors
   Parentheses omitted
   '23' instead of '32'
   'i' instead of '1'
   '·' instead of '·'
   Variable name misspelled

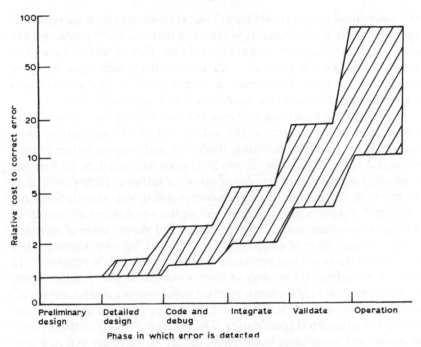

Fig. 2.15. Software error correction cost.

in translation or in coding or typing. Unless special attention is given to each step of the program development to assure reliability, a large number of errors could be introduced. It must be emphasized that it is most cost-effective to detect and eliminate errors as early in the program development process as possible. There is evidence that about 60 % of all software errors are committed *before* coding even begins. It has been established also that the cost of correcting errors increases exponentially as software development progresses and is very high if the fault is diagnosed and corrected during operation. Figure 2.15 shows the relative cost to correct an error with respect to the phase in which the error is detected. The figure indicates, for example, that an error corrected during operation can cost over 75 times more than if it was corrected during preliminary design. Furthermore, not only is the cost of correcting an error much smaller the earlier it is discovered, but the probability of fixing the error correctly is much greater in an early phase than it is if the error is encountered during a later phase.

A requirement of an effective reliability program is the ability to predict and assess total system reliability at critical development, acquisition and operational life cycle phases. It is essential that reliability prediction and assessments account for potential hardware failure modes as well as possible software errors, since each has relatively little system related significance by itself. Developing a total combined hardware/software system hazard of failure rate distribution with age is a difficult task.

Much work has been done with respect to developing the hazard rate age distribution for hardware; i.e. the well-known failure distribution curve consisting of the infant mortality, useful life and wearout failure periods (Section 2.1, Fig. 2.2). However very little work has been done with respect to establishing the hazard rate distribution for software. Hardware failures are due to design, manufacturing or maintenance defects, overstressing and wearout. Software failures are incorrect sections of code which are due to human errors committed during the design and maintenance of software.

Furthermore there is a fundamental difference between hardware and software as far as the interpretation of the time variable is concerned. In hardware reliability the passage of time is associated with a failure event, i.e. an accumulation of stresses is reached which causes a component to fail. An error in software, on the other hand, is made by the programmer or exists because of an incorrect specification. The passage of time is related to error detection and not to error occurrence; the errors which are detected were made previously. The amount of time (labor or machine) involved in error detection and the probability of error detection are a function of test time, type of test and choice of test data because these factors determine which part of a program will be exposed to testing and to the possible detection of errors on a particular program path. Since software failures are mostly entirely related to design and maintenance and not manufacturing or wearout (in a strict sense) it is reasonable to assume that the hazard rate distribution for software, although initially high, will continually decrease with operating time.

Figure 2.16 depicts the combined hazard rate distribution for a system comprised of hardware and software elements when considering an exponential software mortality distribution. The exponential software mortality distribution assumes that errors are corrected as they are detected and that new errors are not introduced by the corrections.

Initially, as shown in the figure, the system exhibits a high hazard rate resulting from the combination of early hardware failures and software errors. The hazard rate stabilizes (at time T_B) when the weaker hardware items have been eliminated but is still decreasing somewhat because of the influence of the software errors. The hazard rate remains at this stabilized,

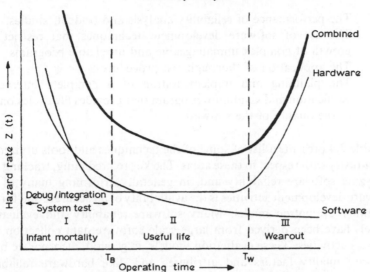

Fig. 2.16. Combined hardware/software hazard rate distribution.

but decreasing, level until it reaches the point T_W where the hazard rate starts to increase noticeably due to the influence of the hardware wearout failure mechanism.

The conventional approach to hardware reliability addresses only the useful life period, where the sum of the distributions of failure result in a constant failure rate that can be described by the exponential failure distribution. Design action is focused on reducing stress-related failures and generally includes efforts to select high-quality, long-life parts that are adequately derated.

The approach to software reliability is to reduce errors through application of structured programming techniques, modularity and through application of detailed, systematic analysis and test during system development and integration.

Achievement of reliable software, like hardware, requires application of highly disciplined and timely engineering tasks and analysis that start during the early system design concept phase and continue through development and plant operation. Some of the more critical elements in the development of reliable software are:

1. The centralization of a software organization that emphasizes program management, thorough documentation and configuration control.

2. The performance of reliability analysis and tradeoff studies.
3. The use of software development techniques that restrict the growth of complex unmanageable and unreliable programs.
4. The application of thorough test procedures.
5. The planning and implementation of a complete interactive verification and validation program that provides life cycle control of the quality of the software.

Table 2.4 presents a list of some of the techniques and tools utilized by the industry with respect to these areas. The key to predicting, tracking and analyzing software reliability and, in general, performing many of the software development activities is the availability of an accurate and formal reliability evaluation model. Many software reliability and evaluation models have been derived from large scale software data collection and analysis activities. These evaluation models, in general, incorporate basic software quality factors and attributes and, like hardware reliability models, are structured to facilitate early development tradeoff decisions. They are used, for example, to provide an early quantitative reliability comparison of alternative software strategies (e.g. programming languages, coding methods, staff organizations, operating systems).

Many of the current software reliability models rely heavily on established hardware reliability techniques. Some significant differences exist, however, between hardware and software from a reliability standpoint and these differences must be taken into account when predicting and evaluating software reliability. For example, software does not degrade due to wear or fatigue. Imperfections or variations are generally not introduced in making additional copies of a piece of software. Software interfaces are conceptual rather than physical. There are many more distinct paths to check in software than in hardware. There is, potentially, an almost infinite number of software entities to check. Any item, including its random value, in a large file, may be a source of error.

In addition, the error modes are generally different. Software errors usually appear with no advance warning, exhibit no period of graceful degradation and more often provide no real time indication of their occurrence. Repair of a hardware fault generally restores the system to its previous condition. Repair of a software fault changes the previous condition of the system. Also, software is difficult to adequately test, due to the high degrees of interdependency within its structure.

Hardware and software are similar in that reliability is a function of complexity, although an acceptable measure of software complexity has yet

TABLE 2.4
Software Reliability Provisions, Techniques and Tools

Software reliability organization
 —Policy
 —Cost and resource estimation and budgeting
 —Staffing
 —Management tools and guidelines
 —Training

System acquisition specification
 —Software performance requirements
 —Software/hardware interface requirements
 —R&QE requirements

System control during development
 —Management plan
 —R analysis (prediction, FMCA, etc.)
 —Maintenance
 —Design review
 —Error analysis
 —Configuration management

Standard and specifications
 —System specification requirements
 —Software performance specification
 —Interface standards
 —Design specification requirements
 —Program specification requirements
 —Coding standards
 —Language standards
 —Reliability analysis methods
 —System control requirements

Software development tools and techniques
 —Structured programming
 —High order language
 —Top down modular design
 —Specified coding structure
 —Program testing
 —Integrated testing
 —Acceptance testing

Reliability analysis methods
 —Software/hardware tradeoffs
 —Prediction and error assessment
 —Failure mode and effect analysis

Error data collection, analysis and feedback

Software reliability improvement

to be found. Also, solid-state electronic devices, e.g. transistors, micro-circuits, if fabricated properly do not have any wearout mechanism that one can see over a long time period. The defects which cause failure (other than obvious misapplication of the device) are built-in during the initial fabrication of the device; the same is true of software. Hardware reliability can be improved by reliability growth testing, e.g. a test-analyze-and-fix program to discover, identify and correct failure modes and mechanisms which cause early equipment failure. This is similar to finding and eliminating 'bugs' in a software program, thus increasing its reliability.

Consequently, attention should focus on the duality that exists between the successful hardware approaches and the emerging software approaches. Once this is accepted, the total problem is simplified because the hardware and software problems can be approached together in a total system concept. The basic difference occurs during full-scale development when hardware is fabricated and tested while software is coded (programmed) and debugged.

FMECA is another system analysis tool that can be applied during software development particularly during the early requirement definition and system analysis phases to assure reliability. FMECA is an iterative and systematic process performed to identify modes of failure and the effect of each mode on system performance. Its effectiveness has been demonstrated during the design of hardware systems to evaluate the potential manner in which failures can occur and to tabulate the results such that designers can formulate compensating corrective actions. Specifically, FMECA has been used to:

1. determine needs for redundancy, fail-safe design features, further derating and/or design simplification;
2. determine the need to select more reliable materials, parts, equipment and/or components;
3. identify single failure points;
4. identify critical items for design review and configuration management;
5. identify safety hazard areas;
6. assure that test program planning is responsive to identified potential failure modes and safety hazards.

The systematic identification of potential failures and their effects using formal FMECA techniques is directly applicable to microcomputing systems.

As depicted in Table 2.4 there are several software tools and techniques that can be applied during development to enhance the reliability as well as the testability and maintainability of the computer programs. The use of high level languages will enhance software reliability by reducing coding errors. High level code is easier to develop, checkout and maintain. A top down modular design strategy can be applied to eliminate many of the software errors particularly errors originating with the specification. Top down programming involves breaking out the program function specification into smaller and simpler statements or functional modules until all statements of the programming language are reached. The context, function and interfaces of the software modules are first completely defined by this process before the details or the coding of the individual modules are specified.

Structured programming techniques are used primarily to minimize coding or branching errors. Structured programs are developed from a small set of standard language programming constructs which prevent undesired skipping to some remote segment of the computational sequence. The decision logic is arranged in a similar manner to that used in design engineering where circuits are configured from a basic set of logic elements. Structured programming avoids the use of the unconditional branch (the 'GO TO' statement) which simplifies the flow of control logic in a program and thus significantly reduces the number of coding errors. Other structured programming characteristics include identification names, module size restrictions and interfacing linkage. A module typically would have about 50 lines of program code (one page) and would have only one link from the preceding module and to the following module.

Consequently, use of top down and structured programming techniques means that program design starts at the system level and gradually becomes more and more detailed until all modules are specified and coded. Software reliability is then dependent on the clarity and quality of the specification and codes for each module. Once coded, a module is tested first by itself, and, if accepted, integrated into the system and finally tested as part of the complete system acceptance test. Thus while program design goes from system to module, acceptance testing goes from module to system.

Software testing consists of exercising different modes of operation through application of various combinations of input data to identify errors. Unfortunately, there is no practical method of testing all modes of operation. The programming and testing of all combinations of program statements even for relatively small programs would be more involved than developing the actual program to be tested and once tested there is no assurance that the test program is correct. Consequently, the best approach for validating the reliability of the software is to develop a complete test plan which stresses testing at the module level where exercising of the modes of operation is a more manageable task.

After module testing and debugging, integration testing and, ultimately, system integration and acceptance testing are performed. Integration testing involves testing each programming unit, sub-programs and routines in groups of increasing size until the entire software package is checked out. Integration assembly and testing is usually done with the aid of a general purpose computer since the operational hardware would not be available until later in development. The test plan should be followed rigorously throughout this process and results thoroughly documented. After completion of the modular integration test (and completion of hardware development) the complete software package is incorporated into the operational hardware and a complete system test is conducted to insure that the hardware and software are compatible and that specified performance requirements are met. Following system integration testing a system acceptance test would be performed against the specified performance requirements.

A configuration management system must be planned and implemented to ensure that the identification and integrity of the operating software is maintained throughout the life of the system. Configuration management is of particular importance for computer-controlled systems because of the abstract nature of software, the volume of documentation and the fact that incorrect program changes can be made relatively easily.

2.6   RELIABILITY ANALYSIS TECHNIQUES

Many methods currently exist for assessing and estimating reliability, each having its area of applicability within the system life cycle. In general, reliability prediction techniques provide an estimate of reliability during the useful life period and involve the following steps:

1.   Decomposing the system into its constituent components or subsystems and performing an analysis (formal or informal) of failure modes, effects (impact on system operation), and criticality (severity of the effect). Based on these results, a logical block diagram is prepared which relates component failure to subsystem and system failure. A strictly in series 'chain' results if all components are critical, and parallel paths result wherever duplicate functions exist or are purposely inserted in the design to enhance reliability.

2. Gathering and analyzing data on the failure rates observed for the same or similar components, subsystems or systems during operational use, life test or from historical data sources.
3. Combining the results of steps 1 and 2 to yield a reliability measure (e.g. MTBF) and computing system reliability.

The specific method and the extent of the analysis that would be applied depend on the system life cycle phase, the analysis objectives and the availability of data.

In general, reliability and data analysis methods can be classified as: *statistical* for the purpose of assessing and analyzing operational reliability and availability from actual empirical data or *predictive* for the purpose of estimating reliability during system development based on historical data derived from similar systems, subsystems and/or components.

Operational reliability assessment generally involves:

1. estimating the parameter(s) or constants (such as failure rate) of a reliability (probability) function (e.g. exponential) and the characteristics that dictate its level from failure data and operating time (or other exposure) information; and
2. applying the probability function (or distribution) to compute system (or subsystem) reliability.

Typically histograms are plotted (e.g. time to failure plots) and statistical techniques used first to test the data and determine the applicable form of the probability distribution. They are then used to identify and evaluate the relationship between the reliability parameter(s), such as failure rate, and the critical hardware characteristics or attributes which impact reliability (such as technology, complexity, application factors, etc.) as defined by the data.

The Kolmogorov–Smirnov test and the chi-square method (in the case of the exponential distribution) are often used to test or verify a hypothesized form of the distribution function. Statistical techniques that are used to estimate and evaluate reliability parameters include:

1. *Chi-square confidence intervals:* to develop a confidence interval around a parametric reliability (e.g., failure rate) estimate; a 60 % confidence interval is that range of values around the estimate that would, with a 60 % probability, include the actual mean of an infinite sample of the device tested.
2. *Stepwise multiple linear regression analysis:* to identify significant characteristics/attributes or variables of a reliability parameter.

3. *Fisher F-test:* to determine if any significant variables have been neglected.
4. *Correlation coefficient (between two variables):* to measure the degree to which one parametric variable may be expressed as a linear function of the other.

There are many probability distributions (both discrete and continuous) that are used to compute reliability. The exponential and Weibull probability distribution functions are two of the most frequently used in reliability analysis. Reliability in the exponential case is completely defined by a single parameter and is given by:

$$R(t) = \exp(-\lambda t)$$

where λ is the failure rate and t takes on any value from 0 to ∞. Its probability density function is

$$f(x) = \lambda \exp(-\lambda t)$$

The Weibull distribution has been used to model reliability performance for certain mechanical parts. The Weibull density function is particularly useful because it can be applied to all phases of a component's life characteristics curve. The exponential density function is a special case of the Weibull.

The Weibull density function is given by:

$$f(t) = \alpha \beta t^{\beta-1} \exp(-\alpha t \beta)$$

where $f(t)$ = Weibull density function, t = the operating time for which we want to know the reliability $R(t)$ of the component, β = parameter of the density function usually referred to as the shape parameter and α = parameter of the density function usually referred to as the scale parameter. The reliability formula associated with the Weibull density function is:

$$R(t) = \exp(-\alpha t \beta)$$

Other distribution functions that have application in reliability analysis are the normal, lognormal and the binomial. The normal density function which has wide use in engineering applications is a perfectly symmetrical distribution; the log normal density function is such that its logarithm has a normal density function; it is frequently used for analyzing wearout failures. The binomial probability function is a discrete distribution associated with repeated trials of the same event. Whatever the event, if the probability of its occurrence in any trial is P, the probability of its non-occurrence is $(1 - P)$. The binomial is used to evaluate the reliability of systems employing redundant components or designed with multipaths to achieve the output function.

Predictive techniques are used to provide a measure of reliability during system development and to show where the design can be improved. The predictive methods primarily deal with the effects of failure and depend upon data emphasizing the attribute characteristics of parts, components, etc. Many formal procedures for their application exist, differing only in the level of data required for their application. There are techniques based on similar equipment, parts count and stress analysis. These methods would be applied according to the life cycle phase of development and utilized according to available information.

The *similar equipment technique* provides a gross estimate of reliability based on data from analogous equipment of known reliability and similar complexity. It is applicable during the early system concept phase.

The *parts count method* provides an estimate of reliability based on a count by part type. This method is applicable during early design studies where the degree of design detail is limited. It involves counting the number of parts of each type, multiplying this number by a generic failure rate of each functional component, subassembly, assembly and/or block depicted in the system block diagram. The advantage of this method is that it allows rapid estimation of reliability in order to quickly determine the feasibility (from the reliability standpoint) of a given design approach. The technique uses information derived from available engineering sources and does not require part-by-part stress and design data. Failure rate data for electronic parts are usually derived from *US MIL-HDBK-217*, while data for non-electronic equipment can be derived from accumulated data of similar equipment, from published sources such as *RADC-TR-75-22*.

The *stress analysis* technique involves the same basic steps as the part count technique. However, the stress analysis technique requires the use of detailed part models plus calculation of component stress values for each part prior to determining its failure rate. Each part is evaluated in its system application based on mechanical, thermal and electrical (if applicable) stress analyses. Once part failure rates are established, a combined failure rate for each functional block in the reliability diagram can be determined. This is the most accurate technique but requires the greatest amount of detail and is applicable during the later development phase.

A new tool that has appeared in the last decade which has proven to be of major aid in reliability analysis and improvement is *fault tree analysis*. Fault tree analysis (FTA) is a structural model for system failure modes composed of logical AND and OR symbols and failure events leading to system failure. Fault trees were initially applied to safety analysis and can be shown to be mathematically equivalent to reliability block diagrams. The main advantage of FTA is that design engineers who are unfamiliar with reliability methods find it easier to use than reliability block diagrams. This is especially important with complex system analysis since the probability of finding an engineer experienced in both system design and reliability analysis may be small.

Bayesian statistics are often used in the performance of reliability assessment. The usefulness of Bayesian statistics is that they allow prior information (e.g. predictive) obtained for a random process to be applied to the analysis of more recent information, such as actual test or experience data. The results thus achieved, therefore, would utilize the widest spectrum of information available. In this fashion, not only would reliability prediction estimates or the latest test information on a system be used exclusively and independently of one another to assess reliability, but also a final assessment can be made that reflects both information sources. It is particularly useful for assessing operational reliability of new complex systems when only limited test data exist.

Engineering procedures for reliability analysis (prediction, FTA, etc.) as well as for implementing an overall reliability program are given in Chapter 6.

To provide cohesiveness and continuity of life cycle reliability analyses it may be advantageous to establish or apply an overall unified analysis process. Figure 2.17 is a flowchart which illustrates the makeup of an overall reliability analysis methodology that can be applied to a nuclear power plant system and components. The interacting process, depicted in the figure, incorporates generic steps present in statistical and predictive analysis methods. It can be applied in conjunction with reported discrepancies, failures and operational data to quantitatively assess reliability throughout the life cycle of a system. Resulting criticality and sensitivity data can be used to provide guidance in the validation of performance requirements, criteria and test methods.

This unique methodology offers a complete and unified analysis process. It provides a means of focusing on areas needing greater reliability and quality control and more careful design considerations. It provides a focal point for reliability allocation, prediction, assessment, data analysis, tracking and for performing life cycle reliability cost tradeoffs. Engineers, manufacturers, operational personnel and others can use the methodology to predict, assess, evaluate and control reliability. It is also a source of continued update and refinement of input data and information. The methodology allows comparison between different system types with inherently different maintenance and repair policy requirements so that a proper decision concerning the most cost effective system can be made.

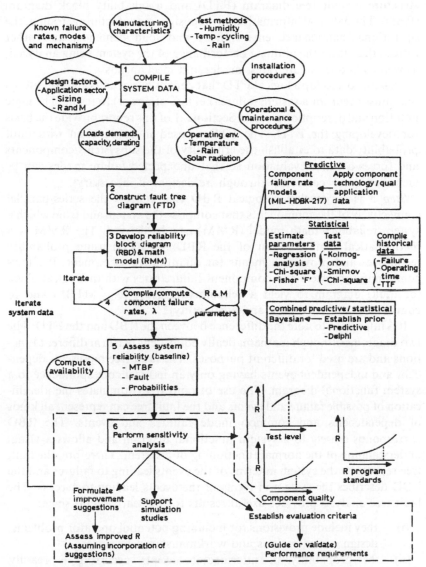

Fig. 2.17. Reliability analysis methodology flowchart.

Step 1 in Fig. 2.17 is the compilation of data and information needed to structure a fault tree diagram (FTD) and a reliability block diagram (RBD). The data and information cover design, manufacturing, testing, the operational/maintenance environment and load demands, and other factors that affect the ultimate performance of the system being analyzed, as well as known failure rates, modes and mechanisms.

Step 2 is to develop a logical FTD that depicts basic faults and conditions that must occur for a system failure (i.e. loss of power) to result. The logic notation and description given in Section 2.1 of this report provide the basis for developing the FTD. The FTD is used in conjunction with fault probability data to establish the criticality of the subsystems/components and forces a closer evaluation so that unexpected failure modes can be identified (and corrected through re-design, if necessary).

Step 3 is to develop a logical RBD that reflects the series/parallel connectively of the components that comprise the system and from which a system reliability math model (RMM) can be derived. The RMM is a mathematical representation of the RBD (or FTD) using probability techniques and provides a means for quantitative assessment. It allows combining subsystem and component failure rates with the logical block (or fault) diagram to yield a reliability measure (e.g. MTBF) and to compute system reliability (and availability).

It is important to note the difference between the RBD and the FTD. The two diagrams although mathematically equivalent perform different functions and are used for different purposes. The fault tree represents dependent and independent events having only an indirect correspondence to a system functional diagram. The use of fault trees stimulates the identification of possible failure and events and the fault tree can represent all kinds of dependencies and common mode failures and events. The RBD corresponds closely to a system functional diagram and allows a visual understanding of the normal functioning of a system. Therefore, the fault tree represents the system in terms of the events leading to failures and the RBD describes the system in terms of the events leading to success. The fault tree analysis provides useful results for nuclear systems since:

(i)   they include provisions for including potential operator problems, design considerations and workmanship deficiencies;

(ii)   the probability of intermediate events occurring is readily observable; and

(iii)   they are easily updated when more detailed data become available.

Step 4 is to compile or compute applicable component failure rates for use as input to the RMM. The component failure rates can be established through statistical, predictive or combined analysis methods, as shown in the figure, depending on system life cycle stage and data availability. Steps 2, 3 and 4 are iterated as necessary until the RBD and RMM reflect subsystem/component detail to a level at which failure rate data exist or can be compiled or developed through the statistical, predictive or combined analysis methods.

Step 5 is to input the failure rate data into the RMM and compute baseline system reliability (and availability) estimates for subsequent sensitivity, system improvement and validation analyses.

Step 6 is to perform a criticality/sensitivity analysis to identify significant areas of unreliability in terms of design, test, application, plant environment and other control attributes or system characteristics for corrective feedback to system design and to the R&D process in general. Criticality is a measure of the relative seriousness of each basic fault (failure mode, operator error, etc.) and can be computed by the following expression:

$$CR = P(x_i)P(H/x_i)$$

where $P(H/x_i)$ is the conditional probability of the overall hazardous conditions given that the basic fault (x_i) has occurred. Once the criticality of a subsystem/component has been established, the technique can be applied to further, and more methodically, list (and review) design, materials, manufacturing processes, operational and maintenance procedures, and particularly the reliability improvement attributes and engineering, and test methods and criteria that are applied.

The sensitivity analysis includes developing histograms, curves and other data that show, for example:

1. the contribution of each major component or generic class of component to the overall failure rate (or MTBF) of the system to provide guidance for subsequent component improvement programs;
2. the relationship of geographical location and operating environmental factors on MTBF and thus provide criteria for thermal and environmental design;
3. the relationship of reliability and quality control levels during manufacturing on MTBF including the application of controlled screening and burn-in.

The baseline estimates and the criticality/sensitivity data are then used to:

1. formulate system improvement suggestions and to reassess system reliability assuming the incorporation of the suggestions;
2. support simulation studies;
3. guide (or validate) the establishment of cost-effective performance criteria and test methods.

The extent of the reliability analysis and the specific areas which are emphasized depend on the system life cycle stage. During system design, emphasis is given to performing a reliability cost tradeoff and establishing specification requirements with respect to reliability. Effort during development includes allocating specified requirements to subsystems and components, assessing the reliability of the design relative to the specified requirements, identifying common cause failures and establishing a basis for a meaningful cost-effective test. Assessments are made during operation to determine achieved levels of reliability and to provide a basis for system improvement programs.

In addition, the technique can be used to monitor the reliability of a system as it progresses through development, installation and operation. It can be used to flag and rank items impacting reliability and this information can be fed back to the R&D process across the industry. It can support the determination of the cost-effectiveness of proposed system changes.

The results of the system reliability analysis, including MTBFs, failure modes, criticalities and their life cycle cost factors, can be incorporated into a dynamic real-time database from which the effectiveness of the overall reliability program can be continually assessed. The database would represent a primary experience pool for designing new systems or improving existing systems with respect to reliability.

Data required as input to life cycle reliability analyses include:

(i) Design:
—configuration
—stress factors
—quality factors
(ii) Manufacturing:
—inspection/process description
—reject rates
—inspection attributes

(iii) Operation:
  —plant environmental factor
  —component deterioration and aging factors
  —inspection and test plans
  —maintenance procedures
  —reject rates

It is recognized that while much of the required data is routinely generated, it is generally not collected and is essentially 'lost' to any centralized effort to utilize it in a long range, disciplined manner. Thus, a permanent, centralized, data system should be established to compile test and failure information, and other reliability relevant data for use in future reliability analyses. The data would be used as direct input to predictions and other analyses as well as provide a basis to track, measure and report reliability results so that corrective action in the hardware or manufacturing/maintenance processes and operational controls can be taken and adjustments made as necessary to minimize cost and maximize effectiveness. The data system should provide the following information:

1. Inspection reject rates.
2. Total operating time for each hardware item.
3. Identification of functional checkout and data recording periods.
4. Identification of discrepancies with reference to test, operating/ non-operating time and conditions.
5. Failure analysis reports of hardware discrepancies, including accumulated operating hours to time of failure, failure modes and cause and type of failure modes.

Cumulative plots of failure events versus time should be prepared and maintained. Also, periodic summary reports should be prepared providing:

1. Failure/reject rates by inspection, test and operating/non-operating time and conditions.
2. Test efficiency factors.
3. Responsible failure and component aging mechanisms.
4. Recommended or accomplished corrective actions.
5. General reliability analysis that correlates design predictions with test results and plant experience.

The specific data that are gathered through dedicated data collection activities can be supplemented with generic data from industry wide databases such as from IEEE, LERs, NERC/GADS, NPRDS, etc.

Chapter 3

Specifying Safe, Reliable and Cost-Effective Equipment

The determination of the appropriate levels of reliability for incorporation into an equipment procurement specification is a complex procedure involving the evaluation of a large number of factors and the use of tradeoff analyses to arrive at an optimum level. It is an iterative process which follows the logic shown in Fig. 1.5. The process is driven by the overall objective to meet safety/plant availability requirements at minimum life cycle cost.

In its simplest form the decision process involves the selection of commercial equipment or components whose inherent reliability easily meets the overall safety/availability requirements. More complex variations arise as availability and safety requirements become more stringent and/or the complexity and life cycle cost of the equipment itself increase.

Once a reliability goal has been established, a life cycle program is then designed to achieve that level. Reliability requirements are formulated covering both quantitative as well as qualitative program provisions and they are incorporated into the equipment procurement specifications which are then used as the basis for governing the entire acquisition and operational process. This chapter first describes the elements that make up a complete life cycle reliability program for equipment used in nuclear power plants and then discusses how to determine the specific requirements to be incorporated into the procurement specifications to reflect the particular characteristics and needs of the applicable hardware items.

3.1 PLANNING AND IMPLEMENTING A LIFE CYCLE RELIABILITY PROGRAM

The purpose of a reliability program is to improve and ensure equipment reliability through application of specific engineering tasks and control elements starting with design and continuing through development/ procurement, production, plant construction and equipment installation (including pre-operational testing) and operations (including maintenance and periodic testing). A detailed program plan is prepared to delineate tasks and responsibilities and used to manage and control all life cycle reliability activities, including engineering analysis to predict frequency of failure and identify areas where improvements can be made most cost-effectively.

The program is structured to maximize the reliability level designed into the equipment and to prevent or minimize degradation of reliability that could result from fabrication, plant integration and operational use. It establishes reliability as an integral part of the system design and specification process, incorporating reliability requirements into plans, drawings and specifications. For existing plants the program deals with reliability problems currently being experienced by defining a disciplined, systematic approach to the implementation of proven reliability improvement techniques without excessive impact upon the current or designed configuration of the plant. It takes into account the reliability implications of the fact that equipments, particularly safety systems, in existing plants are constantly being modified and retrofited.

The reliability program is structured to reflect the characteristics of the equipment used in nuclear power plants. Nuclear equipment, e.g. safety systems, are complex, high technology systems that must operate for long periods of time without serious failure and with a very long total life. A great amount of redundancy and diversity is used in nuclear facilities to ensure the safety of the plant. A large portion of the safety systems operate remotely while depending on human operators for control functions. Maintenance, inspections and overhaul of equipment are usually done at specified time intervals, when the plant is down for nuclear refueling. This process generally follows a pattern of increasing complexity, depending on the operating times accumulated by the systems. The physical environment in which the equipment operates is very severe and can have a serious detrimental effect on the complex mechanical and electronic components of the equipment. High temperatures, vibration, high humidity, the presence of corrosive fluids and gases all take a toll. All of these characteristics affect the structure of the reliability program.

Although the emphasis must be on the reliability of nuclear power plant safety systems, the program is applied to other nuclear plant systems as well. Improving the reliability of nuclear plant systems, in general, will enhance the safety of the plants by reducing the number of 'challenges' to the safety systems. Since the reliability of the safety systems is to some extent on a 'per demand' basis, reducing the number of challenges or demands on the systems per unit time will increase the reliability of the systems on a unit time basis and enhance the overall safety of the plant.

Having a reliability program for all nuclear plant systems has further safety implications in that such a program provides necessary input data for probabilistic risk assessment (PRA). In addition, the program, by providing a complete, disciplined and orderly set of life cycle tasks and elements, would assure that the equipment reliability levels established by PRA to satisfy safety goals are met during operation.

The safety requirements of nuclear power plants are, of course, of paramount concern in the development of a reliability program. However, such a program, whether applied to the safety systems only or to all systems in a plant, has benefits beyond just enhancing safety. Since a nuclear power plant can be forced off-line by outages in the non-power-producing safety systems, improving the reliability of the safety systems will improve plant availability as well. The large cost associated with nuclear power plant downtime (as much as $800 000 per day in power replacement costs alone) provides a strong economic incentive toward improving system reliability. Economic benefits also derive from the reduced operation and maintenance support costs resulting from improved reliability.

In addition, 'backfitting' requirements placed on nuclear power plant licensees by government agencies such as the NRC often result from the unreliability or inability to assure reliability of the existing safety system configuration. Improving the reliability of current systems would reduce the costly manpower and downtime associated with backfitting.

The life cycle cost analysis aspect of reliability engineering can provide valuable information for tradeoff decisions and costs of alternative designs, as shown in Fig. 3.1. By increasing reliability (or availability), the acquisition cost rises and the operation and maintenance support cost falls. At some point within the range of feasible reliability the total life cycle cost is a minimum. This point is shown as R_2. The point R_1 represents the minimum acceptable reliability (derived from PRA safety considerations) and R_3 represents the upper limit of reliability based on the state-of-the-art within a given technology. The shape of the acquisition curve at low reliability levels tends to be relatively flat, being associated with easy-to-do efforts based on simple design changes and procedural and control tasks. The portion between R_1 and R_3 is associated with more complex efforts beginning with component derating and proceeding to the use of higher-quality material, improved fabrication and quality/reliability control

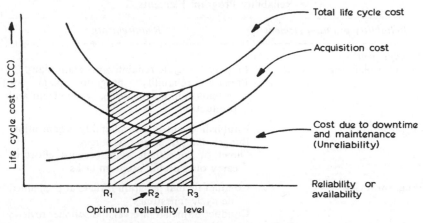

Fig. 3.1. Cost-effective system design.

processes, the employment of diagnostics and the use of redundancy. As a first approximation, costs in this region tend to be roughly proportional to the improvement in reliability. The portion beyond R_3 is associated with completely new systems generally based on the use of advanced/next-generation technology.

An additional consideration which makes it essential that nuclear plant equipment achieve high operating reliability is that failures, particularly in safety systems, are highly visible and if serious enough, or of high frequent occurrence, could have a detrimental public image effect on the whole industry.

Table 3.1 lists a series of reliability program elements and their basic requirements for application during acquisition and operation of nuclear equipment. Many of the elements listed in the table have been developed for, and proven effective, on military and space programs but can be applied either directly to nuclear equipment or with some modification to reflect the unique requirements imposed by the nuclear plant operational environment.

When planning a reliability program based on the elements listed in Table 3.1, one must be cognizant of the dissimilarities in usage and function of nuclear equipment as compared to military/space equipment: among these is the longevity factor. Nuclear power plant equipment are generally designed for a 40 year lifespan. Their design, construction and operation require substantial investment by the utility companies for the purpose of producing revenue and maximizing profit to the stockholders. In contrast, the average weapon system is planned for 10 years of operational use,

TABLE 3.1
Reliability Program Elements

Reliability program element	Requirements
Management	
Program planning	Prepare life cycle reliability program plan
	Prepare reliability program plan for systems/equipment during development and production
Organization	Establish a centralized reliability organization
	Budget manpower and resources
	Direct or coordinate staff/specialist effort to carry out specific program tasks
Program review	Conduct formal program reviews at predefined life cycle milestones
	Conduct informal interval and supplier reviews continuously during acquisition and deployment
Configuration control and management	Define methods for identification and traceability of hardware items
	Establish formal change control techniques and auditing methods
Supplier audit and surveillance	Audit/survey supplier performance
	Maintain supplier performance history
Reliability critical item list (and control)	Develop and structure a reliability critical item list
	Define specific criteria for critical items (equipment considered safety critical from PRA)
Reliability training and indoctrination	Educate appropriate personnel in reliability concepts and fundamentals
Design assurance	
Tradeoff analysis	Perform reliability tradeoff analysis to help evaluate/compare alternate design approaches
Reliability and maintainability specifications	Prepare R & M specifications keyed to criticality of items procured
	Specify quantitative R & M values and program elements
	Establish R & M levels which reflect safety, availability and life cycle cost
Reliability mathematical modeling	Establish mathematical basis for calculation of system reliability values
	Provide detailed models that account for design attributes as well as life cycle degradation factors

TABLE 3.1—*contd.*

Reliability program element	Requirements
Reliability prediction and allocation	Estimate system/component reliability Establish reliability targets for subsystems and components
Maintainability prediction and allocation	Estimate maintenance time factors Establish maintainability targets for subsystems and components
Diagnostics and modularity	Determine extent of diagnostics built-in test modularity and other ease-of-maintenance features Establish key parameters to be monitored Establish level of replacement and level of repair
Human factors reliability analysis	Establish models for human performance/reliability Collect data on human performance and error rates Assess human reliability with respect to operating power plant safety systems
Failure mode and effects analysis; failure mode, effects and criticality analysis; fault tree analysis	Perform FMEA, FMECA, and FTA Establish criticality factors
Common mode failure/common cause failure analysis	Perform CMF/CCF at system and component levels Identify events/conditions that could nullify redundancy
Single failure analysis	Perform single failure analysis
Life cycle cost assessment and analysis	Assess system LCC during concept, development and operational use Perform R & M versus LCC sensitivity studies
Reliability growth and tracking	Track system reliability growth throughout development and plant acceptance Perform reliability growth testing
Reliability demonstration	Perform demonstration test during system development Define test duration, environments and statistical basis for acceptance
Design reviews	Conduct periodic reviews during design and development Establish procedures for design review conduct and follow-up

(*continued*)

TABLE 3.1—*contd.*

Reliability program element	Requirements
Component availability	
Component derating	Establish and implement derating policy for safety related components such that: —equipment reliability meets requirements —equipment is operated within manufacturer's ratings and environmental capability Prepare procedures for derating each component type and category
Component quality level	Establish component quality levels consistent with: —high reliability application (Class 1) —intermediate reliability application (Class 2) —commercial quality application (Class 3) Structure levels around standard component usage
Component failure rate modeling and estimation	Predict component failure rates which account for operating and environmental conditions using modeling and estimation techniques Update predictions as design data becomes available
Component test methods and compliance	Define test methods for classes of tests to cover: —engineering proof tests —qualification tests —production acceptance tests —screening/burn-in tests
Component selection, specification and quality assurance	Establish and implement a formal program of component control with criteria for selection, specification and quality assurance Procure components to definitive specifications
Effects of storage/dormant periods and shelf-life	Establish methods to determine system reliability degradation during idle periods Establish requirements for cyclic/periodic inspection
Maintenance planning (RCM)	Establish a complete maintenance plan for preventive maintenance, overhaul and corrective repair
Spare parts/logistics analysis	Conduct logistics analyses and establish support requirements for the life of equipment

TABLE 3.1—*contd.*

Reliability program element	Requirements
Reliability degradation control	
Controlled screening and burn-in	Design a cost-effective screening and burn-in test program
	Establish screening and burn-in methods
Reliability acceptance	Conduct reliability acceptance testing to determine individual item compliance with reliability criteria
	Define criteria for conduct of tests which account for
	—environmental stresses
	—accept/reject criteria
	—operational verification
	—integrated testing
Degradation control	Establish techniques by which flaws entering through manufacturing processes are controlled and eliminated
	Establish tests to detect latent defects
Operating reliability assurance	
Cyclic inspection	Establish inspection program to identify and eliminate dormancy induced defects
R & M assessment	Establish procedures for periodic R & M assessment which:
	—uncover problems areas and provide bases for product improvement
	—determine effectiveness of R & M programs applied during development
	—track the R & M performance of the operational system
PRA reliability analysis	Establish reliability techniques using PRA to identify critical items
Reliability (or availability) improvement	Establish procedures for reliability improvement through a closed loop process based on data analysis and assessments
Experience feedback	
Failure reporting analysis and corrective action (FRACA)	Establish and implement a closed loop system for reporting, analyzing and providing remedial action for all failures which impact safety and related systems
Data recording and feedback	Establish and implement a program to collect R & M data on operating systems and equipment
	Use data to perform assessment on operating systems

TABLE 3.2

Characteristics of Nuclear Power versus Military/Space Systems and Equipment

Characteristic	Military/space	Nuclear
Size/weight	Small/miniature	Large
Quantities	≥ 100	1–100
System cost	Moderate–high	Very high
Operating environment	Protected	Severe
Mission duration	Minutes/hours	Years
Operational objective	Mission success	Long-term safety
Reliability	Mission oriented	Long-term availability
Maintenance practices	Preventative, corrective	Preventative, corrective
Control mechanism	Contractual	Regulatory

although modification and refurbishment often extend their useful life. Military systems are valued for the operational capability or political posture which attend their employment. Hence, for these systems, the profit motive is not a driving force.

Consider also the conditions under which the respective equipment or systems must operate. Power plant systems operate in fixed ground installations subjected to a severe operating environment. This environment and the stresses to which plant systems are exposed are generated within the plant itself during normal operation and are continuously applied. In addition, nuclear systems must be prepared at all times to respond to even more severe environments and stresses imposed by accident conditions. Military systems and equipment are frequently mobile (i.e. airborne, shipboard, etc.), subjected to atmospheric and operational stresses of wide and often cyclic variation and are frequently operated and maintained by minimally trained personnel. Table 3.2 compares the characteristics of nuclear power plants and military/space systems and equipment.

It must be emphasized that a reliability engineering program, regardless of how well-conceived and proven, will not achieve the desired objectives without strong management support and technical/operating staff enthusiasm. Merely developing a reliability program will not by itself enhance the reliability of equipment in nuclear power plants. Within the nuclear power industry (particularly the utilities/licensees, but also the architect–engineer firms, the nuclear steam supply system vendors and the manufacturers and suppliers) there must be established, within each firm, a specific reliability organization and a reliability program complemented by well-trained reliability engineers who are given sufficient authority to influence design and operation. Aggressive management support in terms of manpower and authority is the key to successful operation of a reliability organization and implementation of an effective program in the nuclear industry. Management must take the initiative to ensure that the reliability program is highly responsive to specific plant safety needs and regulatory requirements, that it is staffed by adequate numbers of well-qualified personnel and that it is conducted as an integral part of the system development, construction, operation and maintenance process. It is the successful implementation of this program which will contribute most to improved safety and operating efficiency in the nuclear industry.

There are many factors that affect reliability program planning. These involve: system function and criticality; development, operation and maintenance life cycle considerations; organizational elements; and funding constraints. The purpose of a reliability program is to assure that new systems comply with safety requirements, achieve plant availability goals and meet regulatory provisions.

Table 3.3 presents a matrix that depicts the program elements keyed to their respective life cycle phase and shows requirements for planning and implementation corresponding to various levels of system/equipment criticality. The program elements are grouped into functional areas of management and control, design assurance, component availability, degradation control, operating assurance and experience feedback. When viewed as an entity, these six areas encompass a complete set of reliability program controls and activities applicable throughout the life cycle phases.

The levels of system/equipment criticality used are:

Critical (CR)—Systems/equipment in which the occurrence of any failure condition or design error would result in a hazard, would prevent the continued safe operation of the plant or would reduce the capability of the operators to cope with adverse conditions.

Essential (E)—Systems/equipment in which the occurrence of any failure condition or design error would reduce the capability (operational ability), but not the safety of the power plant.

Non-essential (NE)—Systems/equipment in which failures or design errors would not degrade plant safety nor significantly reduce operational ability.

This matrix provides a basis for planning and formulating the makeup and provisions of a complete and effective nuclear power plant system reliability program. For example, for systems/equipment in the 'critical' category it may be required that a full complement of reliability program elements be planned and implemented. For systems/equipment in the essential category, the requirements for some of the elements may be eliminated (or relaxed). Equipment/systems in the non-essential category may require only a limited amount of reliability activity. Each individual system reliability program must be structured and tailored to coincide with its specific criticality category and requirements and meet its specified objectives.

Typically, R & M program approaches can be broadly classified according to where the strongest emphasis is placed, such as:

—test, analyze and fix
—statistical analysis
—combination of above

The statistical approach attempts to control and upgrade reliability through analysis of statistical data derived from predictions, assessments, testing and field operations. This latter approach requires statistically meaningful data and, therefore, is most applicable in situations involving relatively inexpensive items produced and deployed in substantial quantities.

On the other hand, the 'test, analyze and fix' approach places major emphasis on stress and environmental testing of a small number (or one) of equipments or systems to identify design and fabrication deficiencies. Subsequent failure and application analyses isolate the root cause of the problem and provide the necessary input for corrective actions. This approach is suitable for low volume, high cost items.

The combination approach employs both of the above in a relatively balanced manner. A thorough, well-organized reliability program typically employs both approaches although, depending upon the system character-istics and application, may still emphasize one over the other in actual implementation.

Table 3.3 includes all elements which comprise a total reliability program and should serve as a conceptual base for adaptation to individual equipment used in nuclear power plants. The reliability plans and implementation procedures that are developed for nuclear systems from this table will most likely favor the 'test, analyze and fix' philosophy. Also, within the US nuclear industry, almost all plant systems (except for cases of backfit), have already been designed, most have already been constructed and many have already been operating for some time. Consequently, the reliability requirements will focus on operation, maintenance and, in the case of backfit, configuration control.

The outputs of each of the program tasks can be defined in terms of an identifiable document or data item. This documentation is a necessary aspect of completely specifying and planning a reliability program and provides evidence that the tasks were adequately implemented. The purpose of the documentation is to disseminate information, record data, document decisions (and the underlying logic) and report program status. By itself the documentation cannot assure system reliability.

Descriptions of standard US military documentation requirements called DIDs (data item descriptions) are available and can be initially used as a basis to completely specify the reliability program. The DoD Acquisition Management Systems and Data Requirements Control List (AMSDL) contains a comprehensive list of DIDs for use by acquisition managers on DoD procurements. These DIDs are identified in the documentation column of Table 3.3. Examples of these DIDs are given in Table 3.4. The specific DID chosen varies with the rigor and scope of the reliability effort applicable during the hardware acquisition process. These military DIDs can be either directly applied or modified to meet specific system requirements and to be compatible with tasks defined in a program plan.

A summary of the characteristics of each of the elements that could comprise a reliability program for nuclear power plant systems is described in subsections 3.2–3.7. These subsections correspond to the six functional areas identified in Table 3.3.

3.2 RELIABILITY MANAGEMENT

Reliability management includes these seven major elements:

1. Program planning
2. Organization
3. Program review
4. Configuration control and management
5. Supplier audit and surveillance
6. Reliability critical item list (and control)
7. Reliability training and indoctrination

While the management aspects of a complete life cycle reliability program are described in terms of the elements listed in Table 3.3, it must be recognized that management pervades each task and control element in all areas of the overall program. As stand-alone items they are of limited value. Each element must be defined relative to its impact on system reliability and ultimately, plant safety. Furthermore, many of the elements provide

TABLE 3.3
Nuclear Systems Reliability Program Requirement Matrix (Industry Requirements)

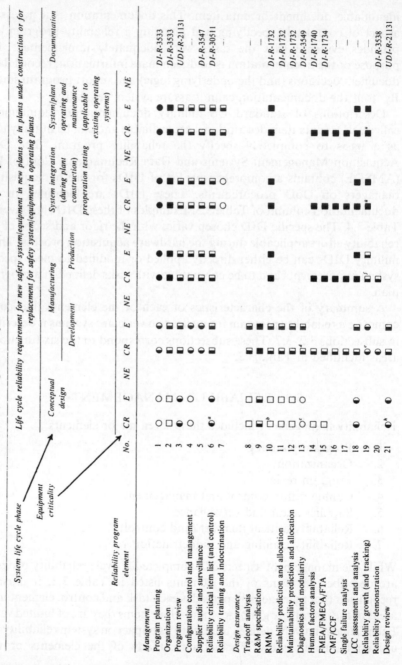

Reliability program element	No.	Conceptual design			Development			Production			System integration (during plant construction) preoperation testing			System/plant operating and maintenance (applicable to existing operating systems)			Documentation
		CR	E	NE	CR	E	NE	CR	E	NE	CR	E	NE	CR	E	NE	
Management																	
Program planning	1																DI-R-3533
Organization	2																DI-R-3533
Program review	3																UDI-R-21131
Configuration control and management	4																—
Supplier audit and surveillance	5																DI-R-3547
Reliability critical item list (and control)	6																DI-R-30511
Reliability training and indoctrination	7																—
Design assurance																	
Tradeoff analysis	8																—
R&M specification	9																DI-R-1732
RMM	10																DI-R-1732
Reliability prediction and allocation	11																DI-R-1732
Maintainability prediction and allocation	12																DI-R-3549
Diagnostics and modularity	13																DI-R-1740
Human factors analysis	14																DI-R-1734
FMEA/FMECA/FTA	15																—
CMF/CCF	16																—
Single failure analysis	17																—
LCC assessment and analysis	18																—
Reliability growth (and tracking)	19																—
Reliability demonstration	20																DI-R-3538
Design review	21																UDI-R-21131

Column headings above the matrix:

System life cycle phase

Equipment criticality

Life cycle reliability requirement for new safety system/equipment in new plants or in plants under construction or for replacement for safety system/equipment in operating plants

Column identifiers: DI-R-1724, DI-R-1732, UDI-R-21135, DI-R-7040, DI-R-1710, DI-R-3535, DI-R-3537A

Component availability
22 Component derating
23 Component quality level
24 Component failure rate modeling and estimation
25 Component testing methods and compliance
26 Component selection, specification and quality (including QPL & STD)
27 Effects of storage/dormant periods and shelf life
28 Maintenance planning (RCM)
29 Spare parts/logistics analysis

Reliability degradation control
30 Controlled screening and burn-in
31 Reliability acceptance
32 Degradation control

Operational reliability assurance
33 Cyclic inspection
34 R&M assessment
35 PRA/reliability analysis
36 Reliability (or availability) improvement

Experimental feedback
37 FRACA
38 Data recording and feedback

Planning: (O) preliminary; (◐) detailed; (●) revise/update. *Implementation*: (□) preliminary (gross); (■) detailed (full); (▉) revise/update.

ᵃ Define critical item criteria, data source and controls.
ᵇ Model(s) developed to accommodate historical data for predictions and actual data for production-operation practical assessments.
ᶜ Determine extent of built-in diagnostic/modularity.
ᵈ Specify diagnostic/modularity level.
ᵉ Develop growth test/tracking model.
ᶠ Assess reliability growth based on development test data.
ᵍ Revise assessment to reflect production/operation experience data.
ʰ Preparation of design review procedure including check list.
ⁱ Prepare component derating guidelines.
ʲ Determine component derating.
ᵏ Procure derated components.
ˡ Prepare component quality guidelines.
ᵐ Determine component quality levels.
ⁿ Procure quality components.
ᵒ Identify effects.
ᵖ Determine effect on component.
ᵠ Apply to dormant equipment.
ʳ Factory acceptance.
ˢ In-plant acceptance.

TABLE 3.4(a)
Data Item Description Example (1)

DATA ITEM DESCRIPTION	2. IDENTIFICATION NUMBER(S)	
	AGENCY	NUMBER
1. TITLE	USAF	DI-R-3533/R-101-2
Reliability/maintainability program plan (example number 1)		
3. DESCRIPTION/PURPOSE	4. APPROVAL DATE	
This plan is used by the procuring activity: (1) to evaluate the contractor's planning for his reliability/maintainability program; (2) to review and approve the contractor's planned reliability/maintainability program; and (3) to monitor and evaluate the contractor's conduct of his reliability/maintainability program.	5. OFFICE OF PRIMARY RESPONSIBILITY AFSC	
	6. DOC REQUIRED	
	8. APPROVAL LIMITATION	
7. APPLICATION/INTERRELATIONSHIP	9. REFERENCES (*mandatory as cited in block 10*) US MIL-STD-470 US MIL-STD-471 US MIL-STD-721 US MIL-STD-785	
This data item description is applicable to systems development contracts during contract validation, full scale development and production phases and equipment development/procurement contracts when contractors are required to conduct a reliability/maintainability program. This plan may be obtained either precontractually, during the RFP/IFP phase, or as a product of the validation phase. When a prior reliability program plan has been proposed and approved by the procuring activity, the specific requirements of this data item description may be satisfied by supplementing the prior plan.		
	MCSL NUMBER(S)	
10. PREPARATION INSTRUCTIONS		
(1) The reliability/maintainability program plan contains the plans for the accomplishment of each reliability/maintainability program task specified by the contract. The plan will provide a cross-index, in accordance with the following outline, which shows the relationships between program tasks and (1) applicable specifications or standards cited by the contract work statement; (2) other reference documents; and (3) contractor policies and standards:		

Format for program plan cross-index

Reference paragraph of US MIL-STD-785 and/or US MIL-STD-470	Applicable task or paragraph no. of program plan	Other reference documents	Company policies, procedures and controls	Estimated manloading for first _____ months

(II) The plan shall also identify and define the following as a minimum:

(A) The work to be accomplished for each applicable task delineated in *US MIL-STD-785* and *US MIL-STD-470.*

(B) The time phasing and manloading involved.

(C) The contractor organizational element assigned responsibility and authority for implementing the reliability/maintainability program.

(D) Lines of communication between the contractor organizational element responsible for implementing the program and other contractor interfacing organizational elements.

(E) Appropriate customer–contractor program milestone review points.

(F) Method of control over subcontractor and vendor reliability/maintainability programs.

(G) The purpose and expected results of each task and the planned methods for monitoring, assessing, reporting and taking appropriate action regarding the status, accomplishments and problems.

(H) Specific techniques for allocating quantitative requirements to lower level functional elements of the system (subsystem, assembly or components).

(I) Specific techniques for making reliability/maintainability predictions.

(J) Proposed methods for demonstrating the achievement for quantitative reliability/maintainability requirements.

(III) The plan shall identify and define interfaces between the reliability/maintainability program and the following closely related programs or elements:

(A) Logistics support evaluations.

(B) Personnel subsystem program.

(C) System engineering.

(D) Systems/cost effectiveness analysis.

(E) System life cycle cost analysis.

(F) Design engineering.

(G) Value engineering.

(H) Data collection and analysis procedures.

TABLE 3.4(b)
Data Item Description Example (2)

DATA ITEM DESCRIPTION	2. IDENTIFICATION NUMBER(S)	
	AGENCY	NUMBER

1. TITLE

Reliability and maintainability allocations, assessments, and analysis report (example number 2)

USAF	DI-R-3535/R-103-2

4. APPROVAL DATE

3. DESCRIPTION/PURPOSE

This report is used: (1) to evaluate the contractor's estimate of reliability and maintainability (the predicted growth, allocation and degree of achievement of these characteristics in the configuration item and its constituent elements); (2) to evaluate the current end potential reliability and maintainability of the configuration item design; (3) to provide information to assist in directing and planning for reliability and maintainability and related program efforts; and (4) to identify design features which are critical to reliability and maintainability.

5. OFFICE OF PRIMARY RESPONSIBILITY

AFSC

6. DOC REQUIRED

8. APPROVAL LIMITATION

7. APPLICATION/INTERRELATIONSHIP

This data item description is applicable to system development during the contract validation and full scale development phases and equipment development contracts for complex equipments through the end of category II tests. It may be applied to appropriate conceptual phase studies, exploratory and advanced equipment developments. It may also be used to define information to be submitted in response to a request for proposal.

9. REFERENCES (*mandatory as cited in block 10*)

US MIL-STD-470
US MIL-STD-499
US MIL-STD-721
US MIL-STD-756
US MIL-STD-785
US MIL-HDBK-217
US MIL-HDBK-472

MCSL NUMBER(S)

10. PREPARATION INSTRUCTIONS

(I) This report shall contain, as a minimum, the following information:

(A) Contractor's analysis of reliability and maintainability potential of the configuration item design, including mathematical models, logic diagrams, functional block diagrams, assumed operating conditions, environmental criteria and other considerations used in the calculations (i.e. combining data and confidence limits when multiple tests are conducted).

(B) Equipment breakdown to the lowest practical level of indenture with associated reliability and maintainability parameters.

(C) Analysis of potential modes of failures; their probable cause and effects on performance, reliability and maintainability. The severity of these effects and the probability of occurrence under applicable operating modes and environments shall be indicated. Definitions of failure must include those expected to be used by maintenance personnel and operators.

(D) Description of the purpose and function of applicable items.

(E) A description of trade studies involving reliability, maintainability and other factors and the resulting effects on overall system effectiveness. Trade studies shall be made available at the request of the procuring activity to substantiate/expand results.

(F) Effects of storage, shelf-life, packaging, transportation, handling and maintenance on the product reliability. Major or critical characteristics of items which deteriorate with age should be included, plus environmental limits, maintenance philosophy, equipment usage, etc.

(G) The contractor's conclusions, identification of problem areas, related actions taken or proposed and a list of further design studies planned as a result of these analyses.

(H) The contractor's allocations of the overall quantitative goals and minimum requirements for configuration item reliability and maintainability as specified by the procuring activity or developed by the contractor. As a general rule, this breakdown should be carried to the level at which failure reports will be submitted.

(I) Current observed achievement of reliability or maintainability of the configuration item and its constituent elements to the lowest practical level of indenture. In each case, the type and units of measurement shall be clearly identified (e.g. the distributions of TBF, CBF, TTR active time, TTR manhours, availability, probability of satisfactory performance, percent successful, etc.). Confidence levels or intervals shall be stated where appropriate. Achieved and predicted reliability growth curves shall be included. A comparison with the analysis and allocation for the configuration item shall be included.

TBF, Time between failures; CBF, cycles between failures; TTR, times of repair.

TABLE 3.4(c)

Data Item Description Example (3)

DATA ITEM DESCRIPTION	2. IDENTIFICATION NUMBER(S)	
	AGENCY	NUMBER
1. TITLE Reliability/maintainability demonstration plan (example number 3)	USAF	DI-R-3538/R-106-2
3. DESCRIPTION/PURPOSE To review, inspect and evaluate the contractor's reliability/maintainability test and demonstration program for selected items.	4. APPROVAL DATE	
	5. OFFICE OF PRIMARY RESPONSIBILITY AFSC	
	6. DOC REQUIRED	
	8. APPROVAL LIMITATION	

7. APPLICATION/INTERRELATIONSHIP
A reliability/maintainability demonstration plan is applicable to contracts where a reliability/maintainability demonstration is required. This plan is not required if a specific test plan or procedure is included in equipment specifications, other contractual documents, or if the test and demonstration program is adequately detailed in the reliability/maintainability program plan (DI-R-3533). Emphasis shall be placed on the integration of the reliability and maintainability demonstrations with the overall test plan.
	9. REFERENCES (*mandatory as cited in block 10*)
	   US MIL-STD-470
	   US MIL-STD-471
	   US MIL-STD-721
	   US MIL-STD-757
	   US MIL-STD-781
	   US MIL-STD-785	

	MCSL NUMBER(S)	

10. PREPARATION INSTRUCTIONS

(I) The reliability/maintainability demonstration plan shall contain the following information as a minimum:

(A) A statement of demonstration objectives including type, unit of measure, and the qualitative/quantitative requirements to be demonstrated.

(B) An identification of all other item tests from which useful reliability/maintainability data can be obtained. This will minimize demonstration requirements.

(C) A description of the demonstration conditions including (as applicable):

   (i) The types and quantities of equipment to be used including:

     (a) support equipment and calibration support requirements;

     (b) tools and test equipment;

     (c) technical publications;

     (d) spares and consumables;

     (e) safety equipment.

   (ii) The data expected from each test along with recording methodology and definition of data elements to be collected.

   (iii) The criteria for classifying test results as success or failure. Definitions of failure must relate to expected symptoms which will be observed by operators and maintenance personnel.

   (iv) The plan of action to be used when test failures occur.

   (v) The principal environmental stress conditions, operating modes, operating time and cycling conditions to be imposed. The relationship of these conditions to eventual use conditions (maximum and minimum) must be specified.

   (vi) A description of the test facilities and instrumentation requirements, including location, insulation, shock mounting, stress mounting and other mechanical factors affecting environmental test levels.

   (vii) The method to be used to report test results.

   (viii) Demonstration constraints, such as personnel skill level, test equipment and their relationship to the eventual use of the items.

   (ix) The maintenance concept.

   (x) The demonstration environment and sites.

   (xi) The levels of maintenance to be demonstrated.

   (xii) The mode of operation during the demonstration considering configuration and mission requirements.

   (xiii) A list of the items to be demonstrated.

   (xiv) The participating agencies including:

     (a) organization;

     (b) degree of participation by each in terms of managerial, technical, maintenance and operating personnel;

     (c) assignment of specific responsibilities;

     (d) qualifications, quantity, sources, training and indoctrination requirements needed for the personnel doing the demonstration (tie in with item C(viii) above).

   (xv) Provisions for a predemonstration phase to prepare facilities, personnel and equipment for the formal demonstration.

   (xvi) Analytical methods and calculation procedures to be used to analyze demonstration data.

   (xvii) A description and schedule of preventive maintenance.

   (xviii) A listing of maintenance tasks to be demonstrated.

(D) A provisional schedule for special or repeat demonstrations to resolve deficiencies and demonstration failures.

output data and information that are necessary input for subsequent reliability tasks. Management is the thread holding each element in its proper perspective. The scope and application-centered details of each element change with each phase within the life cycle.

Current concern centers around systems which impact safety in operating nuclear power plants, i.e. systems which have been operating for 10–20 years and are well along in their life cycle. Thus, structuring a program to meet these needs is of prime concern. Tasks must be planned and implemented to reduce or mitigate aging failure mechanisms after long periods of plant operation, in addition to the reliability tasks applied to the life cycle phases which occur prior to the operational phase. A nuclear power plant reliability program must be structured and organized in recognition of the fact that the operational reliability of a system is a function of its design as well as subsequent life cycle activities in as much as:

(i)   design establishes the 'inherent' reliability potential of a system and is defined by its engineering documentation; and

(ii)   subsequent life cycle activities can only degrade reliability below this inherent design level.

Consequently, assessment and control of operational reliability require deliberate and positive engineering action to be taken throughout a system's life cycle. Reliability must be approached first via design (to establish an upper limit) and then in conjunction with a series of engineering analyses, via tests and controls to eliminate or minimize production and maintenance degradation as well as operational aging.

Within a framework that is consistent with, and meets overall regulatory requirements (e.g. NRC), utility companies provide the overall management, planning and control functions for systems/equipment acquisition, plant integration and operation. During the conceptual phase management tasks are formulated by the utilities/licensees for application during subsequent phases by the architect–engineer firms, the nuclear steam supply system vendors, the manufacturers and suppliers as well as by the nuclear power plants. Also, during the conceptual phase, the feasibility of meeting reliability requirements at reasonable risk with current engineering technology is evaluated. Tradeoff analyses among reliability, system performance, maintenance concepts, development time and life cycle costs are performed to establish an optimum design approach that meets the safety/reliability objectives for the system. In addition, reliability specifications are prepared covering quantitative as well as qualitative design, program, testing and reporting requirements.

An essential aspect of the conceptual phase is the preparation of a complete reliability engineering plan that delineates the sequence of reliability tasks and activities applicable to design, development and production, system integration and plant operation. The plan provides the total life cycle reliability perspective and provides the technical framework for the overall reliability engineering and control program. It reflects changing managerial requirements during subsequent phases and is prepared in a format to accommodate revisions, if and when required.

During development and production, emphasis shifts to the manufacturers, with the utilities acting as monitors to assure compliance to overall reliability objectives. An effective reliability program requires that the manufacturers prepare, during this phase, a comprehensive and detailed program plan that meets the requirements specified in procurement documentation as well as the general engineering and control requirements defined in the overall life cycle reliability system engineering plan prepared during the preceding conceptual phase. The manufacturer's plans identify and tie together all program management and engineering tasks required to accomplish specified reliability design and program requirements applicable to development, production and system/plant integration.

During system/plant integration, emphasis shifts back to the utilities to establish control functions that prevent or minimize reliability degradation resulting from installation and initial operation and test of new (and often unfamiliar) systems and equipment. Reliability and quality controls are applied during this phase to identify and eliminate developing problems prior to on-line operation.

During plant operation reliability is assured by cyclic or periodic inspection and surveillance, startup testing, operating training and qualification, failure reporting and correction action, data recording and feedback and reliability improvement. Scheduled maintenance tasks are performed on critical equipment and components at regular intervals during plant operation to prevent a decrease in reliability or deterioration of safety, or to reduce total life cycle cost. The life cycle reliability plan is updated to reflect operational considerations.

A summary description of each of the reliability management program elements (element numbers 1–7) is presented in Sections 3.2.1–3.2.7.

3.2.1 Element Number 1: Reliability Program Planning
Description
This is the process of developing formal plans (with quantitative goals) for ensuring an 'acceptable' system and component reliability consistent with

plant safety and availability requirements. Complete and detailed program plans are prepared to delineate specific program element responsibilities, interactions and to define a schedule of outputs (and deliverables) relative to system design, development and subsequent life cycle stages. These plans are used to manage and control reliability activities. They serve as a 'road map' for developing reliable systems and components by providing an orderly and systematic approach to reliability engineering. Effective planning is a highly practical management control element that can be accomplished at low cost. Table 3.3 and the descriptions presented in this section can be used as guides in selecting and describing the set of elements that would comprise complete and effective system program plans.

Basic Requirements

Establish and monitor a reliability program planning function and prepare complete life cycle program plans aimed at applying practical proven reliability techniques and control elements during the development and operation of safety critical equipment as well as non-safety critical equipment. These plans are the key management tools for the control and integration of all life cycle reliability activities and requirements. They are prepared and initiated by the utility/licensee during the conceptual phase, further detailed during the manufacturing phase and continually revised and updated as necessary during system/plant integration and plant operation. They ensure actual reliability levels consistent with plant safety and availability objectives during system development or procurement and plant operation. They integrate outputs from all participants and reflect the changing character of individual reliability elements listed in Table 3.3 as emphasis shifts according to system life cycle.

Many of the elements listed in Table 3.3 have been adapted from and can be correlated to *US MIL-STD-785B* which provides guidance in the preparation of reliability program plans for systems and equipment during development and production. It delineates general requirements and defines optional tasks for reliability programs and provides guidelines for the preparation and implementation of a program plan. The procedure for implementing or specifying use of *US MIL-STD-785B* is discussed in paragraphs 1.2 and 4.0 of that standard. It stresses that it is not sufficient to merely list *US MIL-STD-785B* as a reference document 'which forms part of [the system specification] to the extent specified herein' without specific detailed direction as to which paragraphs and sections are applicable. Paragraph 1.2.1 of that standard is quoted here for convenience:

'Tasks described in this standard are to be selectively applied to DoD contract-definitized procurements, requests for proposals, statements of work, and Government in-house developments, requiring reliability programs for the development, production and initial deployment of systems and equipment.'

Table 3.5 lists the elements of a standard hardware reliability program per *US MIL-STD-785B* and shows the importance of each element during the life cycle phases of development and production. However, it must be emphasized that the application of *US MIL-STD-785B* provisions are subject to the discretion of the user activity. The chart given in Table 3.5 is designed to provide the reliability manager with an overview or feeling for the average situation. Since each development program is different, the reliability program is tailored to its specific needs. This tailoring is done by reliability specialists working with the program manager to select *US MIL-STD-785B* tasks and requirements that are most suitable to the specific acquisition and to modify these tasks and requirements where necessary to assure that each tailored task or requirement involved states only the minimum needs of the program. Appendix A of *US MIL-STD-785B* provides tailoring guidance.

Documentation Requirements
Prepare total life cycle system reliability control plans in a form which can be readily modified in recognition of program element emphasis particular to the current phase(s) of system acquisition and plant operation. The planning should contain full and complete descriptions for accomplishing the elements and tasks applicable to each life cycle phase with cross-indexing to show the relationships between the elements and the specifications and standards including development/production plans. Each program element should be identified to the responsible utility, plant, manufacturers and supplier organizational function.

In addition prepare specific program plans covering the design, development and production of new equipment using *US MIL-STD-785B* or other standardized requirements for guidance. These plans can be used by the utility to effect initial review and approval of the manufacturer's reliability program and as the basis for monitoring and evaluating the conduct of the program. They are prepared during the early design phase and updated, as necessary, during the course of the system development and production program. They must meet the requirements of the overall life cycle plan and incorporate any changes to the original planning from

TABLE 3.5

US MIL-STD-785B Application Matrix (taken from Appendix A of US MIL-STD-785B)

Task	Title	Task type	Program phase			
			Concept	Valid	FSED	Prod
101	Reliability program plan	MGT	S	S	G	G
102	Monitor/control of subcontractors and suppliers	MGT	S	S	G	G
103	Program reviews	MGT	S	S(2)	G(2)	G(2)
104	Failure reporting, analysis and corrective action system (FRACAS)	ENG	NA	S	G	G
105	Failure review board (FRB)	MGT	NA	S(2)	G	G
201	Reliability modeling	ENG	S	S(2)	G(2)	GC(2)
202	Reliability allocations	ACC	S	G	G	GC
203	Reliability predictions	ACC	S	S(2)	G(2)	GC(2)
204	Failure modes, effects and criticality analysis (FMECA)	ENG	S	S(1)(2)	G(1)(2)	GC(1)(2)
205	Sneak circuit analysis (SCA)	ENG	NA	NA	G(1)	GC(1)
206	Electronic parts/circuits tolerance analysis	ENG	NA	NA	G	GC
207	Parts program	ENG	S	S(2)	G(2)	G(2)
208	Reliability critical items	MGT	S(1)	S(1)	G	G
209	Effects of functional testing, storage, handling, packaging, transportation and maintenance	ENG	NA	S(1)	G	GC
301	Environmental stress screening (ESS)	ENG	NA	S	G	G
302	Reliability development/growth testing	ENG	NA	S(2)	G(2)	NA
303	Reliability qualification test (RQT) program	ACC	NA	S(2)	G(2)	G(2)
304	Production reliability acceptance test (PRAT) program	ACC	NA	NA	S	G(2)

Code definitions—Task type: ACC, reliability accounting; ENG, reliability engineering; MGT, management. *Program phase:* S, selectively applicable; G, generally applicable; GC, generally applicable to design changes only; NA, not applicable. (1) Requires considerable interpretation of intent to be cost-effective; (2) *MIL-STD-785* is not the primary implementation requirement. Other MIL-STDs, or statement of work requirements must be included to define the requirements.

work accomplished during previous phases of the program. *US MIL-STD-785B* (Task 101) requires that a system reliability plan include the following information:

(a) A description of how the reliability program will be conducted to meet the specified requirements.

(b) A detailed description of how each specified reliability accounting and engineering design task(s) will be performed or complied with.

(c) The procedures (wherever existing procedures are applicable) to evaluate the status and control of each task and identification of the organizational unit with the authority and responsibility for executing each task.

(d) A description of interrelationships of reliability tasks and activities and a description of how reliability tasks will interface with other system oriented tasks.

(e) A schedule with estimated start and completion points for each reliability program activity or task.

(f) The identification of known reliability problems to be solved, assessment of the impact of these problems on meeting specified requirements, and the proposed solutions, or approach, to solve these problems.

(g) The procedures or methods (if procedures do not exist) for recording the status of actions taken to resolve problems.

(h) The designation of reliability milestones (includes design and test).

(i) The method by which the reliability requirements are disseminated to designers and associated personnel.

(j) Identification of key personnel for managing the reliability program.

(k) Description of the management structure, including interrelationships between line, service, staff and policy organizations.

(l) Statement of sources of reliability design guidelines and whether a reliability design review checklist will be utilized.

(m) Description of how reliability contributes to the total design and the level of authority and constraints on this engineering discipline.

(n) Identification of inputs that the contractor needs from operation and support experience with a predecessor item(s). Inputs should include measured basic reliability and operational reliability values, measured environmental stresses, typical failure modes, and critical failure modes.

Applicable Specifications and Texts
AEC, RDT-F2-9T Standard: Reliability Assurance.
British Standard, BS5760: Reliability of Systems, Equipment and Components.
NASA, NHB 5300.4: Reliability Program Provisions for Aeronautical and Space System Contractors.
RADC-TR-79-200: Reliability and Maintainability Management Manual.
US Code of Federal Regulations, Ch. 10, Part B, Appendix B: Quality Assurance.
US MIL-STD-470: Maintainability Program Requirements (for Systems and Equipment).
US MIL-STD-785: Reliability Program for Systems and Equipment Development and Production.

3.2.2 Element Number 2: Organization
Description
The reliability organization is responsible for planning, preparing, as well as implementing, all reliability program plans. The organization plans reliability activities, budgets manpower and resources and directs or coordinates staff specialist effort to carry out specific program elements and tasks. An effective nuclear plant reliability organization would generally maintain a core group of specialists whose backgrounds encompass reliability, maintainability, statistics, physics of failure, data analysis and other technological areas within the reliability discipline. Their exact number, area of expertise and skill level varies with short- and long-term goals and, as situations demand, changes in program emphasis. The group would include specialists, with backgrounds and responsibilities covering reliability management, analysis and control, who are located at the principal utility operation as well as other specialists with backgrounds and responsibilities covering data compilation, failure analysis, plant surveillance and inspection who are located at each operational power plant. This core group can be supplemented by other disciplines, available within the utility operations, such as nuclear engineering, mechanical engineering, electrical engineering, industrial engineering, systems engineering and operations research, as required. Also, at various times during the acquisition/operation of a nuclear power plant, additional outside specialists are used for specific tasks. The reliability organization and its management are reviewed periodically to assure that they provide optimum control of the program, utilize its resources effectively and provide appropriate information to higher management.

Basic Requirements

Establish an organization that combines utility and power plant reliability functions into a centralized coherent program with its own control and management. When established and staffed, the reliability organization functions as follows:

1. It provides detailed structuring of its own organization to implement the program plan, including control of its scope and procedures, as well as those organizations governing manufacturers and suppliers by means of program directives, meetings and other techniques.
2. It budgets manpower and resources and schedules program outputs in accordance with the anticipated best utilization within each of the various life cycle phases.
3. It continuously monitors reliability program costs and effectiveness to ensure that the program is providing maximum support in proportion to the resources available.
4. It plans and directs efforts of internal staff and engineering specialists needed to carry out specific reliability program tasks, i.e. failure analysis, design improvements, etc.

Documentation Requirements

Prepare a chart showing the structure of the reliability organization, names of personnel, functional areas of expertise or responsibility and interfaces with other organization functions such as quality assurance, test laboratories and maintenance. Revise the chart as required to reflect current organization. Prepare procedures and mission/functional descriptions for the overall reliability organization as well as each organizational element.

Applicable Specifications and Texts

British Standard, BS5760: Reliability of Systems Equipment and Components.

NASA, NHB 5300.4: Reliability Program Provisions for Aeronautical and Space System Contractors.

RADC-TR-79-200: Reliability and Maintainability Management Manual.

US MIL-STD-470, Maintainability Program Requirements (for Systems and Equipment).

US MIL-STD-785: Reliability Program for Systems and Equipment Development and Production.

3.2.3 Element Number 3: Program Review
Description
This is the process of conducting formal reviews at predefined system life cycle milestones to assure that the program is proceeding as budgeted and scheduled and that established goals are achieved.

Basic Requirements
Conduct reviews as a self-monitoring function to assure the adequacy and effectiveness of the program plans. Program reviews address the following areas:

1.   Reliability task effectiveness.
2.   Items already reviewed at system manufacturer's site.
3.   Parts, design, reliability and schedule problems.
4.   Results of failure analyses.
5.   Status of assigned action items.
6.   Test schedule: start dates and completion dates.
7.   Other topics and issues as deemed appropriate.

Documentation Requirements
Document the review guidelines and checklists. For each review conducted, document the results in one-to-one correspondence with each item on the list. Distribute the results to key reliability and higher management personnel for appropriate action.

3.2.4 Element Number 4: Configuration Control and Management
Description
This is the discipline of applying technical and administrative direction and surveillance to identify and document the functional and physical characteristics of an item, control changes to those characteristics and record and report change processing and implementation status. Configuration control is applied to all hardware/software changes which occur during a system life cycle to ensure that all changes are included in the technical documentation, e.g. specifications, drawings.

Basic Requirements
Establish a configuration control and management function to identify, control and account for systems design baseline configurations and associated data. This function includes:

1.   Establishment of methods of identification whereby hardware

items, drawings/documentation, software and data can be distinguished from, and related to, other like items.

2. Establishment of a formal configuration and change control procedure to ensure that all proposed changes to approved baseline design configurations, specifications, drawings, parts lists, test plans, etc., are reviewed and evaluated for their effect on functional reliability, maintainability, and other specified characteristics.

3. Definition of a technique by which hardware, documentation, software and data can be readily located, the revision status verified and configuration audits readily accomplished.

Documentation Requirements
Prepare a configuration management plan which incorporates identification and traceability concepts and which recognizes the types, quantities and fabrication of equipment found in nuclear power plants. The plan should also incorporate management provisions and personnel requirements to perform the configuration management function and to ensure its timely implementation.

Applicable Specifications and Texts
US MIL-STD-480: Configuration Control—Engineering Changes, Deviations and Waivers.
US MIL-STD-483: Configuration Management Practices for Systems, Equipment, Munitions and Computer Programs.

3.2.5 Element Number 5: Supplier Audit and Surveillance
Description
This is the process of monitoring the performance of the elements of the suppliers' reliability program plans to ensure that they are performed adequately, in a timely manner and that appropriate corrective action is taken.

Basic Requirements
Establish within the basic planning documents controls applicable to systems/equipment manufacturers and major suppliers which include the following considerations:

1. Facilities surveys prior to procurement.
2. Research into manufacturers/suppliers reliability history.
3. Reliability, maintainability and quality program reviews.

4. Evaluation of cost/technical proposals for reliability.
5. Specification of reliability requirements.
6. Specification of manufacturers'/suppliers' reports and data items.
7. In-plant surveillance.
8. Attendance/participation in manufacturers'/suppliers' design reviews.
9. Maintenance of records of manufacturers'/suppliers' performance.

Documentation Requirements
Establish requirements for manufacturers'/suppliers' deliverable documentation and data items which:

1. are a subset of overall reliability program documentation requirements;
2. are phrased in language applicable to manufacturer and supplier;
3. provide specifications for format content and delivery date.

Applicable Specifications and Texts
British Standard, BS5179 (1, 2 & 3): Guide to the Operations and Evaluation of Quality Assurance Systems.
British Standard, BS5760: Reliability of System, Equipment and Components.
NASA NHB 5300.4: Reliability Program Provisions for Aeronautical and Space System Contractors.
O'Connor, P. D. T., *Practical Reliability Engineering*, Heyden, London, 1981.
RADC-TR-79-200: Reliability and Maintainability Management Manual.
US MIL-STD-470: Maintainability Program Requirements.
US MIL-STD-785: Reliability Program for Systems and Equipment Development and Production.

3.2.6 Element Number 6: Reliability Critical Items List (and Control)
Description
A reliability critical items list identifies and ranks those equipments and components that are most critical in meeting reliability or availability goals. This list is continually updated and published at regular intervals; it provides a basis for prioritizing product improvement programs.

Basic Requirements
Establish criteria for identifying and tabulating items where reliability is critical to the safe performance of power plant functions. The criteria

should encompass items which impact plant safety, availability, downtime and which impact maintenance/logistics support costs adversely. Establish criteria from FMEA, FRACA and operational data analyses, tests (which establish reliability and environmental tolerance) and other program elements performed to assess, control and improve reliability.

Documentation Requirements
Prepare a list of reliability-critical items, their methods of control, current status, problem areas and proposed/implemented solutions and publish at regular intervals.

Applicable Specifications and Texts
AEC, RDT-F2-9T Standard: Reliability Assurance, Quality Assurance.
NASA, NHB5300.4(1A): Reliability Program Provisions for Aeronautical and Space System Contractors.
O'Connor, P. D. T., *Practical Reliability Engineering*, Heyden, London, 1981.
RADC-TR-79-200: Reliability and Maintainability Management Manual.
US MIL-STD-785: Reliability Programs for Systems and Equipment Development and Production.
US MIL-STD-1629: Failure Mode and Effects Analysis.

3.2.7 Element Number 7: Reliability Training and Indoctrination
Description
This is to plan and implement formal training of appropriate industry personnel in basic reliability concepts and fundamentals.

Basic Requirements
The training element of the reliability program is one that falls in the category of a management function in that it is separated from the mainstream of reliability planning and control activities. It is directed toward educating appropriate personnel in special techniques, in the technical disciplines involved in performing reliability tasks and educating other appropriate personnel to recognize the potential benefits of the reliability program to system reliability assurance. It is also directed toward sensitizing program personnel on the importance of designing and manufacturing reliability into the hardware. Because of the varying groups of personnel and disciplines within the nuclear industry, this activity must be approached on a fairly broad front, with each facet of the training program specifically tailored for the group involved.

Documentation Requirements

Prepare appropriate material, handbooks or texts to support reliability training and indoctrination. This material should be geared to accomplish a specific training purpose, such as indoctrination or training of operating, maintenance or management personnel. Material should also contain information of specific interest and value to those charged with the responsibility of utilizing and implementing such information and it should be of sufficient detail to allow implementation or action to be taken with minimal confusion, repetition or error and provide methods of determining individual progress.

3.3   DESIGN ASSURANCE ELEMENTS

Design assurance includes 14 elements (element numbers 8–21):

    8.   Tradeoff analysis
    9.   R & M specification
    10.   Reliability math, modeling
    11.   Reliability prediction and allocation
    12.   Maintainability prediction and allocation
    13.   Diagnostics and modularity
    14.   Human factors analysis
    15.   FMEA/FMECA/FTA
    16.   Common mode failure/common cause failure (CMF/CCF) analyses
    17.   Single failure analysis
    18.   LCC assessment and analysis
    19.   Reliability growth (and tracking)
    20.   Reliability demonstration
    21.   Design review

The area of design assurance represents a set of elements whose function is to improve reliability by providing insights into design parameters and operating characteristics primarily by analytical methods and thus to provide a basis for knowledgeable decision-making. These techniques yield a basis for decision-making prior to: (1) selection and specification of critical in-plant safety and non-safety systems; (2) completion of the design and production of systems and equipment; (3) system integration; and (4) operational use. When applied early in the life cycle, they help establish reliability comparisons of proposed design concepts. When applied during development and early production, they help manufacturers in planning, controlling and assessing reliability and in ensuring that hardware produced will meet specified requirements. When applied during system integration and plant operation, they help to effect the critical transitioning from off-line, conditional and/or intermittent operation to on-line, steady state operation without degrading reliability.

Certain design assurance elements exhibit a dual character. When applied prior to, or during, the early stages of development, these elements provide a forecast or prediction of future events, processes or hardware attributes. When applied to systems already built or installed they represent assessments of results achieved relative to reliability requirements.

A key aspect of the design assurance function is the performance of complete life cycle reliability assessments. This assessment encompasses the inherent reliability of the design as well as any degradation resulting from production operation and maintenance. As was discussed in Chapter 2 (see Fig. 2.1), reliability design and engineering, analysis and control are dependent on the following five major elements:

1. Inherent reliability—represents an upper limit of reliability (e.g. MTBF) established by design. It represents the value of design reliability estimated during development, which may correspond to the value specified in procurement documents.

2. Production degradation—represents the system or equipment reliability resulting from initial or early manufacturing processes before manufacturing learning, inspection efficiency and test procedures have been optimized.

3. Reliability growth during production—represents improvements in the manufacturing process resulting from trained production personnel and well-planned and carefully applied inspection and tests which are designed to force out latent defects for correction.

4. Reliability as released for plant operation—represents the level of reliability of a system or equipment as it leaves the factory prior to operation (or storage). Note that this level, although lower than the inherent level established by design, is nevertheless the highest that can be expected from a mature manufacturing process.

5. Reliability degradation during operation (aging)—represents expected degradation as a result of combined aging and maintenance degradation or degradation as a result of non-operating deterioration factors. The extent of degradation is dependent on the material selected and system design as well as the

effectiveness of the maintenance plan, the inspection and control program and the familiarity and experience of operational and maintenance personnel.

Effective reliability design requires careful planning followed by well-executed engineering tasks that start with design and continue through development, manufacturing and plant service use. A reliability program begins during design with the selection of proven, high quality, well-derated, long life components and the use of adequate design margins. Attention is focused on good construction practices and on the application of preproduction tests to identify potential problems. The process continues during development with the use of well-established, systematic, reliability engineering techniques, supported by tests to stimulate reliability growth and to qualify the design to operational stresses.

To assure that the reliability level designed and developed into a system is not degraded, effort is also applied during production and plant operations. Degradation during production becomes evident when considering that:

1. manufacturing operations can introduce defects into hardware that are not ordinarily accounted for by design engineering efforts; and
2. inspection and test procedures normally interweaved in fabrication processes are imperfect and can allow defects to escape which later result in failures during operation.

Degradation is of particular concern for new fabrication processes where manufacturing learning is not yet complete and a high number of process induced defects can be expected which would, unless controlled, result in a high initial failure rate.

The extent of degradation during production is dependent on the effectiveness of the inspection and quality/reliability engineering control program. Program effectiveness can be measured and evaluated to determine the need for process improvement or corrective changes. The accomplishment of the production analysis and, more importantly, how well subsequent corrective measures are designed and implemented will dictate the rate at which reliability either degrades or grows during production.

In addition, degradation occurs during operation as a result of wearout, with aging as a dominant failure mechanism. Aging is a phenomenon of all components and structures in the presence of environmental stresses and outside influences that interact with the hardware in such a manner as to result in a change in its material properties. Generally, the rate of property changes is influenced by the severity of the environmental conditions, and is deterministic below a certain stress level. Fundamentally then, the design of a component or structure requires the judicious selection of materials and material combinations that are known, or thought, to retain their desired characteristics for extended time periods under the anticipated end use environment.

Critical system or component failure or degradation creates an immediate need for corrective action or product improvement to prevent recurrence. These changes are, generally, implemented by means of plant modification and retrofit change orders. An in-service system modification or repair designed to prevent or mitigate failures or errors can, if not properly evaluated and implemented, compromise the original design integrity and result in a potentially greater hazard. This and other functions establish the need for an effective configuration management system (element number 4) complete with change control and system modification implementation status. Lack of precise configuration identification makes it difficult, if not impossible, to implement changes as well as assess the impact of changes and improvements on the system.

A summary description of each of the design assurance program elements is presented in the remaining pages of this section.

3.3.1 Element Number 8: Tradeoff Analysis
Description
A tradeoff is a rational selection among alternatives in order to optimize a parameter that is a function of two or more variables being compared (traded-off). Generally, system tradeoffs involve varying performance, reliability, maintainability, cost, schedule and risk. These analyses are used to establish reliability and maintainability levels that achieve optimum maximum availability at minimum life-cycle costs while maintaining plant safety. Tradeoff analyses require detailed data on the factors impacting acquisition and operational support costs and their relationship to reliability and maintainability.

Basic Requirements
Perform tradeoff analyses during the early system design and development phase. For this process to be effective, a method is needed for systematically evaluating and comparing performance, reliability, maintainability and life cycle cost of alternate design concepts with respect to prefixed safety objectives needs or constraints. For tradeoff purposes, system performance

is normally evaluated at several discontinuous levels. Each performance level is simply defined relative to functional needs and/or determined through special studies. Reliability, maintainability and cost are the sensitive variables which have continuously varying levels with wide extremes that depend on all facets of a system life cycle. This is an iterative process in which new data is entered until the desired refinements are achieved. The steps involved in performing an R & M tradeoff study are broadly described as follows:

1. *Perform preliminary analysis*: (a) Define appropriate tradeoff measures, i.e. R & M; (b) define the tradeoff criteria; (c) define the level of effort to be applied to the tradeoff process to be consistent with the level of system definition available.

2. *Perform design analysis*: Further define the constraints associated with the system or hardware items and support system characteristics. Define the limitations between which increments in reliability and maintainability may vary.

3. *Define parameters*: Establish the parameters of a standard or 'baseline' design which just meet safety requirements (derived from PRA-safety analysis) and which establish a starting point for all parameters of interest during tradeoff analyses.

4. *Gather data*: Collect, sort and validate system data and, to a lesser extent, component and part-level data from multiple sources including plant operation experiences, system manufacturer's files and industry wide generic databases.

5. *Perform tradeoff studies*: Generate and evaluate design approaches for R & M which satisfy the tradeoff criteria. Generate sensitivity curves which show the breakpoints for R & M with respect to cost, and for given performance inputs.

6. *Refine studies*: Apply design details as they become available for refinement of the tradeoff studies so that the optimum design approach becomes apparent.

Documentation Requirements

Documentation should be prepared relative to each system or equipment for which a tradeoff study was conducted. To be of value for decision making, documentation of tradeoff studies must provide recommendations for hardware acquisition, based on safety, availability and/or economic rationale. Such documentation is generally internal (to the utilities) and the degree of formalism is dictated by internal management policy depending on circulation among top echelons and the leverage they supply in procurement and policy decisions.

Applicable Specifications and Tests
Anderson, R. T., *Reliability Design Handbook* (*RDH 376*), Reliability Analysis Center, USAF/RADC, 1976.
AMCP 706-133: Maintainability Engineering Theory and Practice.
O'Connor, P. D. T., *Practical Reliability Engineering*, Heyden, London, 1981.
US MIL-HDBK-338: Electronic Reliability Design Handbook (ERDH).

3.3.2 Element Number 9: Reliability and Maintainability Specification
Description
This is the process of determining the level of reliability for an equipment that satisfies plant safety and availability requirements. It represents a critical step in assuring that the equipment achieves acceptable reliability during operation. The value of reliability that minimizes total LCC while meeting safety requirements is determined through detailed analysis of attributable cost factors. This value is then incorporated directly (or indirectly through required qualitative design features or attributes) into the hardware procurement specification or possibly used as the basis for establishing a warranty agreement. This process and the resulting optimum reliability level is also used to help compare competing designs, perform design tradeoffs and to evaluate the effect of changes on existing systems.

Basic Requirements
Prepare reliability and maintainability specifications which correspond to the critical, essential or non-essential character of the equipment procured and which reflect specifications prepared from the standpoint of new equipment procurement or replacement of existing equipment. For new equipment classified as critical and essential, prepare specifications which include quantitative reliability and maintainability design requirements, test requirements and program requirements. For replacement of existing critical and essential equipment, prepare specifications which incorporate test/acceptance requirements, failure reporting and reliability assessment provisions. Define levels of reliability for inclusion in procurement specifications which recognize a level of reliability that satisfies safety requirements (derived from safety–PRA considerations) as well as an optimum level of reliability derived from LCC analysis which minimizes overall cost. For essential equipment include only requirements derived

from LCC analysis. Prepare the specification according to standardized format, composition and content and include provisions for changes and revisions. Generation of R & M specifications requires addressing three basic areas:

1. Quantitative R & M design requirements (i.e. MTBF, MTTR availability).
2. R & M program elements (i.e., prediction, FMEA, growth, acceptance).
3. Reporting requirements.

The specifications must recognize the factors shaping the curve in Fig. 3.1 and their applicability for the type of systems/equipment procured. For equipment defined as critical, the specification would incorporate two levels of reliability. The first (shown as R_1 in Fig. 3.1) is that minimum level of reliability necessary to satisfy safety considerations. The second level (shown as R_2 in Fig. 3.1) is the optimum reliability level which minimizes the total life cycle cost of the system/equipment, considering both acquisition cost and maintenance cost. For equipment defined as essential, only reliability level R_2 need be specified. For systems defined as non-essential reliability need not be specified. Standard work statements that define uniform requirements for applicable program elements relative to system criticality classes can be applied and used in the procurement process (see Section 3.8).

Documentation Requirements
Prepare specifications which include the following six sections:

1. Scope.
2. Applicable documents.
3. Requirements.
4. Quality assurance provisions.
5. Preparation for delivery.
6. Notes.

Guidelines for effectively specifying reliability requirements are given in Section 3.8.

Applicable Specifications and Texts
Lakner, A. A. and Anderson, R. T., An analytical approach to determining optimum reliability and maintainability requirements. *The Radio and Electronic Engineer*, **48**, July/August, 1978.

NAVAIR 01-1A-32: Reliability Engineering Handbook.
O'Connor, P. D. T., *Practical Reliability Engineering*, Heyden, London, 1981.
RADC-TR-75-22: Non-Electronic Reliability Notebook.
RADC-TR-79-200: Reliability and Maintainability Management Manual.
US MIL-HDBK-217: Reliability Prediction of Electronic Equipment.
US MIL-STD-785: Reliability Program for Systems and Equipment Development and Production.

3.3.3 Element Number 10: Reliability Mathematical Modeling

Description
Reliability mathematical models are either success-oriented or failure oriented probability diagrams reflecting the logic of the system. They are developed through analysis of the relationships among subsystems and components as represented in functional block diagrams and component drawings. They depict the probability interconnection of the components that make-up the system and provide a basis for performing system–level reliability predictions using individual component or subsystem failure rate estimates.

Basic Requirements
Establish system and component reliability mathematical models which show the consequences of failure on system operation in probabilistic terms. System level modeling includes preparation of reliability block diagrams (RBDs) or fault tree diagrams (FTDs), definition of appropriate statistical distributions and establishment of mathematical relationships by which success/failure rates can be combined. For components and parts, establish modeling techniques which account for operational, environmental, dormancy and other stress factors which determine failure/hazard rates and can be combined to derive reliability values at higher levels of assembly. Establish models in a manner which will accommodate reliability/failure data derived from plant operational experience plus data from existing non-nuclear data bases. These models will serve as the basis for both predicting and assessing reliability. Models should consider the impact of design production, shipment, storage/dormancy, operation and maintenance degradation and aging factors on reliability (see Fig. 2.1). They should account for the unique features, characteristics, stresses and other factors applicable to systems in nuclear power plants and provide a basis for deriving a specific mathematical model that depicts a quantitative representation in terms of

either a RBD or an FTD. The model is an essential element to any quantitative analytical reliability assessment.

Documentation Requirements

Prepare diagrams which include mathematical probability relationships between the items depicted for each critical and essential plant system. The diagrams are constructed through analysis of the functional relationships among items shown by system functional block diagrams and drawings. The interrelation of events is expressed by the way that the blocks are interconnected in the block diagrams. Diagrams should be prepared at system, subsystem and component levels where data can be inputted for later calculation.

Applicable Specifications and Texts

Anderson, R. T., *Reliability Design Handbook* (*RDH 376*), Reliability Analysis Center, USAF/RADC, 1976.
RADC-TR-75-22: Non-Electronic Reliability Notebook.
Review of Equipment Aging Theory and Technology, Franklin Research Center, Philadelphia.
US MIL-HDBK-217: Reliability Prediction of Electronic Equipment.
US MIL-HDBK-338: Electronic Reliability Design Handbook (ERDH).
US MIL-HDBK 472: Maintainability Prediction.

3.3.4 Element Number 11: Reliability Prediction and Allocation

Description

This is the process of quantitatively assessing the reliability of a system or component during design and development—prior to manufacturing and field operation. Predictive methods primarily deal with the effects of failure and depend upon system configuration information, application factors, and data emphasizing the attribute characteristics of parts. They provide a measure of reliability during design with quantitative criteria to identify areas for design improvement or to compare design alternatives. They also provide criteria for testing, maintenance planning, logistics cost studies, and for assessing compliance during later development stages.

Reliability allocation is the process of translating the overall system reliability requirement into reliability requirements for each of the equipments, subsystems, and major components. The apportioned reliability requirements are used to determine if the proposed design will meet system requirements. If the allocated reliability for a specific subsystem cannot be achieved at the current state of technology, then the system design must be modified or the allocation reestablished. This procedure is repeated until an allocation is established that can be achieved and which satisfies the system level requirement.

Basic Requirements

Predict system reliability through use of established models and mathematical relationships (see element number 10) in conjunction with failure frequencies, probabilities or failure rates derived from available data. Compare predicted values to specified values. Review each system and establish methods for allocation of necessary reliability values to lower levels of assembly so that overall system reliability targets can be budgeted and problem areas, if any, identified. Methods should realistically subdivide system reliability according to component/unit complexity, stress or other factors impacting potential reliability. Iterate the allocation process to arrive at realistic component/unit reliability goals which combine to meet system reliability target values. Prepare reliability predictions/allocations as part of the system's planning process using techniques described in *US MIL-HDBK-217*.

In general, there is a hierarchy of reliability prediction techniques available to the designer depending upon (1) the depth of knowledge of the design and (2) the availability of historical data on equipment and component reliabilities. This hierarchy has been developed to accommodate the different reliability study and analysis requirements and the availability of detailed data as the system design progresses. These techniques can be roughly classified in five categories, depending on the type of data or information available for the analysis. The categories are:

1. *Similar equipment techniques*: The equipment under consideration is compared with similar equipment of known reliability in estimating the probable level of achievable reliability.

2. *Similar complexity techniques*: The reliability of a new design is estimated as a function of the relative complexity of the subject item with respect to a 'typical' item of similar type.

3. *Prediction by function techniques*: Previously demonstrated correlations between operational function and reliability are considered in obtaining reliability predictions for a new design.

4. *Part count techniques*: Reliability is estimated by counting the number of parts of each type, multiplying this number by a generic failure rate of each functional component, subassembly, assembly and/or block depicted in the system block diagram.

5.   *Stress analysis techniques*: Reliability is estimated by using detailed failure models plus calculation of component stress values for each part prior to determining its failure rate. Each component is evaluated in its system application based on a mechanical, thermal, and electrical (if applicable) stress analysis. Once component failure rates are established, a combined failure rate for each functional block in the reliability diagram can be determined.

These techniques are further described in Chapter 6.

Documentation Requirements
Document the results of prediction/allocation at predefined system development milestones. Prediction/allocation reports should include the following information:

—Reliability block diagrams that show duty cycle, operating and non-operating failure rates and known applicable failure mode mechanisms.
—Tabulation of failure rates, derating and quality factors, stress and environmental factors and maintenance concepts including inspection periods and repair policies.
—Predictions of specified reliability parameters (i.e. MTBF, system reliability) for system, subsystems and components.
—Consolidation of reliability parameters and comparative analysis between specified/allocated and predicted reliability.

US MIL-STD-785 (Tasks 202 and 203) provides guidelines on the preparation of reliability predictions and allocations based on *US MIL-HDBK-217* techniques.

Applicable Specifications and Texts
Arsenault, J. E. and Roberts, J. A. (eds.), *Reliability and Maintainability of Electronic Systems*, Computer Sciences Press Inc., Rockville, Maryland, 1980.
NAVAIR 01-1A-32: Reliability Engineering Handbook.
O'Connor, P. D. T., *Practical Reliability Engineering*, Heyden, London, 1981.
RADC-TR-75-22: Non-Electronic Reliability Notebook.
US MIL-HDBK-217: Reliability Prediction of Electronic Equipment.
US MIL-HDBK-338: Electronic Reliability Design Handbook (ERDH).
US MIL-STD-756: Reliability Prediction.

US MIL-STD-785: Reliability Program for Systems and Equipment Development and Production.

3.3.5 Element Number 12: Maintainability Prediction and Allocation

Description

This is the process of quantitatively assessing the maintainability of an equipment/system during design and development prior to manufacturing and field operation. Estimates for mean downtime, mean corrective maintenance time and mean preventive maintenance time are developed with quantitative criteria to identify areas for design improvement or to compare design alternatives as well as for maintenance planning, plant availability improvement studies, logistics cost studies and life cycle cost analyses.

Basic Requirements

Predict maintainability of candidate systems based on design characteristics and maintenance features (i.e. test points, self-check features, accessibility, modularization, adjustments, etc.) to determine the ease and speed with which maintenance operations can be performed and failures be diagnosed and corrected. Maintainability should be expressed as a function of time so that repair may be quantitatively evaluated in terms of time required to perform maintenance activities. Maintenance requirements are allocated into repair time elements applicable to lower levels of assembly.

Maintainability predictions are performed based on *US MIL-HDBK-472* techniques. These techniques in general involve the determination of MTTR using failure rate, λ, obtained from reliability prediction studies and maintenance time factors, R_p, derived from a review of system or component design characteristics. Conceptually, the repair of hardware items after the occurrence of a failure necessitates the initiation of a corrective maintenance task which ultimately results in the interchange of a replaceable part or assembly. In order to achieve a complete 'repair', various activities both before and after the actual interchange are necessary. These include activities for diagnosis, alignment and repair verification. The composite time for all repair activities is called the repair time, R_p. In order to provide weighting factors for the expected number of corrective maintenance actions, the failure rate of each replaceable component/part/assembly is used. The failure rate and repair time are combined to arrive at a weighted corrective maintenance action rate. The prediction process also involves preparing a functional-level diagram for the system and determining the repair time for each replaceable item. The functional-level diagram reflects the overall maintenance concept and the complete replacement breakdown for all items that comprise the system.

Documentation Requirements
Document the results of maintainability predictions at pre-defined systems development milestones. Reports should contain the following information as a minimum:

—Maintainability functional level diagrams.
—Tabulation of failure rates, duty cycles, repair policies, maintenance procedures and maintenance conditions applicable to each element.
—Prediction of specified maintainability parameters (i.e. MTTR) for system and appropriate system elements.
—Consolidation of maintainability parameters and comparative analysis of specified and predicted maintainability.

Applicable Specifications and Texts
AMCP 706-133: Maintainability Engineering Theory and Practice.
Anderson, R. T., *Reliability Design Handbook* (*RDH 376*), Reliability Analysis Center, USAF/RADC, 1976.
Arsenault, J. E. and Roberts, J. A. (eds.), *Reliability and Maintainability of Electronic Systems*, Computer Sciences Press Inc., Rockville, Maryland, 1980.
US MIL-HDBK-472: Maintainability Prediction.

3.3.6 Element Number 13: Diagnostics and Modularity
Description
This is the process of determining the degree of built-in-test/fault isolation (diagnostic) self-monitoring and remote monitoring capability required to support specified system maintainability and availability requirements.

Basic Requirements
Establish optimum repair diagnostics and modularization levels through tradeoff studies. Determine, for example, (1) the key parameters which require control room monitoring and display which are indicators of actual or impending failure, (2) the switching circuitry which automatically disconnects faulty equipment and substitutes standby equipment and (3) the levels of automatic or semiautomatic fault isolation for both on-line and off-line diagnosis and repair actions. Also, establish approaches to

modularization consistent with maintainability requirements which simplify training of maintenance personnel, encourage standardization, reduce the number of items in the supply system and permit simplified upgrading of operational equipment. Incorporate requirements for diagnostics and modularity into procurement specifications.

Documentation Requirements
Prepare written requirements which completely describe optimum design requirements for built-in test, modularity and performance monitoring for incorporation into procurement specifications.

3.3.7 Element Number 14: Human Factors Reliability Analysis
Description
This is the process of evaluating the likelihood of human error, the significance of these errors and the effectiveness of corrective action to reduce error probability. Human factors analysis is used for redesigning, relocating, or foolproofing equipment by defining and documenting potential critical problems pertaining to maintenance support, operation and control. It involves studying man–machine interfaces, evaluating personnel task analysis and computing the advantages of automatic versus manual controls. Human factors analysis varies considerably with the significance and magnitude of the function and the nature of the equipment. By identifying and evaluating the effect of human error on the reliability of a man–machine system, significant improvements in reliability can be made through redesign, through elimination of the human element, or through improved training.

Basic Requirements
Establish models for human performance/reliability by which probability of task completion, error rates and other characteristics of operators can be quantified and included in system reliability predictions, failure mode and effects analyses and fault tree analyses. Establish methods for collecting human error data relative to the models established and identifying sources. Sources for human error rate data include nuclear power plants, power plant simulators, process plants and other industries or situations in which tasks, situations and stresses are similar to those within nuclear power plants. Perform human factors reliability analysis in conjunction with the models established, the data collected and the specific man–machine interfacing characteristics of the system/equipment under

consideration. Assess human reliability with respect to operational nuclear power plant systems to determine recommendations for improvement.

The models used to assess human reliability should account for the characteristics of the human working environment, the equipment used by the human operators, the requirements of each task, the stress factors applicable to humans within the work environment and other factors which impact the error rate or the probability of successful task completion. A reliability assessment based on these models should include estimates of: (1) task reliability (i.e. probability of successful completion); (2) the probability of detecting and correcting erroneous performance prior to experiencing undesirable consequences; and (3) the probability that incorrect task performance will result in undesirable consequences.

Documentation Requirements
Describe models used for calculation of human reliability. Prepare formats for the collection of error rate data from plant experience or industry sources. Prepare data summaries through manual or automatic data processing methods which provide the basis for quantifying human reliability. Document the results of periodic assessments performed during plant operation as well as the predefined system development milestones.

Applicable Specifications and Texts
NUREG/CR-1278 (*draft*): Handbook of Human Reliability with Emphasis on Nuclear Power Plant Applications.
US MIL-HDBK-338: Electronic Reliability Design Handbook (ERDH).
US MIL-STD-803A-(1, 2 & 3): Human Engineering Design Criteria for Aerospace Systems and Equipment.

3.3.8 Element Number 15: Failure Mode and Effects Analysis (FMEA), Failure Mode, Effects and Criticality Analysis (FMECA) and Fault Tree Analysis (FTA)

Description
FMEA/FMECA is a systematic process to identify, classify and document failure modes, their effects and criticality. The process is used to evaluate a design and/or process to identify the effects of a failure and to provide criteria for testing, maintenance planning and logistics support. The analysis process is applied at various levels, i.e. component, subsystem or equipment. It is applied early in the design process at the lowest level that can be supported by available data and hardware definition and

subsequently updated to lower and more detailed levels as data become available.

FTA is the systematic process of deductive reasoning that starts with one undesired event and branches off to primary and usually secondary events (failures, human interactions, climatic conditions, etc.) leading to that undesired event. It is also used to evaluate a design and/or process to identify the potential root causes of an undesired event. The technique provides a quantitative measure of reliability or safety. FTA is applied to all undesired events, whose consequences are significant to reliability, safety and cost but whose causes are not readily apparent or are not strictly component related. The top undesired event would normally be uncovered by other tools, e.g. FMEA, design reviews and fault hazard analysis. This technique, like FMEA, is applied at various levels, e.g. components, equipment, subsystem, etc.

Basic Requirements
Perform FMEA/FMECA to systematically evaluate the potential impact of each functional or hardware failure on plant safety, system performance, system reliability, maintainability and maintenance requirements. Establish criticality factors for each potential failure. Rank each potential failure by the severity of its effect in order that appropriate corrective actions may be taken to eliminate or control the high-risk items. Specify failure mode and effects analysis in contractual documents for application during design and development.

Perform FTA on systems and equipment particularly for functional paths of high complexity in which the outcome of one or more combinations of non-critical events may produce an undesirable critical event. Typical candidates for FTA are functional paths or interfaces which could have critical impact on plant safety and, in addition, automated systems in which a multiplicity of redundant and overlapping outputs may be involved. Include FTA requirements in equipment specifications.

Apply FMEA/FMECA and FTA techniques during early design and development of systems and equipment. Prepare revised failure modes and fault tree analyses consistent with the existing design detail. Note that FMECA represents a 'bottom–up' approach, while FTA represents a 'top–down' approach. Both represent analytical approaches for assessing the consequences of failure.

The FMECA can be performed utilizing *US MIL-STD-1629* for guidance. Use either actual failure modes from operational data or hypothesized failure modes derived from design analyses, reliability prediction activities and experiences relative to the manner in which parts fail. In their most complete form, failure modes are identified at the part level, which is usually the lowest level of direct concern to the system designer. FTA involves several steps, among which is the structuring of a highly detailed logic diagram which depicts basic faults and events that can lead to system failure and/or safety hazards. Next is collecting basic fault data and failure probabilities for use in computation. The next step is using computational techniques to analyze the basic faults, determine failure mode probabilities and establish criticalities. The final step involves formulating corrective recommendations for those faults considered critical.

Documentation Requirements
Document the results of FMEA/FMECA and FTA at predefined system development milestones. Resulting reports should provide a complete description of modes of failure in the equipment, the resultant effects on equipment operations and an identification and ranking of critical failure modes. Task 204 of *US MIL-STD-785* can be used as a guide in the preparation of reports. The reports should contain the following information:

(a)   Part/component identification and description (including assembly).
(b)   Failure mode. (List all failure modes associated with the item.)
(c)   Failure effects. (Identify all effects of each failure mode—part, component and system.)
(d)   Failure rate. (List the items failure rates.)
(e)   Probability of occurrence. (List the probability of each item failure rate.)
(f)   Corrective action recommendations. (Identify any recommendations for corrective action.)
(g)   Failure detection method. (Identify any symptoms associated with the failure effect—this information will be useful operational trouble shooting: maintenance analysis report.)
(h)   Criticality analysis. (List the severity level.)

Fault tree diagrams, fault matrices and other back up data should be included in the reports, as applicable.

Applicable Specifications and Texts
Anderson, R. T., *Reliability Design Handbook* (*RDH 376*), Reliability Analysis Center, USAF/RADC, 1976.

IEEE Standard 352-1975: IEEE Guide for General Principles for Reliability Analysis of Nuclear Power Generating Station Protection Systems.

NRC, NUREG-0492: Fault Tree Handbook.

NRC Report WASH-1400 (NUREG-75/014): Reactor Safety Study: An Assessment of Accident Risks in US Commercial Nuclear Power Plants.

US MIL-STD-1629: Failure Mode and Effects Analysis.

3.3.9 Element Number 16: Common Mode Failure (CMF)/Common Cause Failure (CCF) Analyses

Description

This is a technique applied to assess the possibility of a single initiating cause resulting in multiple, cause-dependent failures that negate the benefits of redundant channels of operation. It is used to assess the effects of external events, such as earthquakes, on the equipment. It identifies couplings and dependencies across systems due to human error and testing, maintenance and accident environments. It is also used to assist fault tree development and event tree construction.

Basic Requirements

Perform CMF/CCF analyses on items to determine functionally redundant system susceptibility to failure as a result of single events and to ensure that no single events have been overlooked that could cause any of the system's safety-related functions to fail. Perform initial efforts with respect to historical common failure causes unique to redundant systems. Conduct a systematic search for CCF susceptibility in each redundant item and, in particular, determine if safety-critical functions considered to be redundant are in fact redundant. Perform the CMF/CCF analyses at the component (or lower) level using the results of the FMEA. Review all of the failure modes, failure causes and failure effects identified in FMEA/FTA and classify according to (1) multiple failure modes stemming from a single failure cause and (2) multiple effects stemming from a single failure mode.

Documentation Requirements

Prepare a checklist of intrinsic influences or failure-initiating events based on the review and classification of FMEA/FTA. Incorporate results of CMF/CCF analyses into FMECA/FTA reports.

Applicable Specifications and Texts
AEC Report Wash-1400: Reactor Safety Study, Appendix IV—Common Mode Failures.

IEEE Standard 352-1975: IEEE Guide for General Principles for Reliability Analysis of Nuclear Power Generating Station Protection Systems.

US Code of Regulations, Chapter 10, Part 50, Appendix A: General Design Criteria.

3.3.10 Element Number 17: Single Failure Analysis
Description
This analysis is performed to assess the impact of the failure of series (non-redundant) components identified in reliability models on plant safety.

Basic Requirements
Perform analyses to ensure that single failures are sufficiently remote or that provisions are incorporated to mitigate their effects or consequences. Establish single failure effects and likelihoods as an outgrowth of and in conjunction with FMEA, FTA and CMF/CCF studies.

Documentation Requirements
Prepare documentation to support the conduct of and record the results of single failure analyses. Define format and content requirements in conjunction with FMEA, FTA and CMF/CCF reports.

3.3.11 Element Number 18: Life Cycle Cost (LCC) Assessment and Analysis
Description
This is the process of assessing and analyzing the complete life cycle cost (LCC) of a system and in particular delineating the reliability and maintainability levels that minimize LCC. The LCC of an equipment/system is the total cost of acquisition and ownership of the equipment/system over its useful life. It includes the cost of development, acquisition, operation, maintenance support and, where applicable, disposal. LCC analysis is used to support design tradeoff studies and availability improvement studies. LCC analysis aids in making design, operational, support, budgetary and investment decisions.

Basic Requirements
Establish methods and models for system LCC assessment and analysis.

The models should include submodels for acquisition costs as well as operational and maintenance support costs. Cost elements should take into account factors for replacement power as well as repair/replacement of systems and components over the anticipated 40-year lifespan for such equipment. Factors for reliability and maintainability as an investment during acquisition for which a return is obtained during operational use must be applied within the modeling techniques established. The methods which are established should define cost-estimating relationships which treat current prices, cost escalation and other factors relating cost to the specific elements and characteristics of the reliability program and system reliability performance.

Evaluate modeling factors in terms of areas in which downtime, maintenance and support costs can be effectively reduced by improvements in reliability. The actual levels of reliability achieved during system operation are due to cost incurred to meet design, program test and reporting requirements. These costs can be assessed within the submodel for system acquisition and thus would represent the required investment needed to achieve reliability over the life of the system. This process is then iterated by varying the modeling parameters to achieve an optimized cost over the total life of the system.

Documentation Requirements
Describe models and cost estimating relationships used in LCC assessments and analyses. Provide additional lists for cost data which can be updated as required to track with current costs, prices or interest rates.

Applicable Specifications and Texts
Earles, M. E., *Factors, Formulas and Structures for Life Cycle Costing*, Eddins-Earles, Concord, Massachusetts, 1981.
O'Connor, P. D. T., *Practical Reliability Engineering*, Heyden, London, 1981.
US MIL-HDBK-338: Electronic Reliability Design Handbook (ERDH).

3.3.12 Element Number 19: Reliability Growth (and Tracking)
Description
This is the process of testing prototype hardware to find and remove misapplied parts, to correct design errors, and to eliminate workmanship or manufacturing process induced defects.

Basic Requirements

Plan and apply a structured program of test, analyze and fix (TAAF) during system development to 'force out' design errors and manufacturing flaws. A model should be used to help plan the program and which provides a method by which hardware reliability growth can be dimensioned, disciplined and managed as an integral part of the development or improvement process. The model should provide:

—a technique for extrapolating current reliability status to some future result;
—a method to assess the magnitude of the testing and improvement effort that may be required prior to the actual start of growth testing;
—a means to estimate the total test time needed to grow to a given reliability value with various levels of growth rates depending on the rigor of the test–fix–analyze process thereby allowing tradeoff decisions.

Many overall models for describing the growth process have been established. Some of these models are listed below:

—Duane.
—IBM.
—Exponential–single term power series.
—Lloyd–Lypow.
—Arolf.
—Single exponential.

These models (or curves) are assumed non-decreasing in that once the systems' reliability has reached a certain level, it will not drop below this level for the remainder of the development program.

Establish management control of the growth process through allocation of test resources, scheduling and defining growth targets. Estimate growth rates for systems/equipment in the operational phase based on data derived from plant operation.

Documentation Requirements

Prepare a growth planning document which describes the test, analyze and fix concept and shows how it will be applied to the system or component item under development. An effective plan would include the following information:

—Specified and predicted (inherent) reliabilities plus methods for predicting reliability (model, database, etc.).

—Criteria for reliability starting points, i.e. criteria for estimating the reliability of initial production hardware.

—Test–fix–retest conditions, requirements and criteria, as they relate to, and impact upon, the reliability growth rates.

—Calendar time efficiency factors, which define the relationship of test time, corrective action time and repair time to calendar time.

Prepare reliability growth curves on which actual test data can be plotted and projected relative to the target reliability.

Applicable Specifications and Texts
Anderson, R. T., *Reliability Design Handbook* (*RDH 376*), Reliability Analysis Center, USAF/RADC, 1976.
US MIL-HDBK-189: Reliability Growth Management.
US MIL-STD-781: Reliability Design Qualification and Production Acceptance Tests: Exponential Distribution.
US MIL-STD-1635: Reliability Growth Testing.

3.3.13 Element Number 20: Reliability Demonstration
Description
Reliability demonstration provides decision information prior to the production or deployment of a system or equipment based on realistic test and evaluation, of equipment performance and reliability under given environmental conditions. It provides a measure of confidence that the equipment will perform as specified prior to actual operation.

Basic Requirements
Perform reliability demonstration tests on certain select production hardware to verify conformance to reliability requirements. It should be noted that reliability demonstration requires extensive test data and therefore is judged most applicable in situations where a quantity of items is produced and tested. On the other hand the test analyze and fix approach, as previously described, places emphasis on stress and environmental testing of a small number of (or even one) item(s) to identify design and fabrication deficiencies.

Plans for reliability demonstration should be established during development for subsequent implementation during production and should include specific criteria covering:

—actual environmental condition and operational stresses to be applied (based on actual operational plant use conditions) and their variation within the test period;

—test periods and durations which assure valid results when considering test decision risks and failure expectations.

The plans should reflect an integrated approach to testing that includes experience and takes into account results from previous test efforts so that test costs and time are conserved. The test plan should contain information regarding how the system/equipment will be tested, the test duration and location, and other details for the management and control of the effort. For test plans applicable to equipment whose distribution of times-to-failure is exponential (or can be assumed so), *US MIL-STD-781* provides appropriate guidance. For equipment whose distribution of times to failure is non-exponential, specific test plans must be formulated. Test plans can be classified within several categories:

1. Attributes demonstration tests:
   (a) plans for small lots;
   (b) plans for large lots (including the Poisson approximation method);
   (c) attributes sampling using *US MIL-STD-105*;
   (d) sequential binomial.
2. Variables demonstration tests:
   (a) time truncated—
       (i) exponential distribution,
       (ii) normal distribution,
       (iii) Weibull distribution;
   (b) failure-truncated tests—
       (i) exponential distribution,
       (ii) normal distribution (known),
       (iii) normal distribution (unknown),
       (iv) Weibull distribution;
   (c) sequential tests—
       (i) exponential distribution,
       (ii) normal distribution;
   (d) interference demonstration tests;
   (e) Bayes sequential tests.

Documentation Requirements
A detailed test plan should be prepared which delineates the following:

—System configuration to be tested.
—Condition of test.

—Statistical background.
—Accept/reject criteria.
—Treatment of failed items.
—Definition of failure.
—Number of units to be tested and test duration.

The document should be prepared in sufficient detail with step-by-step procedures to allow complete control, implementation and monitoring of the demonstration test effort.

Applicable Specifications and Texts
AMCP-706-198: Reliability Measurement.
Anderson, R. T., *Reliability Design Handbook* (*RDH 376*), Reliability Analysis Center, USAF/RADC, 1976.
Ireson, W. G., *Reliability Handbook*, McGraw-Hill, New York, 1966.
US MIL-STD-781: Reliability Design Qualification and Production Acceptance Tests: Exponential Distribution.

3.3.14 Element Number 21: Design Reviews
Description
Scheduled reviews are conducted to assure that the design is meeting specified requirements and is proceeding in accordance with contractual milestones and in particular that the system, subsystem, equipment or component reliability requirements will be achieved. These reviews are a very effective method of assuring that the requirements are being met during the procurement phases. Normally, a preliminary design review (PDR), a critical design review (CDR), a test readiness review and a production readiness review are scheduled and held as the system moves from design through production.

Basic Requirements
Conduct periodic reviews during the acquisition of new systems and equipment to evaluate the technical adequacy of the design, approaches to testing and demonstration, reliability performance and other areas of concern. Conduct reviews to evaluate initial or early design progress as well as to evaluate the final design prior to fabrication. Prepare procedures for the conduct of the reviews including detailed checklists that identify the areas to be covered and which are designed to extract maximum information from the review process. Specify requirements for design reviews in procurement documents. Document the results of each design

review conducted during the development effort. Each procedure (and checklist) must be tailored and directed toward the specific review to be conducted and the amount of detail which will be available at that time. Design review procedures must contain provisions for effective follow-up of corrective and/or managerial actions which result from the review process. Procedures and checklists should be designed to:

—detect conditions that could degrade equipment reliability.
—provide assurance of equipment conformance to applicable specifications.
—assure the use of preferred or standard components as far as practical.
—assure the use of preferred materials/processes as far as possible.
—evaluate the electrical, mechanical and thermal aspects of the design.
—provide stress analysis to assure adequate component derating.
—assure accessibility of all components that are subject to adjustment.
—assure interchangeability of similar subsystems' circuits, modules, and subassemblies.
—assure that adequate attention is given to all human factors aspects of the design.
—assure that the quality control effort will be effective.

Documentation Requirements

Prepare detailed design review procedures/checklists for use during actual conduct of each review. The format should provide, in addition to the checklist, spacing for recording the result relative to each item or question. Results of each review as embodied in the checklist item should be formalized and maintained on-file for action and/or future reference. Following each review, a report should be prepared. The report can be organized to monitor and evaluate the effectiveness of each design review and, in particular, the associated deficiency follow-up control activities. The report should contain the following information:

(a)   The checklist and criteria used in performing the review.
(b)   A summary description of the design review data package including (as applicable):
    —Program plans.
    —Critical parts list.
    —Component selection and application data.
    —Component derating application data.
    —Component failure rate data and sources.
    —R & M allocations/predictions.

—FMECA.

—FTA.

—Maintenance concept.

—Test plans (i.e. growth, demonstration, screening and acceptance).

—Special studies (e.g. R & M and cost tradeoffs).

—Technical configuration data including block diagrams, schematics, detailed drawings, etc.

(c) A list of discrepancies which are identified during the review (particularly interface problems).

(d) Techniques by which the discrepancies are followed up and corrected.

(e) A specific accounting of each design review action item including a description of action to be taken by both the manufacturer and the utility.

(f) A summary discussion of the status and resolution of the discrepancies which were identified and the action items remaining open from previous review meetings.

(g) A record (and description) of decisions made (e.g. project go-ahead).

Applicable Specifications and Texts
AEC, RDT-F2-9T Standard: Reliability Quality Assurance.
British Standard, BS5760: Reliability of Systems, Equipment and Components.
RADC-TR-79-200: Reliability and Maintainability Management Manual.
US MIL-HDBK-338: Electronic Reliability Design Handbook (ERDH).
US MIL-STD-785: Reliability Program for Systems and Equipment Development and Production.

3.4 COMPONENT AVAILABILITY

Component availability includes eight elements (element numbers 22–29).

22. Component derating
23. Component quality level
24. Component failure rate modeling and estimation
25. Component testing methods and compliances
26. Component selection, specification and qualification (including QPL and STD)

27. Effects of storage/dormant periods and shelf life
28. Maintenance planning (RCM)
29. Spare part/logistics analysis

Activities to assure component reliability and availability include selection of standard components and the control of critical components.

A general rule is that, wherever possible, standard components with extended life and proven reliability characteristics are to be selected. Standard components may be defined as those which, by virtue of systematic testing programs and a history of successful use in equipment, have demonstrated their ability to consistently function within certain specific electrical, mechanical and environmental limits and, as a result, are considered to be of established quality. For electronic parts, *US MIL* specifications exist which thoroughly delineate a part's substance, form and operating characteristics for practically every known electronic component type.

A well-controlled component program involves establishing a vendor control program, conducting audits of vendor processes, establishing source inspection where applicable and preparing associated documentation. The component control efforts include identification of critical components from the standpoint of safety, reliability, replacement life, cost, procurement lead time, etc. Critical component control includes establishing provisions for special handling, identifying the critical characteristics to be inspected or measured, preparing material review procedures, defining quality assurance and traceability criteria and conducting periodic audits. Detailed documentation is prepared that describes procedures, tests, test results and efforts to reduce the degree of criticality of each component.

A summary of each of the component availability elements is presented in Sections 3.4.1–3.4.8.

3.4.1 Element Number 22: Component Derating
Description
Derating is the process of reducing the stress/strength ratio in the application of components for the purpose of decreasing the occurrence of stress-related failures.

Basic Requirements
Establish and implement a derating policy and procedure for components and material to assure that under expected plant operating conditions (1)

the equipment reliability meets stated requirements and (2) the components are, as a minimum, operating within the manufacturer's specified ratings and environmental capability. The derating procedures should include specific rules for derating of each component type and category taking into consideration those stress and environmental factors that are significant contributors to degradation and failure of the component.

The derating procedures and guidelines are to be followed by design engineers as part of the component selection task. *US MIL-HDBK-217* provides derating curves and failure rate versus stress data for most electronic devices. For mechanical items, failure rate versus stress data may be obtainable from manufacturers and from historical experience.

Develop a uniform derating policy and guidebooks/handbooks specifically for the classes and categories of safety-related components utilized in nuclear power plants. The procedures should be founded upon accepted principles and practices but also take into account actual component experience accumulated over the years of nuclear plant operation, the unique materials, design features and configurations of nuclear plant safety-related components and the total normal and abnormal operating environment (climate, physical, electrical, maintenance/repair, etc.).

Documentation Requirements
Document the derating policy and procedural recommendations. For critical and essential equipment, a log itemizing each component along with its rated value, applied stress and derating level should be maintained as part of the design documentation. This log should be updated as necessary to reflect design changes affecting safety-related component choice or applied stress. These logs are to be made available for formal design reviews and for review by key reliability personnel as required.

Applicable Specifications and Texts
Anderson, R. T., *Reliability Design Handbook* (*RDH 376*), Reliability Analysis Center, USAF/RADC, 1976.
US MIL-HDBK-338: Electronic Reliability Design Handbook (ERDH).

3.4.2 Element Number 23: Component Quality Level
Description
This is the process of selecting and specifying component quality levels that meet system reliability requirements.

Basic Requirements

Establish various levels of component quality to meet a range of equipment reliability needs, cost-effectiveness and manufacturing control. Within each level identify appropriate screening/burn-in tests, performance tests, allowable parameter variation, statistical sampling/acceptance criteria, materials selection, mechanical tolerance and other characteristics through which quality levels can be structured. Establish quality levels consistent with the levels of system criticality (Table 3.3):

  —Critical (CR)—high reliability application
  —Essential (E)—intermediate reliability applications
  —Non-essential (NE)—commercial quality application

These levels are structured to encourage use of standard components having favorable history for high reliability applications and to discourage use of non-standard components.

Prepare component quality guidelines and determine the specific details for each quality level. Establish a component control function which maintains standardization criteria at each quality level. Prepare specifications and procure components in accordance thereto.

Documentation Requirements

Prepare lists of standard components for each quality level. Maintain a literature file on the capabilities/attributes of components which can be used as a basis for component selection.

Applicable Specifications and Texts

British Standard, BS9000: Electronic Components of Assessed Quality. O'Connor, P. D. T., *Practical Reliability Engineering*, Heyden, London, 1981.
US MIL-HDBK-338: Electronic Reliability Design Handbook (ERDH).

3.4.3 Element Number 24: Component Failure Rate Modeling and Estimation

Description

This is the process of estimating component failure rates using modeling and estimation techniques that account for the materials and design characteristics of the components, time–stress failure physics and the operating and environmental application conditions.

Basic Requirements

Utilize existing failure rate prediction models and estimation methods such as provided in *US MIL-HDBK-217* and *NPRD-2*. *US MIL-HDBK-217* covers most electronic and electrical and certain electromechanical component categories. It also provides failure rate models as a function of component material, physical, functional, quality factors and application considerations including applied load stresses and operating environment. In the absence of existing models for specific components estimation methods based on such factors as similarity, maturity, expert opinion (Delphi technique), etc., should be employed. The models and estimation methods should be exercised early in the design as major safety-related components are identified (generically) and updated and refined periodically as specific components are selected and design data becomes available.

Documentation Requirements
Component failure rates should be documented at key intervals during the system/equipment conceptual design phase and updated as necessary during the development and production phases for review and action (if deficiencies exist) by reliability and utility management personnel. The rationale used to derive failure rate for each component shall be stipulated in detail, denoting the assumptions made, the failure rate source (models, other estimation means or observed/historical) and the values assigned to the modeling factors (e.g. component quality, stress level and environment class).

Applicable Specifications and Texts
Arsenault, J. E., and Roberts, J. A. (eds.), *Reliability and Maintainability of Electronic Systems*, Computer Sciences Press Inc., Rockville, Maryland, 1980.
British Standard, BS9000: Electronic Components of Assessed Quality.
NPRD-2, Non-Electronic Failure Rate Data.
RADC-TR-75-22: Non-Electronic Reliability Notebook.
US MIL-HDBK-217: Reliability Prediction for Electronic Systems.

3.4.4 Element Number 25: Component Testing Methods and Compliance
Description
This is the process of selecting or designing appropriate testing methods to accept and qualify components produced or procured for use in nuclear plant safety systems.

Basic Requirements
Apply tests with pre-defined acceptance criteria, to sample production components to determine their acceptability for fabrication in equipment.

The following classes of tests may be performed:

1. Engineering proof tests—performed on engineering samples in order to determine the general capability of the design relative to the application requirements. Since specifications are generally not frozen at this stage, design changes can be readily accommodated.

2. Qualification—a formal testing program, including environmental and life/reliability tests, conducted on production (or pre-production) components to establish qualification compliance of the supplier to specified requirements.

3. Lot (or production acceptance) tests—formal tests conducted on components intended for fabrication into safety-related equipment to assure compliance with specifications and conformance with qualifications test specimens.

4. Screening/burn-in tests—post-production stress tests for the purpose of uncovering defects, including latent defects introduced during production, so that inferior units can be isolated and removed prior to fabrication into equipment. Screening normally includes short-term physical, thermal and mechanical stresses, whereas burn-in denotes longer-term application of thermal and electrical stresses.

Current testing methods directly applicable to electronic and electrical components are:

1. *US MIL-STD-202*—establishes uniform test methods for active and passive components, covering environmental stresses, physical characterization tests and electrical performance tests.

2. *US MIL-STD-750*—covers performance and environmental testing of discrete transistor and diode components, except screening and burn-in procedures which are treated in *US MIL-S-19500*.

3. *US MIL-STD-883*—comprehensive testing procedures for analog and digital microcircuits with heavy emphasis on reliability tests including operating and storage life testing, screening procedures and burn-in tests.

Prepare procedures for components not covered by standards based on manufacturers' recommendations, equipment performance and reliability requirements, and the nuclear plant environment. These tests should include provisions for each category described above. Provisions for tracking defect and reject rates, and requirements for effecting corrective actions for non-complying components should be provided.

Develop testing methods for component configurations unique to the nuclear industry and the specialized application environment of the industry with the emphasis on safety. The methods should incorporate complete test design provisions and acceptance criteria. Also, methods for handling non-complying components and manufacturing lots must be included. Special emphasis should be given to screening and burn-in methods as a means for removing marginal units.

Documentation Requirements
Test reporting formats should be developed and employed for documenting and certifying the results of each test program for all qualification lot acceptance, and screening/burn-in tests performed. The reports should contain test conditions applied, failure (acceptance) criteria, number tested, number failed and failure modes of non-complying units. Additionally, periodic summary reports should give management and reliability personnel an overview of specification non-compliance by vendor, component type, quantity failed and status of corrective actions.

Applicable Specifications and Texts
British Standard, BS9000: Electronic Components of Assessed Quality.
US MIL-STD-202: Test Methods for Electronic and Electrical Component Parts.
US MIL-STD-750: Test Methods for Semiconductor Devices.
US MIL-STD-883: Test Methods for Semiconductor Devices.

3.4.5 Element Number 26: Component Selection, Specification and Quality Assurance
Description
This process is to establish and implement a formal program of component control, including selection, specification and quality-assurance criteria. The program should require the use of 'standard' components whenever possible and require formal justification for using any 'non-standard' components. Each safety-related component should be procured to a definitive procurement specification which has explicit provisions covering functional, performance and reliability requirements plus quality assurance testing.

Basic Requirements
Provide guidelines for formal selection, specification and quality assurance programs applicable to equipment development programs. Practices defined in *US MIL-STD-965* can be applied.

Develop component selection and application criteria and methodologies specifically for nuclear plant safety-related components, including a formal program which defines requirements and component review provisions. Prepare criteria for establishing standard components and implement methods for qualifying components as standard items. Maintain current listings of standard components. Consideration should be given to the development of preferred component lists (PCL) for those items used in quantity or used in critical applications which require extensive evaluation for qualification. Prepare selection rationale and justification requirements for the use of non-standard parts which include the reason(s) for selecting the non-standard component and detailed consideration of its capability and suitability for the intended application along with test and/or operational reliability, or reliability physics analysis data as supporting evidence.

Documentation Requirements
Consideration should be given to the preparation of component selection lists for reliability and management review. The lists should indicate for each component its set of identifiers (manufacturer, type number, rating, etc.) and its qualification status relative to standard parts list and preferred parts lists. Complete backup documentation is to be furnished for each safety-related component which does not appear on either of these listings.

Applicable Specifications and Texts
US MIL-STD-965: Parts Control Program.

3.4.6 Element Number 27: Effects of Storage/Dormant Periods and Shelf Life
Description
This is the process of determining the effects of long non-operating/idle periods on reliability and to establish the need for periodic inspection and maintenance.

Basic Requirements
Establish methods to determine system reliability degradation during non-operational/idle periods. In order to determine if, when and what maintenance should be performed, system testing may be required. Maintenance in this instance is the inspection and reconditioning performed on an idle (but operationally capable) system to prevent

reliability degradation. It involves defining requirements and intervals for cyclic or periodic inspection during non-operational/idle periods.

Conduct surveys and measurements of the ambient environmental conditions surrounding equipment during its idle periods. Relate these conditions to known degradation effects upon the materials, structural elements and parts which compose the system/equipment. Prepare inspection requirements which focus on these effects as well as on the intervals at which they are performed. Operate systems and equipment off-line at periodic intervals, where possible, to burn off residues of moisture, dust or contaminants and to establish operational readiness.

Documentation Requirements
Prepare written inspection procedures to ascertain possible degradative effects or reduction in operational capability.

3.4.7 Element Number 28: Maintenance Planning (RCM)
Description
This is the process of establishing cost effective maintenance requirements to prevent a decrease in hardware reliability and to maintain safety at acceptable levels.

Basic Requirements
Develop system maintenance plans including graphical and pictorial data in a form that can be used by plant maintenance personnel to accomplish necessary preventive maintenance/overhaul tasks and corrective repair. The plans should be consistent with the maintenance and repair policies established for the nuclear power plant. They should be completely integrated documents providing specific criteria and clear-cut direction for accomplishing maintenance tasks for physical inspections, calibration and alignment, replacement of short-life items and component overhaul as well as for performing fault isolation, repair and repair verification tasks. Criteria for maintenance and overhaul should reflect the fact that a system's condition after restoration would be less than the 'as new' condition. Consequently, criteria shall be incorporated into the plans reflecting the number of overhaul/repair cycles that can be accomplished while still maintaining satisfactory conditions within the plant construction/operation-life period. Task scheduling should be based on failure mode and failure frequency data derived from reliability prediction and assessment studies. The maintenance plans should be initially prepared during

development, further detailed during plant integration/construction and revised as necessary during plant operation.

Apply reliability-centered maintenance concepts (RCM) to develop and maintain system maintenance plans. RCM is a relatively new program first initiated by the airlines and now applied by DoD for deriving optimum maintenance plans. RCM is a concept which uses an analytical methodology, or logic, for influencing a design's maintainability and reliability, and for establishing specific maintenance tasks for complex systems or equipment. RCM concepts are described in Section 6.2.

Documentation Requirements
Prepare maintenance requirement plans that identify the reliability critical components requiring maintenance and classify them according to (1) Hard time (2) On condition or (3) Condition monitoring. The plans should also define the intervals, identify responsibilities and specify the required records to be maintained.

Applicable Specifications and Texts
AMCP 750-16: AMC Guide to Logistic Support Analysis.
DAP 750-40: Guide to Reliability Centered Maintenance for Fielded Equipment.
DARCOM Final Report DAG39-77-C-0169: Reliability Centered Maintenance Study.
MSG-2, Airline Manufacturers' Maintenance Program and Planning Document.

3.4.8 Element Number 29: Spare Parts/Logistics Analysis
Description
This analysis is performed to evaluate the failure rates of all replaceable components in an operating system and to provide necessary data for determining stockage requirements. Failure rate estimates or assessments are made throughout the life of the system as new data become available. Component failure rates are used to establish the number of spare parts required to assure the availability of the equipment at minimum cost. Spare parts/logistics analysis is performed as an integral part of the overall maintenance engineering program.

Basic Requirements
Conduct logistics analyses which account for all facets of system support from initial installation and operation through its useful life. Base these

analyses on reliability centered maintenance considerations, as reflected in the maintenance plan, to determine scheduled and unscheduled maintenance tasks and the resources to be acquired. Establish logistic support requirements to be specified during system procurement which acknowledge the long-life support requirements for power plant systems. Perform analyses which provide information in the following areas:

—Numbers and types of spare parts required to support the system over its anticipated lifespan and stock levels to be maintained
—Requirements for tools and test equipment for system maintenance
—Maintenance/overhaul intervals including time required to accomplish each schedule maintenance action
—List of consumable items required to support the system (e.g. lubricants, fluids, filters)
—Maintenance data (parts lists, schematics and instructions)
—Skill levels of maintenance personnel

Documentation Requirements
Prepare provisioning documentation which specifies the total support requirements for the system or equipment under consideration. A total support package establishes numbers and types of spares, maintenance facilities/equipment, technical documentation and personnel requirements to maintain the system over its operating life.

Applicable Specifications and Texts
Arsenault, J. E. and Roberts, J. A. *Reliability and Maintainability of Electronic Systems*, Computer Sciences Press Inc., Rockville, Maryland, 1980.

3.5 RELIABILITY DEGRADATION CONTROL

Reliability degradation control includes three elements (element numbers 30–32).

30. Controlled screening and burn-in
31. Reliability acceptance
32. Degradation control.

Once the inherent or designed-in reliability is established, engineering efforts focus on the prevention or reduction of degradation that may result from manufacturing or from system/plan integration. Process defects can be introduced and escape inspection which later result in operational failures. Special inspections, screen tests and reliability/quality control methods are designed and applied during production and plant/system integration to eliminate these defects and minimize possible degradation.

This section discusses reliability degradation and its control. Basic requirements and approaches are described that can be used to plan reliability control measures during production, acceptance and plant integration.

3.5.1 Element Number 30: Controlled Screening and Burn-in

Description

Controlled screening and burn-in involves the application of selected stresses during hardware tests on a 100% basis for the purpose of revealing inherent, as well as workmanship and fabrication process induced, defects without weakening or destroying the product. The application of stress serves to reveal defects which ordinarily would not be apparent during normal quality inspection and testing.

Basic Requirements

Establish methods and techniques to reduce reliability degradation by proper use of screening and burn-in the purpose of which is to compress (timewise) a system's early mortality period and reduce its failure rate to an acceptable level as quickly as possible. Define the rigor of the applied tests and the subsequent failure analysis and corrective action efforts to minimize reliability degradation and preserve inherent reliability. Identify appropriate screen tests, their effectiveness and limitations in order to plan and implement an optimized production screening and burn-in program.

A screen/burn-in test program should be designed and optimized relative to the individual hardware technology, complexity and end-item application characteristics as well as the production volume and cost constraints of the particular system or component being manufactured. Develop and structure an effective screening/burn-in program by first identifying the kinds of failure modes that can occur and then the assembly level at which they may be induced. Guidelines to aid in planning and specifying a screening program for electronic control equipment are presented in Section 5.2. The planning and specification process involves exercising various options and comparing expected reject or fall-out data to the number of possible defectives that may escape as estimated from an analysis of production data on a similar system and equipment. The process permits rapid convergence on the selection of test levels, test duration and number of cycles, failure-free criteria, and other control parameters. The process is carried out in the context of measurable costs and cost-avoidance data, and overall system effectiveness.

The rejects that result from screening provide an indication as to whether or not the test is effective and that potential field failures are being removed from the system. If the screen test shows very few defects, it is either insufficiently severe, or the system or subsystem items being screened are already reliable. The 'reject rate' is a measure of the number of items deviating from specification.

Documentation Requirements
Prepare a complete screen test specification that encompasses:

—Test sequence and application levels
—Test conditions including test duration, number of cycles, failure-free criteria, cumulative operating time and critical electrical parameters
—Expected reject or fall-out rates
—Test facilities
—Special test equipment
—Data recording requirements and methods
—Failure reporting analysis and corrective action procedures
—Manpower and training requirements

The specification should be prepared in a form that allows the established screening tests to be effectively transferred to production.

Applicable Specifications and Texts
Anderson, R. T., *Reliability Design Handbook* (*RDH 376*), Reliability Analysis Center, USAF/RADC, 1976.
ESSEH, *Environmental Stress Screening Guidelines for Electronic Hardware*, Institute of Environmental Sciences, 1981.
IEC Publication 605: Equipment Reliability Testing.
NASA, NHB 5300.4 (3D): Test Methods and Procedures for Microcircuit Line Certification.
US MIL-STD-175: Microelectronic Device Data Handbook.
US MIL-STD-750: Test Methods for Semiconductor Devices.
US MIL-STD-785: Reliability Program for Systems and Equipment Development and Production.
US MIL-STD-883: Test Methods and Procedures for Microelectronics.

3.5.2 Element Number 31: Reliability Acceptance

Description

Reliability acceptance tests are performed on production systems/equipment to determine compliance to reliability criteria. These can be described as a periodic series of tests to indicate continuing production of acceptable equipment and are used to indicate individual item compliance with reliability criteria. The tests are intended to simulate in-service operation of the item and to provide verification of reliability parameters.

Basic Requirements

Plan and implement reliability acceptance tests using *US MIL-STD-781* for guidance. This standard provides guidelines for test conditions applicable to production reliability acceptance testing with provisions for developing acceptance criteria. Test procedures with complete acceptance criteria should be prepared well in advance of the actual testing. All required operational modes should be exercised as part of the acceptance tests including verification of any functional redundancy or backup.

Documentation Requirements

Prepare written detailed test plans and procedures which describe:

—System/equipment configuration and its control
—Parameters to be measured and recorded as the basis for acceptance under the conditions specified
—Test conditions, equipment set-up and measurement techniques
—Statistical considerations as required
—Accept/reject criteria
—Reporting, analysis and corrective action for test failures
—Detailed definitions of failure
—Test duration

The plans and procedures should be written in sufficient detail to provide step-by-step instructions to personnel conducting the test and should include data recording, control and monitoring of the test effort.

Applicable Specifications and Texts

British Standard, BS5179 (1, 2 & 3): Guide to the Operation and Evaluation of Quality Assurance Systems.
IEC Publication 605: Equipment Reliability Testing.

US MIL-STD-781: Reliability Qualification and Production Acceptance Test—Exponential Distribution.

3.5.3 Element Number 32: Degradation Control

Description

These are the procedures and steps to be taken to ensure that the inherent reliability of an equipment/system is not degraded during production, and plant operation including long non-operating periods. Once the inherent or design-in reliability is established, engineering efforts focus on the prevention or reduction of degradation. Well-planned and carefully executed inspections, tests and reliability/quality control methods are applied during production (as well as during plant operation) to eliminate defects that could lead to failure. Environmental stresses as well as inspection methods and operational/maintenance procedures are continually assessed to determine the need for better inspection, screening and control provisions to improve reliability.

Basic Requirements

Establish and implement a program to ensure that the inherent design reliability of a nuclear system is not degraded during production, installation and operation. The control program should include efforts that start with design by specifying materials that account for manufacturing variations and continue through manufacturing, installation and operation with the application of special inspections and test specifically designed to detect and eliminate workmanship induced defects. Degradation control includes measuring the reliability of a system as it is released for operation, by quantitatively taking into account potential defects induced by production as well as the installation process and the inspections tests and other measures applied to remove the defects. Actual reliability should be continually assessed through detailed analysis of process flow, reject rate statistics and estimates of inspection efficiency factors. The intent is to identify areas where better inspection, screening and control provisions can be applied to reduce degradation and to ultimately improve reliability.

    An essential aspect of the degradation control function is the application of screen tests specifically designed to detect and remove latent reliability defects. These tests involve the application of stresses or screens to convert latent (or unobservable) defects such that they may be removed by conventional testing methods (see element number 30). A complete interactive life cycle reliability analysis can be applied to plan, assess, control and improve post design reliability control measures based on the above requirements. This would provide:

—A means by which the inherent reliability embodied in the design can be retained during manufacturing and operating/non-operating periods
—A means to determine the need for additional screen tests or better inspection
—A technique to assure a smooth transition from design to production to installation and operation
—The capability to assess, grow and control actual reliability.

Application of a complete life cycle methodology would be a logical extension of proven accepted reliability and statistical analysis practices, such as that given in *US MIL-HDBK-217*. Submodels can be developed that account for potential production, installation and dormant deterioration mechanisms and degradation factors as they apply to nuclear systems, equipment and components.

Documentation Requirements
Prepare control plans that define methods to assure that the inherent reliability designed into systems and equipment are not degraded. The plans should include: identification of the processes involved in the production–installation cycle; establishment of special test techniques to supplement conventional inspection methods, where required; calculation of inspection/test efficiencies; estimation of defect percentages introduced at each phase or process in the manufacturing/installation cycle and estimation of the outgoing levels of reliability. Plans should define the effort required in the areas of test, fabrication and inspection and methods of handling. The plans should include provisions for upgrading various processes and/or inspections responsive to the need for improvement in the manufacturing/installation operation. Process flow charts, efficiency report formats and degradation control worksheets would form a part of the finalized control plan.

Applicable Specifications and Texts
US MIL-HDBK-338: Electronic Reliability Design Handbook (ERDH).
US MIL-Q-9858: Quality Program Requirements.

3.6 OPERATIONAL RELIABILITY ASSURANCE

Operational reliability assurance includes four elements (element numbers 33–36).

33. Cyclic inspection
34. R & M assessment
35. PRA/R analysis
36. Reliability (or availability) improvement.

The transition of a system or equipment from a production environment to an operational environment can result in a decline in system reliability. This potential decline stems from several sources. One source is from installation errors or from damage which can occur during integration of the system/equipment at the plant site; another results from the degradative effects of dust, moisture or corrosion during idle periods usually after installation but prior to full operation and, generally, while some construction within the plant is still in progress.

In addition to these effects, normal operation over extended time periods give rise to aging effects and subsequent reliability degradation. These effects take their toll on materials and components due to heat, pressure, vibration, dust, condensed containments and other environmental stresses incident to the nuclear power plant operating environment.

A summary of each of the operational reliability assurance elements is presented in this section.

3.6.1 Element Number 33: Cyclic Inspection
Description
Cyclic inspection is performed to ascertain system/equipment reliability and readiness for use, to detect deterioration while idle and to furnish data for any necessary reconditioning or overhaul action. A knowledge of the system or equipment item, particularly its deterioration properties and their effects is necessary to plan and specify optimum cyclic inspection requirements.

Basic Requirements
Establish an inspection program whose objective is the elimination of defects brought about by non-operation (dormancy) while installed in the operational site. Perform periodic inspections of system/equipment during idle or non-operating periods to provide data for assessment of system

reliability and operational readiness, detect deterioration and other deficiencies caused by aging and long-term non-operating or dormant stresses and to provide a basis for hardware reconditioning or overhaul. The type, extent and frequency of the cyclic inspections are dependent on the deterioration properties of the hardware and parameters of unreliability as defined by FMECA.

An effective inspection program requires the performance of thorough and detailed visual inspections to identify non-operating defects, deterioration and aging mechanisms followed by a functional checkout to verify operability. Emphasis is placed on inspecting for evidence of deterioration, contamination and corrosion, as well as for loose or frozen parts, damaged parts, leakage or excessive moisture. The functional test is defined such that it can be applied simply and quickly on sensitive and critical hardware items to assure that they perform satisfactorily and are ready for use. The functional test is not intended to represent a complete and detailed inspection of the hardware to determine compliance to requirements but is designed to verify operability fully utilizing end item functions to indicate readiness. The tests can range from a relatively simple checkout of a critical component to perhaps a full functional test of a complete system.

Documentation Requirements
Cyclic inspection procedures should be prepared that define the frequencies of inspection, methods of inspection and acceptance criteria. Included should be step-by-step instructions for performing each inspection and test specified. Criteria should also be established and reflected in the inspection procedures that recognize the fact that hardware deteriorates due to age and plant environment and when deterioration begins to take effect its condition will decline from the 'as new' condition. Although the effectiveness of the inspections and the adequacy of the reconditioning operations and controls will minimize the decline, the resultant hardware reliability would be somewhat lower than that initially installed. Forms should be designed for recording the results of each inspection for permanent filing and storage of inspection data.

3.6.2 Element Number 34: Reliability and Maintainability (R & M) Assessment
Description
Operational, failure and repair time data is collected and then employed to assess, quantitatively, in-service reliability, availability and maintainability. These assessments are used to: (1) perform component-ranking

analyses to identify candidates for reliability improvement; (2) identify and evaluate improved maintenance procedures; (3) evaluate the effect of proposed changes on system availability; and (4) perform cost/benefit analysis of availability improvement programs.

Basic Requirements
Assess systems reliability based on statistical analysis of operational test and actual experience data. The operational data should first be reduced and statistically evaluated. The reduced data can then be directly applied or it can be inputed into reliability math models (element number 10) to assess system and component reliability in lieu of analytical data, such as from *US MIL-HDBK-217*. Statistical techniques such as multiple regression analysis, *F*-test and correlation coefficient analysis can be applied to the operation data to identify those variables that significantly impact system and component failure rate. The assessments are to provide consistent and accurate data and information which can be fed back into the plant's system improvement process as well as to provide an 'experience' information base for future developments. The specific objectives are to:

—Uncover problem areas, effect timely corrective action and provide a basis for reliability and maintainability improvement programs
—Determine the effectiveness of design, test and program reliability concepts
—Track the reliability performance of the operational system.

In addition Bayesian statistics can be applied to the assessment of system reliability. The usefulness of Bayesian statistics is that it allows prior information obtained for a random process to be applied to the analysis of more recent information, such as actual experience data. This prior information would include all pertinent data, such as analytical results (e.g. predictions), development test data (e.g. reliability growth), demonstrations, production and operational data. The results thus achieved, therefore, would utilize the widest spectrum of information available. In this fashion, not only can reliability prediction estimates and the latest test information on a system be used independently of one another to assess reliability, but a final assessment can also be made that takes into account both information sources.

Perform R & M assessments on those critical systems and equipment as dictated by actual operational experience data. Actual or operational reliability and maintainability data should be reviewed in detail as part of the assessment effort to assess compliance with requirements, to compare with predictions, to identify trends and problem areas and to set corrective action or improvement priorities. For example, the assessment data can be reviewed to identify specific areas for reliability design improvement. This would include evaluating:

—The reduction of failure rates by operating components at reduced (derated) stress levels. This is accomplished by selecting components which have ratings well in excess of those required for their system application.
—The use of improved components for which reliability has been significantly increased through special manufacturing techniques, quality control procedures and testing methods.
—Design simplification to eliminate parts or components.
—The substitution of functionally equivalent items with higher reliability.
—The overall reduction of failure rate through increased control of the internal and external system environment—e.g. through reduction of ambient temperature, isolation from handling effects and protection from dust.
—The provision of features which enable prediction of incipient failures, and permit remedial action to be taken before an operational failure occurs.
—The provision of design features which reduce the probability of human initiated errors.
—The provision of multiple, identical parts, paths or higher functional levels (redundancy) in order to prevent a system failure in the event that one element fails.

Similarly, the assessment data can be reviewed to identify specific areas for maintainability design improvement. This would include evaluation, for example, of the need for: (1) improved maintenance practices; (2) higher quality technical manuals and maintenance aids; or (3) better training to improve the skill level of technicians.

Documentation Requirements
Prepare assessment procedures, data recording forms, survey questionnaires and other documents through which an R & M assessment can be effectively conducted. The procedures should delineate specified, predicted and tested performance parameters for specific systems and equipment found in nuclear power plants for comparison to current performance.

Documents containing the results of each assessment performed should be maintained on file.

Applicable Specifications and Texts
AMCP 706-133: Maintainability Engineering Theory and Practice.
NAVAIR 01-1A-33: Maintainability Engineering Handbook.
US MIL-HDBK-338: Electronic Reliability Design Handbook (ERDH).

3.6.3 Element Number 35: Probabilistic Risk Analysis/Reliability Analysis
Description
Reliability engineering tasks are performed to support probabilistic risk assessment (PRA) as well as to analyze resulting data. This includes establishing component-failure and human-error rates, performing FTA, common cause analyses, and human reliability analyses, and performing component sensitivity studies.

Basic Requirements
Establish an overall methodology for providing reliability data for input to PRAs as well as for analyzing subsequent output data. The methodology should delineate how results from PRAs can be used to identify hardware items whose potential failure and criticality to plant safety warrant further action to preclude failure effects. It should form a closed-loop effort which complements early tasks for new designs to define reliability critical items and form the basis for on-going review.

Perform a sensitivity analysis of PRAs to identify and prioritize those system/equipment items whose failure or degradation has potential hazardous impact on plant operation. Compare these critical hardware items derived from PRAs with items derived from operational experience or industry data sources. Compile a list of items based on these reviews and comparisons. Iterate this process as new PRAs and other information from plant operation/industry data sources becomes available. Prepare requirements for corrective action or improvement of critical equipment items.

Documentation Requirements
Prepare lists of items judged critical based on PRA and reliability analysis. In conjunction with each item on the list prepare or identify corrective action plans for actual implementation.

Applicable Specifications and Texts
NRC, NUREG/CR-2300: PRA Procedures Guide.

3.6.4 Element Number 36: Reliability (or Availability) Improvement

Description

This is a logical process for analyzing system failure and repair data to determine those elements or procedures which can be modified to improve system availability or lower life cycle cost while maintaining plant safety. The reliability improvement program is applicable to operating systems where data is available to ascertain those components or subsystems which are limiting system availability through high failure rate or long repair times.

Basic Requirements

Establish procedures whereby improvement of operational systems and equipment can be achieved through a closed-loop process of positive action based on operational data analysis and assessments. An essential facet of a reliability improvement effort is determining a method for identifying the most beneficial improvements to be incorporated. Candidates for improvement must be selected based on expected future benefit or current reliability values substantially lower than established during design, development or early operational use. Improvement programs must work in conjunction with data recording and assessment efforts to establish an integrated, cost-effective process.

Implement reliability or availability improvements into systems or equipment based on the following process:

Step 1. Review available data that describe operational and failure characteristics of the system as well as maintenance and logistic cost data to assess the need for product improvement. If indications are that improvement in reliability is needed for economic reasons, then a detailed quantitative cost–benefit analysis should be performed.

Step 2. Assess the reliability of the item as it is currently configured within its operational environment to determine quantitatively the extent of improvement considered necessary. Close coordination with appropriate operation and maintenance personnel are required in order to assure that improvement goals can be achieved with minimum risk to plant operations.

Step 3. Review backup data in detail to identify areas and criteria for improvement and formulate recommendations that would meet the criteria.

Step 4. Compute the impact of the recommended improvements which appear most useful for cost tradeoff consideration, on MTBF,

MTTR, overall downtime and system performance and determine the total cost for their implementation.

Step 5. Compute cost–benefit factors—develop a numeric for each R & M recommendation which reflects the total cost of the change and its impact on system performance. This will allow the determination of those change recommendations which have maximum effectiveness. The recommended changes can then be presented in an improvement plan in decreasing order of cost effectiveness as defined by the computed cost–benefit factors.

Documentation Requirements
Prepare a written improvement plan which identifies the process by which improvements are formulated, evaluated and incorporated into operational hardware systems. The plan must call out data sources, techniques for determining cost versus benefits and methods to incorporate changes to minimize equipment outages.

Applicable Specifications and Texts
AMCP 706-133: Maintainability Engineering Theory and Practice.
NAVAIR 01-1A-33: Maintainability Engineering Handbook.
US MIL-HDBK-338: Electronic Reliability Design Handbook.
Workshop on performing power plant reliability and availability analysis, *Ninth Annual Engineering Conference on Reliability for the Electric Power Industry*, 16–18 June, 1982.

3.7 EXPERIENCE FEEDBACK

Experience feedback includes two elements (element numbers 37 and 38).

37.   Failure reporting and corrective action (FRACA)
38.   Reliability data recording and feedback

Experience data are the foundation upon which most reliability concepts, techniques and decisions are based. Failures provide the vehicle for understanding degradation mechanisms, stress influences and other phenomena leading to improvements in materials, designs and stress control features. Often, this valuable knowledge is lost through inadequate or uncoordinated data recovery procedures or absence of any data collection mechanism.

A meaningful reliability data collection program consists of reporting both failures and the duration of successful operation of the monitored systems and equipments and should include supporting information relative to the on–off–standby timing profile of the individual equipments and the surrounding operational environment, including unusual occurrences.

Reliability data serves a number of useful purposes including:

1. Verifying that the equipment is meeting its reliability requirements.
2. Discovering deficiencies in the equipment and providing a basis for corrective actions and product improvement programs.
3. Establishing failure histories and rates for comparison and for use in predictions.
4. Establishing accurate means for determining spares requirements.
5. Estimating the degradation and wearout characteristics of components for determining preventive maintenance schedules.
6. Estimating the qualifications of maintenance personnel and number of manhours required to achieve a desired availability level.

Experience data should be collected from all available sources such as:

—Component development and manufacture
—Reliability tests
—Subcontractors and vendors
—Plant operations

Each person provides data with unique characteristics and value. In combination they can provide an integrated insight into the reliability capabilities of the components being tracked. The usefulness of experience data is directly keyed to the thoroughness and accuracy of the reported data. Deficiencies in recording at the data source inevitably lead to incorrect conclusions, erroneous decisions and inappropriate actions.

A summary of each of the experience feedback elements is presented in Sections 3.7.1 and 3.7.2.

3.7.1 Element Number 37: Failure Reporting Analysis and Corrective Action (FRACA)

Description

The FRACA is a closed-loop, real-time system, the primary purpose of which is to identify failures/problem areas leading to initiation of aggressive follow-on activities to effect and verify corrective action to eliminate or minimize recurrence of the problem. It is evident that the effectiveness of correcting observed deficiencies requires rigorous and persistent tracking and then coordination of actions pursued by different organizations and disciplines. This responsibility is assumed by the reliability organization, which must, coincidentally, report the incident status periodically to program management and obtain their concurrence on recommendations and proposed corrective measures.

Basic Requirements
Implement a closed-loop failure reporting and corrective action system which systematically reports all failures observed during development, installation, manufacture, testing, construction installation and checkout to assure that failures adversely affecting reliability of safety systems are identified and eliminated.

The program will include:

1. Complete documentation of the observed failure.
2. Investigation and analysis of all failures regardless of how trivial they may appear.
3. Implementation of corrective actions.
4. Evaluation of the effectiveness of the corrective action.
5. Management visibility of active failure/problem investigations.

Pattern failure reporting and corrective action program procedures after *US MIL-STD-785* (Task 104), which requires that the program be capable of collecting, analyzing and reporting failure for the purpose of determining failure cause and recording the corrective actions taken. Develop uniform reporting forms, delineate responsibilities and points of interface between the various organizations involved and stipulate clear lines of authority for establishing priorities and obligating resources. Identify qualified laboratories that have the necessary facilities and skilled staff to perform failure analyses on components normally used in nuclear plant equipment.

Documentation Requirements
Document the FRACA procedure and reporting format. Produce a completed FRACA reporting form for each observed failure containing actions, dates and authorized signature for the step in the process. Publish a biweekly (or monthly) summary of outstanding failure reports denoting the investigation priority and status of each step in the process.

Applicable Specifications and Texts
US MIL-STD-785: Reliability Program for Systems and Equipment—
Development and Production.

3.7.2 Element Number 38: Data Recording and Feedback

Description
This task is to collect reliability and maintainability data during system operation. The data consists of reports of failures and repairs and reports of duration of successful operating of the monitored equipment/system. The data is valuable in providing inputs to many program elements such as reliability improvement, spare parts, maintenance scheduling and logistics, etc. In contrast to FRACA, reliability data recording and feedback applies primarily to the collection, reduction and reporting of data from operating systems. This feedback does not provide closed-loop corrective action control but rather is intended to serve management personnel with a broader perspective of productivity, cost and other problem areas in order to isolate major areas of concern and establish priorities for corrective investigations. A number of utilities are utilizing computerized data systems of this type. Data residing in these databases also represent valuable historical resources, which if combined would greatly increase the confidence of statistical inferences made from the data. The INPO–NPRDS presently consolidates data acquired from individual utilities including LERs.

Basic Requirements
Establish and implement a program for collecting reliability and maintainability data on operating systems and equipment. Reporting is not limited to safety-related equipment and is to consist of reports of failures and repairs and reports of successful operating time periods (or cycles) of the monitored equipment/system. Collected data should be utilized in reliability assessments (see element number 34) to determine compliance with requirements, to identify trends and problem areas, to aid in formulating maintenance and logistics plans and then retained as corporate memory for future reference and for application to new designs.

Prepare data collection plans and procedures consistent with internal data requirements, the NRC, LER requirements and present centralized industry data systems such as NPRDS and GADS. Implement data collection on operating plants/equipments and monitor closely to assure integrity and completeness of reporting. Collected data should be made available to the centralized data system as a means of providing industry wide measures of reliability, availability and maintainability performance.

Documentation Requirements
Produce collection and data processing procedures, tailored to the needs of individual utilities. For each failure event or operating period prepare reports describing the events as observed. Publish analyzed data summary reports at designated intervals to effect essential management visibility and control. Submit event reports to NPRDS, GADS and NRC (LERs) as appropriate.

Applicable Specifications and Texts
NUREG-0161: Licensee Event Report.
NUREG-0020: Operating Unit Status Report.
IEEE STD-500-1977: Nuclear Reliability Data Manual.

3.8 DETERMINING RELIABILITY SPECIFICATION REQUIREMENTS

Determining reliability specification requirements for complex nuclear power equipment involves balancing many interrelated variables and factors. Some of the considerations that must be stressed when formulating a specification for a given procurement are as follows:

—Application of detailed analyses and cost tradeoff studies during the early design phase to aid in the achievement of safe and reliable operations at minimum life cycle cost.
—Definition and implementation of an effective management and control program. This program would directly enable reliability personnel to influence design, provide timely outputs consistent with major design and program decision points and, in general, provide the means to develop a system that meets cost-effective objectives and requirements.
—Continuous application of systematic and highly disciplined engineering tasks during the design phase. Their purpose is early identification and correction of problems which will force the design to be iterated as necessary, prior to the build-up of hardware.
—Early procurement, build-up and reliability growth testing of critical components.
—Performance of equipment reliability growth testing that emphasizes failure analysis and corrective action and provides a test cycle that reflects the application environment.

—Implementation of a reliability assurance program that allows a smooth transition from design and development to production and to system/plant integration and operation, without degrading reliability, and which emphasizes screening tests at critical stages in the equipment fabrication process.

—Implementation of an integrated test and plant acceptance program instead of separately managed testing programs. This requirement prevents both duplications and omissions in testing and provides a single test baseline in parallel with a closely interrelated program of reliability assessment. This approach emphasizes the intimate relationship of the testing effort with equipment requirements and underscores its role as an input to the various project decision points.

The equipment function and criticality, its plant location (sheltered or unsheltered), complexity, state-of-the-art and how close the hardware is to existing or similar hardware systems must be considered in determining the most appropriate reliability level to specify. In general the process involves first reviewing the hardware to be procured in terms of its performance, location and other system parameters; next, several reliability and LCC options or point estimates are developed based on selected application of program elements and design attributes at levels consistent with the hardware performance, location and criticality; finally, from a review of the point estimates the reliability value that is most cost effective while satisfying plant safety and availability requirements is determined.

The designer must make tradeoffs between the quantitative reliability levels and program elements with projected LCC estimates of the system while meeting plant safety/availability requirements. Since acquisition cost and operational and maintenance support cost are driven by the reliability characteristics of a system, it is essential that cost-effective levels are specified. A low reliability specification level will reduce acquisition cost but will result in high operational and maintenance costs. Conversely, a high reliability level will increase the initial cost but reduce maintenance cost. In between there is an optimum point that minimizes LCC; this level is compared with the level necessary to satisfy plant safety (risk) availability requirements and tradeoffs performed as necessary (see Fig. 1.5).

Determining the optimum reliability specification level requires the identification and definition of quantifiable reliability attributes and the use of cost-estimating relationships that relate reliability and, in particular, MTBF to acquisition cost. A key system reliability program activity is the determination of the MTBF value that minimizes LCCs and meets the minimum criteria determined from availability/risk analysis and thus represents the specification requirement to be exercised over the acquisition process.

The relationship between reliability and cost is given in Fig. 3.2. The figure shows that as a system is made more reliable (all other factors held constant) the support cost will decrease since there are fewer failures. At the same time, acquisition cost (both development and production) must be increased to attain the improved reliability. At a given point, the amount of money (investment) spent on increasing reliability will, because of the amount saved in support cost, represent the reliability for which the total cost is a minimum. Consequently, reliability can be viewed as an investment during acquisition for which the return on investment (ROI) is a substantial reduction of maintenance support.

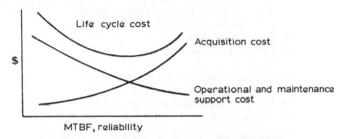

Fig. 3.2.   Life cycle costs versus reliability.

The total LCC has two principal cost components: those costs incurred before a system is operational (acquisition) and those costs incurred after a system becomes operational (operational and maintenance support logistics). Some of the major elements contributing to these costs are shown in Table 3.6. Of the LCC factors, the cost for reliability design and engineering have a most significant impact. If high quality, long life and established reliability components are used with stringent design safety factors to enhance equipment reliability, cost will increase. Redundancy and diversity also add substantially to cost. On the other hand, costs associated with production test failures will be reduced because rework, scrap rate and quality control inspection become lower as a result of increased reliability. More important, however, during plant operation, the need for replacement power because of plant shutdown for repair would be substantially less and logistics and maintenance support costs (which are highly sensitive to system reliability characteristics) will be significantly

TABLE 3.6
Life Cycle Cost Factors

—*Acquisition*
1. Basic engineering design
    (i)   Redundancy, diversity
    (ii)   Derating
    (iii) Part selection
    (iv)   Modularity
    (v)   Diagnostics
2. Test and evaluation
    (i)   Development
    (ii)   Reliability growth and screening
    (iii) Reliability acceptance
3. Program management and control
4. Manufacturing and quality engineering
    (i)   Process planning
    (ii)   Configuration and change control
    (iii) QA planning, audits, liaison, etc.
5. Parts and materials
6. Fabrication, assembly
7. Quality control
8. Inspection and acceptance testing
9. Packaging and shipping
10. Test equipment and tooling
11. Documentation

—*Operation cost*
1. Operators
2. Materials, excluding those for maintenance
3. Replacement power
4. Component spares
5. Maintenance, repair and overhaul
6. Inventory entry and supply management
7. Support equipment
8. Personnel training and training equipment
9. Technical data and documentation
    logistics management
10. Configuration and change control

lowered. High reliability levels will materially reduce corrective maintenance, preventive maintenance, plant retrofits and spares requirements.

The logistics plan, including staffing patterns and spares provisioning, as well as the maintainability concept currently in practice, strongly influence LCC. Such features as automatic built-in test, fault isolation and automatic remote maintenance monitoring will fundamentally alter equipment requirements and costs. The precise effects of higher acquisition cost on operational support cost can only be determined by detailed analysis using established LCC models. Staffing and the degree of training can also materially affect support cost by modifying the actual effective utilization of maintenance and support labor. However, to apply these costs in analysis, one must be sensitive to the total process.

The establishment of LCC models that reflect exact relationships between reliability and cost in the context of total LCC requires detailed evaluation of the factors that constitute acquisition, operational support costs. These relationships must be quantitative and accurately defined in order to perform a total LCC analysis that allows effective tradeoffs and the determination of optimum parameters for specification.

In determining which costs to include for tradeoffs and which to exclude, one should employ the concept of variable costs. Those costs that are fixed regardless of the decision choices are excluded; e.g. basic design engineering, tooling, production and general operational costs. Those costs that vary with the reliability level are included; e.g. part quality level, derating, reliability testing, replacement power, repair and component spares.

Figure 3.2 shows that as the level of reliability increases, the operational and support (O & S) costs decrease. The shape of the curve in the high reliability portions is relatively flat, being associated with the application of more involved and higher cost improvement attributes and the more steeply sloped areas are associated with the application of lower, less complex and lower cost attributes. Thus, as a system is made more reliable (and maintainable), everything else being equal, the O & S costs will decrease due to fewer failures which, in turn, will decrease the amount of labor needed for maintenance (and its associated cost) and also the cost for replacement of components and materials and for power.

The logistic and maintenance support cost factors may be divided into three major categories that are impacted by the system reliability performance level:

1. the cost of corrective maintenance (CM) actions throughout the life of the system;
2. the cost of preventive maintenance (PM) actions throughout the life of the system;
3. the cost of recurring staff maintenance training.

With the exception of the cost of replacement power due to plant shutdown the operational cost of the system is generally independent of reliability performance.

The cost of corrective maintenance actions is the total cost of failed components including labor and cost consumables. This cost is a function of the number of expected failures over the life of the system, the repair time and various other reliability and maintainability factors.

The cost of preventive maintenance actions includes the cost of replacement of components of both labor and consumables. This cost is a function of the time to perform the actions and the interval between the actions which are dependent on the design-in maintenance features and the actual reliability performance level. Preventive maintenance includes efforts to:

(a)   Restore or replace parts exhibiting aging or wearout—The effects of wearout on critical equipment can be minimized through replacement at predetermined time periods. Since wearout failures increase with time, a suitable time period for replacement of parts can be defined through detailed age exploration studies based upon the distribution of failure and the acceptable probability of failure before replacement.

(b)   Align and checkout—The determination of gradual drifts in various operating parameters of an equipment may aid in the identification of the degradation of various components. Through alignment and checkout of an equipment, peak performance may be retained or, when the limit of acceptable performance is reached, may also permit the preventive replacement of a component before a failure condition occurs.

(c)   Clean, lubricate and service—The nature of an equipment and the environment in which it is operating to a large extent determine the frequency of this type of preventive maintenance action.

The cost of recurring training is directly proportional to the number of personnel required to maintain the baseline system at the plant, the labor turnover rate and the cost of training of a single repairman to proficiency in the system. This cost is also dependent on the design-in reliability level and the maintainability features of the system.

Figure 3.2 shows that acquisition cost increases with reliability. Improved equipment reliability is the result of the application of a well-planned reliability program during acquisition. The reliability program defines the extent and effectiveness of the controls, disciplines and provisions employed during system development. In general two major reliability program levels can be defined as follows:

1. High reliability program—applied to systems having high criticality and/or because unscheduled maintenance actions are very difficult and expensive. Stringent specifications and extensive reliability provisions are required consistent with a well-defined, tightly-regulated product.

2. Normal reliability program—applied to systems having lower criticality and maintenance procedures can be applied relatively easily and are inexpensive. Downtime is not critical.

Specific provisions and controls for each level are determined by considering each of the elements described in Sections 3.2–3.7, that would constitute the program relative to the needs and constraints of the equipment to be procured. For example, the quality level of the parts selected (element number 23) has a major impact on reliability and cost. To meet the full spectrum of reliability and cost needs, varying levels of part screens, burn-in tests and manufacturing control provisions have been defined and applied. Table 3.7 depicts these levels as they relate to microcircuits, semiconductors, resistors and capacitors and can be used as a guide in selecting the most cost-effective level to be applied in a given equipment application.

Derating (element number 22) also has a major impact on part

TABLE 3.7
Component Control Levels

Microcircuits:	Class S	Classes B, B1, B2	Class C	Commercial
Semiconductors:	JAN TXV	JANTX	JAN	Commercial
Resistors:	S	R	M, P	Non-established reliability
Capacitors:	T, S	R, P	L, M	Non-established reliability

US MIL-STD-1562 provides equipment designers and manufacturers with lists of microcircuits considered most acceptable for various reliability level applications.
US MIL-STD-701 provides device characteristics, ratings and other parameters of standard semiconductors.
US MIL-STD-199 provides equipment designers with a selection of standard resistors for use with various reliability level applications.
US MIL-STD-198 provides the equipment designer with a selection of standard capacitors for use with various reliability level applications.

reliability and, ultimately, equipment/system reliability. Derating can be defined as the operation of a part at less severe stresses than those for which it is rated. In practice, derating can be accomplished by either reducing stresses or by increasing the strength of the part. Selecting a part of greater strength is usually the most practical approach. Tables 5.6–5.9 depict recommended derating levels for various part classifications as they would apply to electronic equipment at 25 °C (ambient).

Modularity (element number 13) refers to the separation of the components of a system into physically and functionally distinct groups or entities to facilitate fault isolation, removal and replacement. Modularized systems provide simplicity, uniformity and flexibility of configuration which results in reduced training of maintenance personnel, reduced skill levels, standardization and a smaller number of replaceable items. The extent of built-in diagnostics will determine how rapid and economical the maintenance functions can be performed.

Reliability growth testing (element number 19) will significantly improve reliability. Achievement of reliability goals is largely dependent on the extent to which testing has been used during development to 'force out' design and fabrication flaws, and on the rigor with which these flaws are analyzed and corrected. A primary objective of growth testing is to provide methods by which hardware development can be dimensioned, disciplined and managed as an integral part of overall development. Reliability growth testing also provides a technique for extrapolating the current reliability status (at any point during the test) to some future result. In addition, it provides a method for assessing the magnitude of the test–fix–retest effort prior to the start of development, thus facilitating the making of tradeoff decisions.

The model used most often for reliability growth processes, in particular reliability growth testing, is one originally published by J. I. Duane (see Fig. 6.26). Essentially, this model provides a deterministic approach to reliability growth such that the system MTBF versus operating hours falls along a straight line when plotted on log-log paper. That is, the change in MTBF during development is proportional to t^α where t is the cumulative operating time and α is the rate of growth corresponding to the rapidity with which faults are found and changes made to permanently eliminate the basic causes of the faults observed.

Production reliability screen tests (element number 30) are employed to eliminate incipient failures during the manufacturing process. A screen is designed to remove inferior devices from production and, thus, reduce the product failure rate by methods of stress application. The application of a stress test, or a sequence of tests, to a component or subsystem will reveal inherent weaknesses (and thus incipient failures) without destroying the integrity of the item. The purpose of screening is to compress the early failure period and reduce the failure rate to acceptable levels as quickly as possible. A thorough knowledge of the item to be screened and the effectiveness and limitations of the various tests is necessary to establish a useful and reliable screening process. Included among the screening tests for electronic assemblies are fixed time tests, fixed time–temperature tests, time–temperature cycling tests, and time–temperature–power cycling tests.

Supplier audit and surveillance (element number 5) will also bring about significant reliability improvement. An effective surveillance program involves a comprehensive evaluation and continuous monitoring of contractor facilities and program activities (from a reliability and quality assurance standpoint) beginning during contractor selection and continuing through development and production.

Figure 3.3 illustrates, conceptually, some of the general relationships between this application of key reliability elements and acquisition cost. Cost-estimating relationships such as that shown in the figure can be established for the actual equipment to be acquired and then used to produce rough estimates of the extent and cost of reliability improvement. These cost-estimating relationships are established for each applicable reliability element or design attribute by considering that the other elements or attributes are constant and at the level generally considered most appropriate for the parameter being varied. The relationships shown in Fig. 3.3 provide a basis for determining the reliability levels that are most cost-effective and thus are incorporated into the hardware procurement specifications.

The specification of proper reliability requirements is essential in acquiring reliable and maintainable hardware systems and equipment. The specification requirements and in particular the reliability attribute levels govern the entire acquisition process. They provide a basis for allocating reliability design and cost targets as well as structuring and scheduling the overall acquisition development program including the reliability growth and tracking process.

Determining the reliability levels that are cost-effective is a task that involves engineering LCC tradeoff analyses based on the attributes and relationships given in the previous paragraphs of this chapter. It is an iterated analysis process which may, in order to be most cost-effective, require some revision of the initial system performance requirements.

Once the reliability level has been established and the program

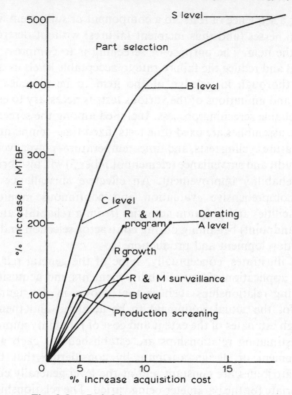

Fig. 3.3. R & M cost estimating relationships.

requirements structured, effort then focuses on preparation of the requirements for inclusion into the actual procurement specification. Generally there are six sections of a procurement specification:

(1) Scope
(2) Applicable documents
(3) Requirements
(4) Quality assurance provisions
(5) Preparation for delivery
(6) Notes

The following guidelines identify some of the necessary elements for effectively specifying reliability requirements:

Section (1), Scope, should include a clear, concise abstract of specification coverage with a description of the item(s) in sufficient detail to preclude misinterpretation of the extent of the coverage intended by the specification.

Section (2), Applicable documents, lists only those specifications and documents that are referenced in the body of the specification. Documents referenced must be in approved form at the time of specification issue.

Section (3), Requirements, includes clearly expressed quantifiable requirements which reflect minimum acceptable operational demands, definitions of satisfactory performance and criteria for success or failure by mode, function and degree. Also included are time frames of interest, environmental and special field conditions, program requirements specifically applicable to system development and/or production, references to appropriate general specifications, reporting requirements, including submission dates for special reports required by specifications or referenced documents called out, and date of submission of detailed acceptance test plans for approval.

Section (4), Quality assurance provisions, includes general test or inspection conditions (e.g. duty cycles), description of item(s) to be accepted under tests if different from the total system as defined under 'Scope', and the number and sampling plan for selection of items to be tested together with the estimated test duration. Also included are success and failure criteria related to test conditions and the accept/reject criteria of the test plan.

Section (5), Preparation for delivery, includes special packaging and bracing requirements and may even specify the type of carrier. Also includes disposition of test items.

Section (6), Notes, includes unique or changed definitions of terms in addition to explanatory information to assure clarity of the previous sections.

All reliability factors applicable to the item to be procured are contained in the requirements section of the specification document. (Note: Only sections (3) and (4) of the specification document are contractually obligating.)

The task of specifying reliability program elements to support the procurement of systems and equipment is performed early during the hardware acquisition process. As previously indicated there are two major classes of procurements:

1. Development: Provides for the development of a new system. Such a contract may, but does not necessarily, include provisions for the fabrication of test hardware models. Also included is the design of

new equipment in order to achieve additional or improved system capabilities.

2. Commercial (off-the-shelf): Provides for the procurement of existing hardware for particular applications without special design adaptations for integration with (or support of) current plant equipment. Section 3.9 discusses methods for achieving reliability in commercial equipment.

Also, there are procurements involving continuations. For example, a manufacturer may be awarded a contract for the development of new hardware with a follow-on contract to fabricate and install at the plant. Reliability requirements are dependent upon the particular procurement type combination.

Development of reliability specifications for new systems requires knowledge of:

1. Quantitative design requirements (e.g. MTBF, MTTR).
2. Qualitative program elements (e.g. prediction, FMEA, etc., as described in Sections 3.2–3.7).

The purpose of a system reliability program is to provide an organized method of assuring that the requirements and objectives are properly considered during acquisition and maintained during plant operation. The depth and level of a program is governed by the safety/availability requirements for the equipment function, the configuration constraints of the function itself, the state-of-the-art (in view of the functional requirements), the maintenance concept and cost limitations. Many times it may be cost-effective to apply military reliability standards to the procurement of 'high rel' power plant equipment. Following is a description of some of the MIL-STDs that may be applicable to power plant hardware procurements:

US MIL-STD-785—establishes criteria for a reliability program and provides guidelines for the preparation and implementation of a reliability program plan. Although the standard was established for the US DoD procurements, it is also applicable to other procurements where system reliability is an important consideration.

US MIL-STD-756—establishes uniform procedures for predicting the quantitative reliability of electronic equipment throughout the development phases to reveal design weaknesses and to form a basis for apportionment of reliability requirements to the various sub-divisions of the product.

US MIL-HDBK-217—Provides essential failure rate data for electronic parts and indicates how *US MIL-STD-756* may be implemented using this data. The handbook is designed to improve prediction accuracy. Application factors are included to account for the severity of the use environment.

US MIL-HDBK-472—provides information on current maintainability prediction procedures and valuable information and guidance to personnel concerned with the design, development and production of equipment and systems requiring a high order of maintainability.

Determining the appropriate program elements calls for reviewing the type of contract in view of the reliability design requirements and attributes. The nature of the procurement (e.g. development or commercial) will to a large extent dictate the requirements. If the hardware to be procured is an off-the-shelf, commercial product, no reliability prediction or reliability growth and demonstration testing would be included. However, depending upon the required reliability level, acceptance tests may be required (see Section 3.9). The relationship of the specified reliability level to the state-of-the-art will also dictate the extent of the program activities. If the specified MTBF, for example, is close to the maximum that can be achieved within the state-of-the-art (that is, if there is little room for reliability improvement) then a very rigorous and intensive program would be required. However, if the specified value is not stringent and there is ample room for reliability improvement, then the program would not have to be extensive.

Table 3.3 presents a reliability program matrix that can be used to aid in determining program requirements. This matrix identifies the essential reliability program elements relative to each life cycle phase and to the levels of system/equipment criticality. This matrix provides a basis for selecting applicable program elements and ultimately formulating R & M specifications.

A reliability specification checklist is presented in Fig. 3.4. This checklist is intended to aid in preparing hardware specifications and to ensure that appropriate requirements are imposed. In addition, the completed checklist provides a brief abstract on the nature of the reliability requirements called out in the procurement of hardware systems. This overview of the R & M requirements provides assurance that basic cost-effective requirements have been thoroughly considered, reviewed and included in the specification.

Standard requirements for each of the program elements provide an initial basis for structuring a specification. The information compiled in

R&M Specification Checklist

I. General Information:
Specification number _____
Title _____
Program manager _____
Equipment description _____

II. Reliability Design Requirements and Attributes:
MTBF _____ Service life _____
MTTR _____ Availability _____
Forced outage rate _____ Planned outage rate _____
Startup reliability _____

III. Program Elements:
(1) Program planning ☐
(2) Organization ☐
(3) Program review ☐
(4) Configuration control and management ☐
(5) Supplier audit and surveillance ☐
(6) Reliability critical item list (and control) ☐
(7) Reliability training and indoctrination ☐
(8) Tradeoff analysis ☐
(9) R&M specification ☐
(10) Reliability math, modeling ☐
(11) Reliability prediction and allocation ☐
(12) Maintainability prediction and allocation ☐
(13) Diagnostics and modularity ☐
(14) Human factors analysis ☐
(15) FMEA/FMECA/FTA ☐
(16) CMF/CCF ☐
(17) Single failure analysis ☐
(18) LCC assessment and analysis ☐
(19) Reliability growth (and tracking) ☐
(20) Reliability demonstration ☐
(21) Design review ☐
(22) Component derating ☐
(23) Component quality level ☐
(24) Component failure rate modeling and estimation ☐
(25) Component testing methods and compliances ☐
(26) Component selection, specification and qualification (including QPL and STD) ☐
(27) Effects of storage/dormant periods and shelf life ☐
(28) Maintenance planning (RCM) ☐
(29) Spare parts/logistics analysis ☐
(30) Controlled screening and burn-in ☐
(31) Reliability acceptance ☐
(32) Degradation control ☐
(33) Cyclic inspection ☐
(34) R&M assessment ☐
(35) PRA/R analysis ☐
(36) Reliability (or availability) improvement ☐
(37) Failure reporting and corrective action ☐
(38) Reliability data recording and feedback ☐

Fig. 3.4. Reliability specification checklist.

part 3 of the checklist for a given system can then be applied to tailor the standard requirements. The design requirements (reliability level and attributes) can be obtained from part 2 of the checklist. It must be emphasized, however, that the program requirements for a given procurement must be tailored to meet its need and constraints. Boiler plate requirements must not be used *per se*. The actual requirements for a given procurement are prepared by incorporating carefully selected reliability design requirements, and program elements with scheduling and submittal information based on the methods discussed in this section.

3.9 ACHIEVING RELIABILITY IN COMMERCIAL EQUIPMENT

Readily available commercial equipment and components are used extensively in nuclear power plants. The advantages of using commercial hardware include: proven design, reduced lead time and low cost (because of minimal development expense or none at all). Possible disadvantages associated with commercial equipment include inability to meet specific reliability requirements, limited performance, parts availability, reduced control of changes and increased logistic support requirements.

Commercial equipment, in general, is used when criticality in terms of operational requirements is normally low and when tradeoff studies indicate that this acquisition approach is optimal because no new development is required.

Procurement of commercial equipment requires effort to select items with 'as is' suitability and demonstrated acceptability to meet the needs of the operating plant. Specification efforts are restricted to describing only those performance requirements necessary to assure hardware acceptability. Design requirements are specified only to the extent necessary and essential to satisfy operational needs. No detailed reliability program elements are specified.

Ensuring adequate reliability of a commercial hardware item starts with the performance of an initial reliability assessment based on an analysis of all available data. Many sources of data may be used in the initial assessment. These include: plant performance/reliability data, industry-wide general reliability/outage data, and R & M estimates from formal prediction procedures. Adjustments are then made to the initial assessment, based on judgement factors, to arrive at the actual level to specify in order to account for the impact of technology differences and/or changes in production methods applicable to the new procurement. Because of field data limitations, predictive procedures are often applied to determine the reliability of commercial equipment during acquisition. Among the reliability prediction procedures are part count and stress analysis techniques; both of which use failure rate data available in *US MIL-HDBK-217*.

The part count technique provides an estimate of reliability based on a count by part type (resistor, capacitor, integrated circuit, transistor, etc.). The method allows rapid estimates of reliability in order to quickly determine the feasibility of a given design approach. The stress analysis technique provides a more accurate estimate of reliability based on the use of detailed part models plus a calculation of application stress values for each part. Procedures for performing reliability predictions based on part count and stress analysis techniques are given in Chapter 6.

Comparative evaluations may be performed to provide supplementary part reliability data. This involves the extrapolation of failure data from well documented parts (i.e. *US MIL-HDBK-217*) to parts having little or no failure data, provided similarity exists. Similarity refers to those performance type, class, construction, material, or rating parameters by which the comparison can be made. To remain valid, extrapolation must account for the differences between the parts compared as well as their similarities. Also, regression analyses may be applied in lieu of, or to supplement, *US MIL-HDBK-217* estimates in those cases where a strong correlation exists between performance variables and reliability. Regression analysis is a means, using analysis of failure data, to relate reliability (i.e. MTBF) directly to major performance parameters.

In addition, several qualified suppliers for the equipment should be identified if possible. Also, the application of a reliability improvement warranty (RIW) should be given consideration as a means for further committing a supplier to a specified actual reliability and for reducing life cycle costs. The thrust of a warranty is to achieve acceptable reliability through an improvement profit incentive. The supplier warrants the product to be free from defects in design, material and workmanship and to operate in its intended environment as required for a specified period. The warranty commits the supplier to replace failed items and/or perform repair services for the specified duration. The supplier prices the warranty to cover his expected repair cost which must be consistent with a specified reliability level. If the actual reliability exceeds the specified level the supplier gains. Conversely, if the actual reliability is less, the supplier loses.

Also, it is particularly important in the case of commercial equipment that cost effective maintenance requirements be determined and applied to assure reliability during plant operation. RCM concepts as described in Section 3.4.7 (element number 28), should be applied to determine optimum maintenance requirements. This involves application of a logical analysis process on critical failure modes identified by the manufacturer or through an engineering analysis conducted during equipment selection to determine those maintenance tasks that should be performed to prevent a decrease in reliability and/or deterioration of safety to unacceptable levels. Specific requirements for on-condition inspection, hardtime replacement and condition monitoring should be determined and applied as part of the reliability program. Procedures for planning and implementing RCM principles are given in Chapter 6.

Chapter 4

Evaluating Reliability Programs for Nuclear Power Plants

Compliance to specified requirements during system acquisition is assured by first auditing and then monitoring the supplier's reliability plans, organizational functions, their interfaces and engineering analyses as well as by participating in design reviews, witnessing formal tests and conducting source inspections. Results of an audit and subsequent monitoring evaluations are used to track the implementation status and effects of the reliability program and to rate and compare different programs and the elements within them. This chapter discusses reliability program evaluation, provides a series of questions to be used in conducting an effective audit and provides extensive guidelines for program monitoring.

The effectiveness of any reliability program is related to the extent to which it is implemented and to the rigor with which it is pursued by the system designer, the manufacturer and the utility. Individual program plans may vary in emphasis and scope yet, when properly weighted and implemented, have the same effect of achieving operating reliability. A program should be dynamic and flexible enough to accommodate changes as the system progresses through its life cycle. Evaluating and monitoring the program is an essential element of the total reliability effort. The utility (or system designer), in addition to preparing reliability requirements that are integrated into hardware system specifications, the statement of work and other contractual documentation, also evaluates proposals, reviews data items (i.e. program plans, predictions, analyses, etc.), participates in design reviews and, in general, continually evaluates and monitors the outputs of the program throughout development, production and plant operation.

The reliability program is evaluated during source selection by performing a detailed review of proposed plans. An example of criteria that can be used to evaluate the reliability aspects of proposed programs is presented in Table 4.1. Effective proposal evaluation will not only lead to the selection of the most qualified supplier with respect to reliability, but will also establish the course for subsequent audits and evaluations during development and production.

Evaluation continues, after source selection, by performing a detailed audit to determine the adequacy of a supplier's overall reliability program and organizational structure and then by monitoring the reliability tasks (including data items, test results and modifications thereto), continually, throughout the development, production and acceptance phases. The auditing and monitoring evaluation process for a given equipment involves, first, developing criteria from a review of the contractual requirements, including the specification, the statement of work, the

TABLE 4.1
Program Evaluation Criteria (Source Selection)

Compliance with requirements
—Meeting R & M parameter values defined in the equipment specification.
—Complying with applicable program specifications and data requirements.
—Demonstrating of R & M values (without minimizing performance capability or incurring excessive cost).

Understanding of the problem
—The scope or range of tasks that make up the R & M effort.
—The R & M technologies (mathematical, statistical, etc.).
—Other advanced, yet proven, methods for R & M programming.
—The interaction between various R & M elements and the system design and development process, including the interface aspects of R & M with development milestones.

Soundness of approach
—Adequate manpower, facilities and other resources to implement the described approach.
—Sufficient flexibility to accommodate design changes, program delays or extension of R & M elements.
—Ability to meet the objectives of the R & M program within the scheduled time period.

Technical expertise
—Background or prior experience in R & M and related areas.

Management
—The R & M organizational structure and its functioning within the overall corporate and program management, including personnel assigned, their technical expertise, management techniques and lines of communication.

---

documentation requirements, and other supplemental information to form a detailed program evaluation basis. The evaluation criteria are then applied to determine the adequacy of the reliability activity.

The reliability program considerations that are stressed when formulating an equipment specification can also be used to formulate criteria for the subsequent program evaluations. Some of the other considerations that should be emphasized and reflected into the evaluation criteria are:

—The specific relationship of the reliability program to other development activities. For example, the reliability program should be a formally organized entity with central management, a documented program plan and separate accountability for program resources. It should be negotiated together with the overall contract (rather than after contract execution), delineating the scope and cost of all reliability efforts.

—The performance of program reviews (and revisions of the program plan, if necessary, depending on the results of the reviews). The reviews should serve as a means of implementing the recommendations of the reliability program evaluation effort.

—The control of subcontractor and supplier reliability programs and the assessment of their effect on the reliability of the overall system.

—Project data including independent reliability assessments of purchased equipment and components.

4.1   RELIABILITY PROGRAM AUDIT

A complete reliability program audit would encompass the following four areas:

(I)   The structure and function of the overall *reliability organization.*
(II)   The *practices and control elements* applied during acquisition and operation.
(III) The *effectiveness* of the reliability program.
(IV)   The *achieved operation and reliability performance* of the equipment.

Each of these areas is addressed in this section in terms of a series of questions that can be incorporated into a given program audit. It must be emphasized, however, that a program audit must be tailored to meet the needs and requirements of the equipment based on criteria developed from the equipment specifications as well as other considerations previously discussed. Guidelines are presented following the audit questions (Section 4.2) to help in the monitoring process through evaluation of specific aspects of a given equipment reliability program to determine compliance with specified requirements.

(I)   Audit of the overall *reliability organization* involves addressing such questions as:

1.   Is a formal reliability organization maintained with a permanent in-house staff?
     What is the background and experience of the person responsible for reliability?
     What is his position in the overall company organization?
     What are the names of other key people in the organization?

2.   What is the relationship of reliability to other organizational functions?
     Do they interface directly with:

Project management?	Design?
Quality assurance?	Other?

2.   What is the size of the reliability staff?
     Is staff composed of experienced reliability engineers?
     Number of professional? Degrees? Average experience?
     Number of non-professional?
     Are supplementary analysts used?

4.   What are the general functions of the reliability organization?
     Specification? (Prepare, review from reliability standpoint.)
     Supplier reliability audit and surveillance?
     Reliability (availability) analysis?
     Design drawing and specification review (for adequacy of reliability requirements)?
     Component engineering?
     Design review?
     Acceptance test monitoring?
     Failure analysis?
     Data recording and analysis?
     Operational reliability assessment?
     Product improvement programs?

5.   Are other organizations involved in reliability (or related) functions?
     Test laboratories?

Quality assurance?

Failure analysis?

6. Are formal reliability training programs conducted?

Reliability engineering training?

Operational and maintenance training?

(IIa) Auditing the effectiveness of the *practices and control elements* applied during *acquisition* involves addressing such questions as:

1. What is the basic approach to assuring system/equipment reliability?

What reliability indices are employed?

Are there formal policies/procedures covering:

determination of reliability requirements?

are requirements derived from system risk analysis and optimized with respect to cost?

acquisition of new (or replacement) system/equipment to meet the requirements?

operation and maintenance to assure that reliability is maintained?

2. Are reliability requirements applied contractually to hardware suppliers?

Design?

MTBF?	Service life?	Forced outage rate?
MTTR?	Availability?	Planned outage rate?
Start-up reliability?		

Component quality?

Redundancy or back-up modes of operation? Fail safe design?

Ease of maintenance? Diagnostics? Modularity? Test?

Development/reliability growth?   Screening?

Demonstration/qualification?   Acceptance?

Management and control?

Reliability program plan?	Failure analysis and reporting?
Parts control?	Data reporting, analyses and feedback?
Critical item control?	Reliability assessment?
Configuration control?	Subcontractor/supplier control?
Program/design review?	

3. Are standard reliability program requirements imposed (either as a guide or contractually, fully or limited)?

US MIL-STD-785?

US MIL-HDBK-217?
US MIL-STD-781?
4. Are *system* reliability program elements planned and applied?
(A) Reliability prediction/apportionment/assessment?
    Allocation techniques?
    Numerical evaluation/prediction analysis?
    Redundancy aspects of passive components?
    Verification test methods?
    SFP detection methods?
The objectives of the reliability methods should be addressed, e.g. are they used to: (i) Establish sub-system/component reliability goals? (ii) Provide quantitative measures of reliability during the development process? (iii) Determine where the design can be improved? (iv) Aid design tradeoff decisions? (v) Provide criteria for planning (and verifying) reliability growth and demonstration tests? (vi) Establish the need for redundancy? (vii) Identify single failure points and provide quantitative input for early spare provision plans? Information on the extent and effectiveness of system models and how they are applied should be obtained. For example: how do the models account for the reliability interconnection of the subsystems and components? What failure rates are used to support reliability predictions? What are the uncertainties in the data and what impact do the uncertainties have on the assessments? What techniques are used to apportion reliability requirements and to set design goals for subsystems and components? Are actual test/field data used to assess achieved reliability?
(B) Maintainability prediction/assessment?
    Allocation techniques?
    Numerical evaluation/prediction?
    Verification methods?
The objectives of the maintainability methods should be addressed, e.g. are they used to: (i) Establish subsystem/component maintainability goals? (ii) Provide a quantitative measure of how easily a design can be maintained? (iii) Determine where the design can be improved? Information should be obtained on the effectiveness of the methods to aid design tradeoff decision, plan and verify test methods and provide input for early spare provisioning plans. Maintenance level diagrams, work factors, repair time data (e.g. determined

via maintenance analysis and which account for human factors and maintenance errors), repair frequencies (e.g. based on component failure rates for maintainability predictions), should be obtained, as well as information on techniques to apportion maintainability requirements and set maintainability goals among subsystems and components.

(C) System interface?
  Compatibility?
  Malfunction effects analysis?
  Detection/evaluation methods?
  Remedial action?
  Consequence analysis?
  FMEA/FMECA?

The effectiveness of these analysis techniques in evaluating system interfaces and compatibility between subsystems/components should be evaluated particularly to determine the consequences of failure or malfunction on overall system reliability. Information should be obtained on how the methods are applied to determine, for example, the need for redundancy and fail safe design features. Identify single failure points, and critical items, and how to assure subsystem/component compatibility. Also, the audit should address the effectiveness of these methods in providing input to R & M models/predictions, identifying critical items and their priorities, defining techniques for failure detection and providing key inputs for developing maintenance strategies and plans. The extent, depth and rigor of the techniques and, in particular, the uncertainties of their results should be considered including: (i) FMECA procedures and the basic data and information (e.g. from design configurations, component engineering and part failure rates resulting from prediction studies) used to support the process; (ii) structural models (logical 'and' and 'or' symbols and failure events) used to support FTA and (iii) other analysis techniques that are applied to further locate and ultimately force out potential malfunctions.

(D) Tradeoff study?
  Improvement techniques?
  Cost-effectiveness?
  Improvement evaluation?

Techniques used to help make R & M tradeoff decisions involving the evaluation of design alternatives as well as the determination of program/test requirements should be evaluated. Information on how tradeoff studies are performed to determine, for example, the optimum MTBF/MTTR mix that would maximize availability should be obtained including sensitivity curves and other data which would show the relationship of R & M parameters controls and engineering tasks to availability and cost.

(E) Design review?
    Techniques applied?
    Reliability impact?
    Remedial action?
    SFP assessment?
    Techniques to surface hidden system faults?

The audit should address procedures applied to systematically review performance, reliability, maintainability and various other system characteristics at major design and testing decision points. The audit should obtain information on the adequacy and completeness of checklists developed to support design reviews. Criteria applied to determining conformance or adequacy should be obtained covering such design/program items as:

—Program plans
—R & M allocation, predictions and assessments
—Identification and evaluation of critical components
—Test plans and procedures
—Maintenance concepts
—Substrate and component specifications
—Remedial actions
—Single point failure (SPF) assessment
—FMEA/FMECA/FTA or other techniques to surface hidden system faults
—Failure analysis reports
—Growth test data
—Production reliability assurance plans
—Supplier control methods
—Configuration management
—Documentation and reports

Also, the methods employed to control design reviews,

including the thoroughness of deficiency follow-up control procedures, should be evaluated.

5. How do data uncertainties effect the system reliability program elements?
6. How are the effects of operator and maintenance actions taken into account on safety system reliability analysis?
7. Are hardware reliability program elements planned and applied?
    (A) Hardware specification?
        Reliability requirements (MTBF, MTTR)?
        Application techniques?

The audit should address techniques used to determine hardware R & M specifications and, in general, how requirements are established that satisfy safety requirements and, operational availability needs, and that are, at the same time, within the state-of-the-art. Information on how quantitative requirements are established and how the requirements are formulated into a hardware specification that reflect an effective balance of the various demands should be obtained.

    (B) Hardware selection?
        Reliability experience?
        Failure rate considerations?
        Interface considerations?
        Hardware maintainability considerations?

Basic procedures and criteria applied to select system hardware (e.g. based on proven R & M and long-life characteristics and demonstrated acceptability to meet system needs) should be evaluated. Information on how failure rate and mode experience data, subsystem/component interfaces (particularly between R & M parameters and the system design and development process), logistic factors and the supplier's background or prior experience in the R & M and related areas are considered in selecting critical hardware items should be obtained.

    (C) Component derating?
        Policy?
        Techniques?
        Are guidelines applied?

Are derating guidelines used in the design of hardware items to assure that all components are operated well within recommended stress limits? Are techniques applied to reduce the probability of hardware-induced failures and to allow the components to realize the full extent of their inherent reliability?

(D) Screening?
　　Burn-in techniques?
　　Testing criteria?
　　Selection (part) approval?

Methods, techniques and guidelines used to plan and implement hardware screening and burn-in programs should be evaluated. Information on the application of stress screening during hardware production on a 100% basis for the purpose of revealing inherent, as well as workmanship and process-induced, defects without weakening or destroying the hardware, should be obtained including screen test profiles, time duration, acceptance criteria and other elements and controls. The methodologies and techniques used to plan optimum screen programs and to determine the most effective burn-in time periods should be examined.

(E) Production degradation control?
　　Method of control?
　　Preventive action?
　　Acceptance?
　　Failure reporting analysis and corrective action?

Techniques used to control reliability during manufacturing, to minimize degradation of intrinsic or designed-in reliability and to accelerate reliability growth should be evaluated. Information on methods of control, including techniques to isolate intrinsic and induced defects in a manner such that special inspections or screens can be applied to eliminate the defects, should be obtained. Information on failure analysis and data collection programs covering failures reported during manufacturing and actual experience during operation and how the data is applied to modify and improve the manufacturing process should also be obtained.

8. Describe any other reliability practices and control elements that are applied during acquisition?
9. How is component reliability preserved in storage, and during installation and construction?

(IIb) Auditing the effectiveness of the *practices and control elements* applied during *operation* involve addressing such questions as:

1. How is reliability assured during operation and maintenance? Is there an operating philosophy? (e.g. minimum number of hours per start)?
   How is the operational staff organized?
   How is the maintenance staff organized?
2. How are operational/failure data collected and analyzed? Is the data system computerized?
   What computer codes are used?
3. Are operational failures reported, analyzed and fed back to system designers and manufacturers?
4. Is available operational reliability data sufficient?
   If, not, how can it be improved?
   At what cost?
5. Are operational reliability (and availability) assessments performed periodically and reports prepared and issued?
   How do data uncertainties affect the reliability assessments?
6. Have product improvement programs been initiated?
7. What reliability program elements are planned and applied?
   (A) Maintenance policy/practices and strategy?
   Preventive maintenance?
   Corrective maintenance?
   Fault detection/isolation?
   Logistics?
   Downtime control?
   Service (life) time?
   Basic maintenance concepts should be evaluated including required maintenance personnel skill levels, support equipment requirements, logistics training repair management, maintenance manuals and support data and other maintenance parameters. Information should be obtained on the rationale and benefits of the maintenance parameters and techniques for establishing maintenance frequencies, corrective maintenance procedures, fault detection/isolation methods and on techniques for controlling downtime over the entire service life of the system/equipment (40 years).
   (B) Replacement strategy?
   Time constraints (frequency)?
   Replacement criteria?
   Verification?

Various replacement strategies including those based on time constraints (or number of cycles) as well as those based on the operational condition of the hardware should be evaluated. Information should be obtained on methods for establishing replacement criteria (time or condition factors), throwaway concepts and techniques for verification that replacement was accomplished properly and that the hardware is restored to full operational integrity.

(C) Reliability growth program?
     Failure analysis procedures?
     Reliability improvement techniques?
     Diagnostic activity?
     Requirements verification MTBF/MTTR)?
     Immaturity failures versus random failures analysis?
     Detection of latent defects during test, method?
     Wearout failure problem, solution?

Methods applied to analyze, correct, improve and, in general, grow reliability should be evaluated. Information on automatic monitoring function designed to survey selected system performance parameters or operating conditioning (such as temperature) in order to detect impending system/component malfunction and to make (or allow) compensating adjustments or corrections should be obtained.

The extent and depth of built-in hardware diagnostics, the application of end-to-end verification testing and the rigor and thoroughness of failure/data analysis procedures should be evaluated. Also failure/data analysis procedures and activities should be evaluated with respect to determining the extent and effectiveness of: (i) analysis techniques to determine root causes as they relate to various hardware technologies; (ii) statistical techniques to isolate infant mortality, random and wearout failures and to establish trends; and (iii) control methods to define personnel responsibilities, scheduling requirement depth of analysis activities, reporting forms, feedback mechanisms and output requirements particularly relative to assessing achieved R & M parameters such as MTBF and MTTR.

(III) Auditing the *effectiveness* of the reliability program involves addressing such questions as:

1. During system/equipment acquisition, were all of the original

reliability program requirements (system/hardware elements) completed in their entirety?

2. As the programs progressed, did the attention to the reliability requirements increase, decrease or stay the same?

3. Which requirements or program elements are considered most cost-effective in detecting and correcting failures prior to plant operation?

4. Did significant management changes or organizational changes occur during the programs affecting the manufacturer, system designer or the utility?

   If so, did this change the attitude regarding the reliability requirements?

5. Were there major changes in the course of the programs such as program stretchouts, performance definition changes, etc?

   Did the program changes affect the reliability requirements?

6. Did major engineering/design changes occur as a result of reliability deficiencies uncovered during:

   —Design (reliability prediction, FMEA, etc.)?
   —Development/reliability growth tests?
   —Screening?
   —Acceptance?
   —Plant operation?

7. Were there major problem areas uncovered during the design reviews?

   If so, were these resolved satisfactory and in a timely manner?

8. Did significant cost overruns occur during acquisition?

   To what were these attributed?

   Were reliability deficiencies significant contributors to any cost overruns?

(IV) Auditing the actual *reliability performance* of equipment used in nuclear power plants involves addressing the following questions:

1. In general are the system/equipment used in nuclear power plants considered:

   —Reliable (performs without problems)?
   —Satisfactory (performs in spite of minor problems, requires maintenance, but easy to maintain)?
   —Poor (performs but fails often and requires extensive maintenance)?
   —Unsatisfactory (fails often and requires extensive and difficult maintenance)?

2. What are the principal reliability problems?
   Design?   Workmanship?
   Operational software procedures?   Human factors?
   Maintenance procedures?
   Environmental considerations?
3. Does actual operational reliability generally agree with what was predicted and measured during acquisition?
4. What subsystems, equipment or components fail most often?
   Mechanical?   Outage rate?
   Electrical?   Outage rate?

In addition to auditing and evaluating the reliability aspects of the program, the effectiveness of the configuration management function must also be assessed. Configuration management is a discipline in which technical and administrative direction and surveillance are applied to identify and document the functional and physical characteristics of a hardware or software item that satisfies an end use function, to control changes to those characteristics and to record and report implementation status on approved changes. It is used to establish and maintain a formal set of procedures by which a uniform system of change identification, coordination, approval and status accounting is accomplished. The procedures are planned early during development and implemented later during acquisition when the system is first placed under configuration management control. Controls are defined for formal documentation (specification, plans, drawings, manuals, etc.) related to facilities, systems and equipment items and computer programs. The implementation of a well-planned properly executed configuration management function is essential to the performance of many of the elements of a reliability program (e.g. assessment, failure analysis and data recording) as well as to the effectiveness of the overall program.

4.2 RELIABILITY PROGRAM MONITORING GUIDELINES

Guidelines that can be used in determining the effectiveness of specific reliability program tasks performed during system development and plant operation are presented in the following pages of this chapter. These guidelines are intended to provide a complete self-contained basis for monitoring an equipment reliability program. Because of the similarity of the audit and program monitoring functions, the guidelines address many of the same topics covered in the audit questions; however, in this case they

TABLE 4.2
Reliability Documentation Requirements Schedule

Required document	Schedule
(1) Reliability program plan	Shortly after award of contract (prior to PDR)
(2) R & D status report	Monthly
(3) Reliability apportionment report	Shortly after award of development contract (prior to PDR)
(4) Reliability prediction and analysis report	At PDR and CDR
(5) FMECA report	At PDR and CDR
(6) Maintenance plan	At PDR
(7) Maintenance allocation report	Shortly after award of development contract (prior to PDR)
(8) Maintenance prediction report	At PDR and CDR
(9) Parts reliability control plan	At PDR
Maximum stress criteria	
Component specifications	
Vendor controls	
(10) Critical item control	Prior to PDR (update quarterly)
(11) Subcontractor R & M control plan	Covered in R & M program plan
Schedule of deliverables	At PDR
Data and reports	As required
(12) Design review plan	Covered in program plan
Agenda	Prior to CDR and PDR
Checklist	Prior to CDR and PDR
(13) Integrated test plan	At CDR
Growth tests	
Screening tests	
Acceptance tests	
(14) Test procedures	At CDR updated 30 days prior to start of testing
(15) Test reports; analysis and corrective action	Within 30 days of test completion
Growth tests	
Screening tests	
Acceptance tests	
(16) Failure summaries	Within monthly status reports
(17) Reliability data collection plan	Described in R & M program plan
(18) Reliability assessment reports	As required

focus on the monitoring of reliability engineering tasks performed on specific equipment.

Guideline 1 and Table 4.2 address project reliability organization, methods of control, planning, and reporting activities. Guidelines 2–18 address individual reliability engineering and control activities. In addition to the technical criteria associated with each task, certain aspects associated with management and control are covered. The reliability program must be evaluated with repect to management relative to each task. In addition, the managerial aspects and basic premises embedded within these tasks must show the interaction of each with all of the others within the framework of the overall reliability plan, and how each task impacts detailed design activities.

Guideline 1: Reliability Project Organization and Control Criteria
—The project reliability organization shall consist of an identifiable group, separate from design, QC, etc., whose manager has direct access to program management and who reports at the same level as design.
—The organization shall be defined with respect to its own critical reliability functions as well as with respect to allied functions (e.g., QC, manufacturing).
—The organization shall consist of a team of specialists having expertise in all pertinent reliability areas (e.g. statistics, physics of failure, component engineering).
—The reliability project manager shall possess sign-off authority on all decision efforts with respect to reliability.
—The overall guiding philosophy of the program shall be defined and the impact on the design effort established (i.e. define fully the tie-in with early design results and describe the interaction of all tasks).
—A schedule shall be provided showing all tasks as well as the interaction of each task with other reliability tasks and task timeliness relative to design and other efforts. Program and hardware milestones shall identify applicable reliability constraints.
—A list of deliverable items and delivery dates shall be provided (see Table 4.2).
—Program plans shall state the intended methods of control (meeting, PERT, reviews, audits, etc.), that include discussions of policy formulation and information dissemination and status reporting.
—Plans shall define status reporting methods including format, scheduling, and delivery.

Guideline 2: Reliability Allocation Criteria
—Overall allocation methodology shall be based on criticality, complexity of design and function, operational use environment, previous experience with similar equipment and relation to the state-of-the-art.
—Specific allocations shall be based on conceptual goals and prediction and shall, if possible, include a further improvement factor that challenges designers; for example, the improvement factor could be 125% of the predicted value.
—Allocations shall be made to lower levels within the system (i.e. subsystem, equipment, assembly or subassembly to be designed or purchased as an entity) and provide design goals for subassemblies and higher level equipment items.
—Allocations shall be completed shortly after the start of the design phase; submittal should be well in advance of PDR.

Guideline 3: Reliability Prediction Criteria
—Effort shall consist of analytical estimates of system reliability (or MTBF) based on mathematical models, failure rates and stress/environmental factors and on the underlying statistical distribution of failures.
—Predictions shall include factors for equipment function, duty cycle, operating and non-operating failure rates, and known applicable failure modes and mechanisms.
—Prediction efforts shall establish inherent reliability to aid in design based tradeoff decisions, provide criteria for the starting point of reliability growth testing, and foster elimination of design flaws.
—Prediction efforts shall support the design of a system that exhibits the inherent MTBF resulting from conceptual design tradeoff studies. Further criteria are:
  —System (or operating mode) failure is a direct reflection of part failure.
  —Reliability is determined from a series arrangement of parts or components (except where redundant design has been employed, or where engineering analysis shows that failure does not degrade performance beyond acceptable limits).
—Data sources for the prediction effort shall consist of historical data derived from plant specific data systems, industry wide databases such as NPRDS, NERC/CADS, IEEE 500, LERs, *US MIL-HDBK-217* and the *RADC-TR-75-22* (see Appendix B).

—Prediction is an iterated process—initially based on gross part counts and subsequently based on detailed stress analysis.

—Scheduling should show prediction as a continuous effort during detail design with predictions updated periodically; submittals prior to preliminary design review (PDR) and the critical design review (CDR).

Guideline 4: Failure Mode Analysis Criteria

—Analysis shall be done part-by-part (and, if possible, failure-mode-by-failure-mode) to determine the consequences of failure on system reliability and planned functions, particularly safety functions, in a way that relates parts, components and functions to their failure effects.

—Analysis shall be based on data and information from design configurations, components engineering and part failure rates resulting from prediction studies, relevant historical information and earlier analyses.

—Analysis shall quantitatively determine the probability of modal failure for each mode identified.

—Results of analysis shall be used to accomplish the following:

  —Provide input to reliability predictions and aid in defining corrective action priorities.

  —Identify critical parts, assemblies, parameters and chracteristics that can be used as basic criteria for inspection.

  —Establish corrective action criteria in advance of equipment fabrication without early large scale testing, to aid in the generation of test plans and procedures.

  —Provide failure-rate-by-mode distributions.

—Analysis shall be updated periodically, based on data from failure analysis and other data collection activities.

—Effort is performed continuously during design iterations; submittals correspond to PDR and CDR.

Guideline 5: Maintenance Plan Criteria

—Plans shall provide definition as to what consitutes a repair action and the scope of maintenance activities planned for execution at site, off-line and at maintenance centers. The approach to periodic or scheduled maintenance activities should be included.

—The maintenance plan shall state the scope and character of fault isolation and post-repair checkout activities including the following:

  —Requirements needed to support the system at each level of repair.

—Amount of operating time needed to perform post-repair checkouts.
—Personnel skill level requirements.

—Plans shall describe the methods and criteria established by which the maintenance concept is translated into hardware design features.
—Scheduling shall show the finalization of the maintenance plan during the early stages of the detailed design effort.
—Definitions of the maintenance plan are submitted and finalized at the PDR.

Guideline 6: Maintainability Allocation Criteria
—Plans shall indicate how repair times (or MTTR) are allocated to systems, components and levels of assembly corresponding to the repair activities performed at the operating plants.
—Each repair time assigned shall include an improvement factor over and above a strict subdivision of system MTTR requirements, which forces emphasis and provides goals during detail design activities. (Improvement factors could possibly be based on a 25 % reduction in MTTR.)
—The results of the allocation shall be used to generate maintainability demonstration and test plans, provide design goals, and indicate marginal areas requiring concentrated effort.
—Specific allocations of MTTR shall account for anticipated repair frequency based on system and component failure rates.
—Allocations shall be completed shortly after the start of the detailed design phase; submittal should be well in advance of PDR.

Guideline 7: Maintainability Prediction Criteria
—Predictions should provide a quantitative evaluation of the design in terms of MTTR, repair rates and other statistical parameters for each level of repair.
—Predictions shall indicate the feasibility of meeting system MTTR objectives and shall provide an assessment of the probability of correct failure indication.
—Predictions shall be supported by maintenance level diagrams, work factors, and other data determined via maintenance analysis.
—Analysis shall identify areas requiring periodic cleaning, adjustment or replacement.
—Predictions shall be used to define preventive maintenance intervals,

to identify time replaceable items and to aid in logistics/supply provisioning.

—Results of predictions shall be submitted to correspond with major review points—PDR and CDR.

Guideline 8: Critical Item Control Criteria

—Plans shall list initial critical items and include parts, equipment/components and other items considered critical from any of the following standpoints:

   —Are reliability sensitive (from early reliability studies, apportionments, etc.).
   —Have limited life.
   —Are high cost items.
   —Have long procurement leadtimes.
   —Require formal demonstration testing.

—Plans shall provide for critical item identification, control and special handling, and shall identify critical item characteristics to be inspected or measured during incoming inspection. Methods include MRB procedures, traceability of material and periodic audits.

—Plans shall cover rules for early procurement of critical parts as well as early buildup and reliability growth testing of critical components as deemed necessary. Specific supplier controls of test methods, which indicate how defects are forced out and reliability growth is achieved, shall be identified.

—Items identified as critical shall be documented. Procedures, tests, results, growth status and efforts to reduce the degree of criticality of each item shall be included in the documentation.

—Documentation for critical items shall be submitted initially prior to PDR and updated quarterly.

Guideline 9: Component Control and Standardization Criteria

—Component control and standardization effort shall be applied to select, specify and control all critical mechanical, electronic and electromechanical parts. A part approval cycle shall be established with a continuous effort to minimize numbers and types of parts and components used.

—The selection process shall include design evaluation, reliability history review, construction analysis, FMEA and cost-effectiveness studies, as necessary.

—This control effort should include the development of meaningful procurement specifications which, when completed, reflect a balance between design requirements, QA and reliability needs consistent with apportionment studies and vendor capabilities. The procurement specification should cover:

—Lot acceptance testing.
—QA provisions (including incoming inspection).
—Qualification testing, if required.

—The component qualification scheme should include detailed and formal submittal of data to support approval requests (data to be either statistical test data or analytical data for components where similarity exists, or a combination of these two types). Note that those components that require formal statistical test data for qualification should be entered under critical item control (Guideline 8).
—The maximum allowable (design application) stress levels for each component type shall be defined.
—A vendor control program with audits of vendor processes, associated documentation and source inspection, where applicable, shall be established.
—A continuous component improvement effort shall be provided based on failure analysis, controlled testing programs and operational reliability assessment.

Guideline 10: Supplier Reliability Control Criteria
—Plans shall show approaches and methods by which they intend to control subcontracted material, including the imposition of requirements on subcontractors in accordance with *US MIL-STD-785* and *US MIL-STD-470*.
—Programs shall include:

—Analytical tasks, such as apportionment, prediction, FMECA and FRACA, which shall be performed with the same degree of rigor as contractor efforts.
—A component control and standardization effort, which interrelates with the control program (especially in the areas of commonality of critical component approval, maximum stress criteria and qualification rationale).
—Growth tests, demonstration tests and qualification tests on subcontracted items.

—Supplier's documentation shall include a reliability program plan, a schedule for accomplishing reliability tasks, and a list of deliverable documentation.

—Submittals of supplier data and reports shall be timed to fit logically into the overall development schedule.

Guideline 11: Design Review Criteria
—Reviews shall be performed against a comprehensive checklist and criteria for reliability and shall provide the means for formal assessment of design effort.

—Review procedures shall provide for formal reviews as well as informal reviews conducted internally.

—Formal reviews shall include PDR, CDR, reliability, program managers, and mock-up reviews. Specific checklists shall be prepared for each review and shall cover the items shown in Table 4.3.

—Review procedures shall contain provisions for deficiency follow-up control.

—A detailed checklist and agenda shall be submitted prior to formal review (prior to PDR and CDR).

Guideline 12: Reliability Growth Tests Criteria
—The test plan for reliability growth testing shall show a vigorous test, fix and retest program, which emphasizes comprehensive and detailed failure analysis activity, and shows relationship between various time factors, growth rates and starting/end points.

—Specific growth test plans shall be formulated as part of the integrated test program and shall show:

—Predicted MTBF.
—Demonstrated MTBF.
—Starting point.
—Growth rate.

—Growth plans shall include the cumulative test time required to grow, to the specified MTBF, the number of test units subjected to growth tests, and the anticipated test time per unit. In addition:

—The growth plans shall indicate realistic time factors which recognize that, in order to grow under a constant level of corrective action, sufficient downtime must be allowed for adequate implementation of corrective action before restarting the growth tests.

TABLE 4.3
Design Review Checklist Considerations

Preliminary design review (PDR)
Identification of critical components
Program plans
Preliminary test plans
Design progress
Reliability allocations and predictions
Maintenance concept
Special studies (e.g. detailed tradeoffs)

Critical design review (CDR)
Subsystem and component specifications
Test plans and procedures
Critical component evaluations
Final design configuration
Reliability analyses
Test results

R & M design review
Reliability allocations
Reliability predictions
FMECA
Failure data
Growth test data
Production reliability assurance

Program managers review
Supplier control
Configuration management
Cost/schedule
Documentation and reports
Status of overall project

—Plans shall include:

Calendar time/month available
Test time/calendar time
Description of test cycle (environment on/off time)

—The plan shall be submitted as part of an overall integrated test plan at CDR.

—Progress of growth testing shall be tracked and logs and data forms maintained to record the number of units on test, test times accumulated, failures, corrective actions and level of reliability of MTBF achieved during the time period.

—Final growth test report shall be submitted after completion of test (within 30 days).

Guideline 13: Reliability Demonstration Test Criteria
—The test plan shall indicate tests to be conducted per *US MIL-STD-781*.
—Plans shall indicate reliability level (MTBF) to be demonstrated and the associated confidence level, and shall show the relationship between demonstrated MTBF, confidence and test times.
—Plans shall show the number of units for test, expected test time, calendar time factors and scheduling of effort.
—The plan shall indicate the kinds of data to be gathered during the test and relationship to maintainability tests.
—The reliability demonstration plan shall be submitted as part of the integrated test plan at CDR.
—The progress of demonstration testing shall be tracked and logs/data forms maintained that record the number of units on test, test time accumulated, failures, corrective action, statistical decision factor and accept/reject criteria.
—Monthly summary reports shall be submitted and final report shall be prepared, per *US MIL-STD-781*, and submitted within 30 days after completion of test.

Guideline 14: Maintainability Demonstration Test Criteria
—The plans shall indicate tests to be conducted per *US MIL-STD-471*. They should include:

  —Parameters to be demonstrated.
  —Confidence associated with demonstration (that is, relationship of the number of failure events (trials) to the total potential failure modes from FMEA studies).
  —Number of units (or systems) involved.
  —Repair levels.
—Representatives from the utility shall be involved in the selection of simulated maintenance trials (failures) to be induced into the system.
—Test plans shall specify scheduling of demonstration and duration of each effort and shall indicate data to be recorded during testing.
—Plans shall be submitted as part of the integrated test plan in time for CDR.
—Progress of demonstration testing shall be tracked and logs/data

forms maintained that record number of trials, nature of repair, repair time, statistical decision factors and criteria for success.

—Summary reports shall be prepared that indicate the test status. A final report shall be prepared within 30 days after completion of test.

Guideline 15: Failure Reporting, Analysis and Corrective Action Criteria (FRACA)

—Plans shall describe methods for reporting, analysis and corrective action of all failures regardless of their apparent magnitude through a formal 'closed loop' failure analysis function.

—Plans shall indicate that activities are to be controlled by a formal written procedure which describes methods, personnel responsibilities, forms, documentation submittals and scheduling of effort. Plans shall indicate specific failure recurrence control procedures and include the following:

—Basic failure analysis approach.
—Failure analysis procedures.
—Depth of analysis.
—Forms and reporting formats.
—Corrective action follow-up procedures.

—Plans shall indicate the applicability of FRACA activities with regard to all development, qualification, pre-qualification, acceptance, growth, demonstration, critical item and other test activities, and their extension through development, production, plant integration and operation of the hardware system. Plans shall contain sufficient detail to describe the sequence of events which occur upon detection of a failure including methods for failure verification and classification.

—Failure analysis methods shall be described which indicate the physical analysis techniques and controlled testing efforts currently used to determine the causes of failure.

—Plans shall describe corrective measures/techniques that can be applied to eliminate (or minimize) the basic failure mechanism. These measures involve (as applicable):

—Component selection criteria.
—Special non-destructive tests to weed out specific failure mechanisms.
—Qualification requirements.
—Special in-process fabrication inspections and tests.

—Component stress/strength criteria.
—Special reliability assurance provisions.

Guideline 16: Reliability Data Collection Criteria
—The data collection effort shall provide management information suitable for long range planning for future needs and shall include experience information in the following categories:

—Resource requirements.
—Logistics requirements.
—Training requirements.
—Overhaul programs.
—System improvement.

and provide the basis for accurate field assessment of equipment reliability.
—Plans shall provide specific mechanisms for collecting operational, maintenance and installation data at plant sites, maintenance areas and during factory test for feedback.
—Data collection shall utilize, where practicable, existing procedures, forms and methods of collection.

Guideline 17: Production Reliability Assurance Criteria
—Plans shall indicate the methods used to assure that the inherent reliability designed into equipment is not degraded during production. Plans shall describe methods for incoming inspection, in-process and final (acceptance) testing. Plans shall show effort in the areas of test, fabrication and inspection procedures and methods of handling/storing components, subassemblies, and other production items.
—A statistically derived quality control plan shall be designed and implemented to achieve maximum control at minimum cost. It shall include increased and more comprehensive inspection at all levels of assembly.
—Plans shall show methods by which stress/screening tests are applied at various levels of assembly.
—Reliability shall be continually assessed during production through detailed analysis of production process flow, actual reject rate statistics and estimates of inspection efficiency factors.
—Scheduling shall show production reliability procedures to be prepared during design with initial submittal at CDR and updated as

required prior to full scale production. Summary reports indicating current production reliability shall be submitted continually during full scale production.

Guideline 18: Reliability Assessment Criteria

—Plans shall indicate how actual achieved reliability will be assessed from system test data and actual plant operating/failure experience.

—Assessments shall compare predicted values and achieved values.

—Assessments shall be performed during development, production and operation. Bayesian statistics could be used to combine the results of theoretical considerations, engineering analysis and test results to yield assessments that use the widest possible range of available data and information.

—Plans shall indicate sources of data, the data reduction method to be used and a system for feedback of these results via an assessment report.

—Assessments should include all pertinent data, such as analytical results (e.g. predictions), development test data (e.g. reliability growth), demonstrations, production and plant operating data.

Chapter 5

Designing for Reliability

This chapter discusses how reliability specifications can be translated into equipment designs and describes those reliability improvement attributes which facilitate this translation into acceptable hardware. The reliability specifications, described in Chapter 3, are established to meet operational needs and to minimize total LCC while maintaining plant safety levels.

The achievement of reliability begins in the early design stage by selecting the quality level of the parts, the extent of derating and the degree of redundancy as well as by selecting the level of built-in test, fault isolation, modularity and other ease of maintenance features to incorporate into the design. Selecting the extent and level of these improvement attributes establishes the inherent reliability and maintainability of the equipment. Rough reliability models are applied during early design to establish the relationship between the known failure rates of individual components and the reliability of the equipment as a whole. The maximum load or stresses to which the components may be subjected are determined and ample safety factors are provided in their specification.

The operational life characteristic curve, presented in Chapter 2, distinguishes between three failure periods in the life cycle of an equipment. The effects of these periods have special importance during design. Chapter 2 discussed some of the failure causes and suggested methods to reduce the failure rate during each of the three periods including the use of high quality parts and the application of screening and derating. In addition to reducing failure rates the components or subassemblies must be accessible for ease of replacement. Less reliable components requiring more service should be more accessible than high reliability components or modules requiring less service. Downtime for repair can be greatly reduced by minimizing the time required for fault isolation through modularization. There is an expanding field of expertise in built-in test and fault isolation

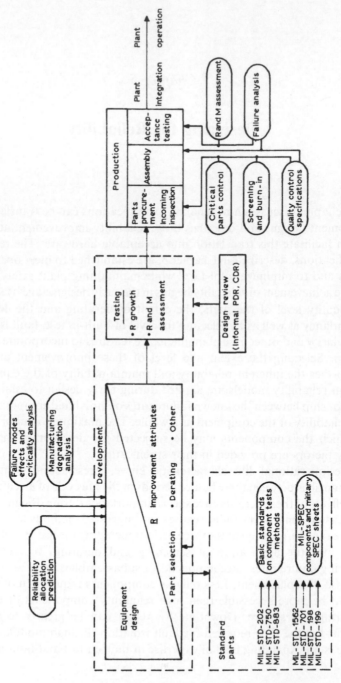

Fig. 5.1. Design reliability activities during equipment acquisition.

techniques from which diagnostic options can be evaluated to select an optimum approach. Redundant or standby subsystems can also be designed into the equipment to reduce lost time and to avoid system failure. Of special importance is the performance of reliability growth testing on prototype hardware in order to refine a design and eliminate latent design shortcomings. Also, various screening procedures, including burn-in, can be planned and then applied to initial hardware to reduce 'infant mortality' and thus further establish the reliability potential of the equipment early in its life cycle.

A balance between reliability and maintainability should be maintained while remaining within the specified requirements. High reliability is required for safety critical equipment and components or to meet plant availability requirements. High reliability is also required when maintainability is expensive and/or difficult to achieve due to unique equipment functions, poor plant location or other design factors. The continuous application of LCC analysis during the equipment design and development phase is important in order to keep the acquisition cost in balance with operating cost.

As the equipment progresses through development, tradeoff studies of alternative design approaches are performed and reliability is continually assessed and evaluated using established models and analysis techniques. An overall life cycle methodology can be implemented (see Fig. 2.17) that consolidates established reliability models for prediction, FMECA and other design improvement techniques for application during development (as well as operation). The methodology would provide baseline estimates of the inherent reliability of progressive levels of design definition for comparison with the level specified and to provide a basis for subsequent reliability improvement during development prior to plant installation and operation.

This chapter describes the reliability improvement attributes (i.e. part selection, derating, etc.) that can be incorporated into the design of an equipment. Figure 5.1 indicates the relationship of the reliability activities during design to the overall equipment acquisition process. Chapter 6 describes the reliability engineering procedures that can be applied to ensure that reliability is designed and built into the equipment.

5.1 PART SELECTION

The parts used in systems and equipment are selected and specified during design. This involves selecting standard high quality parts with proven reliability and long life characteristics wherever possible. Many standard parts are available for application to modern control equipment. These parts constitute the building blocks from which equipment is developed and, as such, greatly impact the actual reliability of the hardware. Since the reliability of the end item is largely dependent upon these building blocks, the importance of selecting and applying high quality, cost effective parts cannot be overemphasized.

Part selection is of particular importance for complex microcircuits, semiconductors and other electronic parts used in computerized control systems. These parts are susceptible to a variety of failure modes and mechanisms and their failure rates are dependent on the part technologies, how they are processed and the quality controls and screen tests applied. Table 5.1 presents a list of some of the factors which must be considered in the selection of cost-effective parts.

The following part selection guidelines cover relative qualification and screening requirements, failure rates and cost-effectiveness trends for microcircuits, semiconductors, resistors and capacitors. These guidelines show the relative difference between the reliability/quality levels and, consequently, are intended to provide an initial basis for part selection. The

TABLE 5.1
Part Selection and Control Factors

Does the part failure mode impact plant safety (from failure mode studies)?

Does the part have a short replacement life?

Does the part have a long procurement lead time?

Does the part require qualification testing?

Is the part a high cost item?

What is the required part failure rate?

What is the derating factor required (based on reliability allocation studies)?

Is burn-in (or other screening) needed to achieve the failure rate?

Is a standard MIL-STD item available from a qualified vendor?

What is the normal delivery cycle?

Will the part be available throughout the life of equipment?

Is there an acceptable standard procurement document?

Are existing procurement specifications available?

Is a part procurement specification necessary?

Are multiple sources available?

---

guidelines are not intended to be used for specifying part screening tests or for establishing part failure rates. Specification of part screen tests are discussed in Section 5.2. Part failure rates are determined from a review of their specific application in view of the technologies, stresses and other factors using *US MIL-HDBK-217* or other equivalent data source (see Section 6.1).

5.1.1 Microcircuit Selection Guidelines (Fig. 5.2)

Microcircuits can be procured to specified levels of quality as defined by *US MIL-M-38510* and *US MIL-STD-883* (and *British Standard, BS 9000*)

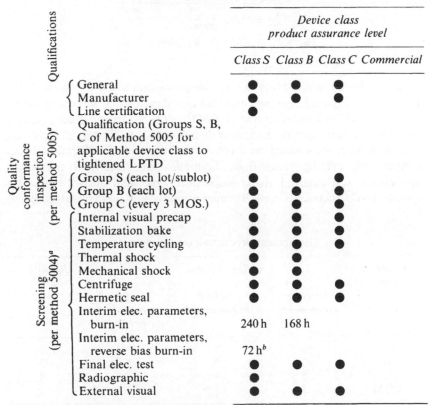

Qualifications and Screening Requirements
for *US MIL-M-38510* Microcircuits

	Device class product assurance level			
	Class S	Class B	Class C	Commercial
Qualifications				
General	●	●	●	
Manufacturer	●	●	●	
Line certification	●			
Qualification (Groups S, B, C of Method 5005 for applicable device class to tightened LPTD)				
Quality conformance inspection (per method 5005)				
Group S (each lot/sublot)	●	●	●	
Group B (each lot)	●	●	●	
Group C (every 3 MOS.)	●	●	●	
Screening (per method 5004)				
Internal visual precap	●	●	●	
Stabilization bake	●	●	●	
Temperature cycling	●	●	●	
Thermal shock	●	●		
Mechanical shock	●	●		
Centrifuge	●	●	●	
Hermetic seal	●	●	●	
Interim elec. parameters, burn-in	240 h	168 h		
Interim elec. parameters, reverse bias burn-in	72 h[b]			
Final elec. test	●	●	●	
Radiographic	●			
External visual	●	●	●	

[a] Per *US MIL-STD 883*.   [b] When specified.

Fig. 5.2. Microcircuits. (a) Screening requirements.

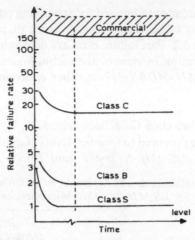

Fig. 5.2.—*contd.*   (b) Reliability levels.

which are keyed to reliability failure rate prediction models given in MIL-HDBK-217. Figure 5.2(a) indicates that qualification and screening for all class S, B and C microcircuits (per *US MIL-STD-883*) first includes general and manufacturer certification, and then proceeds to the use of screening and inspection tests based on internal visual precap, stabilization bake, temperature cycling, centrifuge, hermetic seal, electrical parameter verification and external visual inspection. Over and above these high quality levels (classes S and B), microcircuits are further subjected to

Equipment production costs (relative)

Control level	Relative cost			Relative reliability
	Discrete components	*Hybrid microcircuits*	*Complex monolithic circuits*	
S	50	[a]	[a]	150
B	12	9	5	75
C	5	4	2·5	9
COM	3	2·5	1	1

[a] Not readily procurable.

Fig. 5.2.—*contd.*   (c) Cost trends.

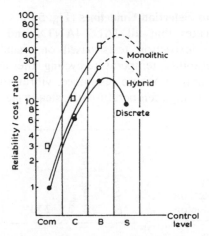

Fig. 5.2.—*contd.*   (d) Cost-effectiveness trends.

screening based on thermal and mechanical shock. Beyond this, class B microcircuits are subjected to 160 h burn-in. Class S microcircuits undergo a longer, 240 h, burn-in and then proceed to a special 72 h reverse bias burn -in, followed by a radiographic examination. Commercial type microcircuits can also be procured but, generally, they are not subjected to special manufacturing procedures, inspections or burn-in. However, as a minimum, they generally undergo some form of visual and electrical parameter testing.

Figure 5.2(b) indicates the resulting differences in failure rate that are observed due to the special procedures and processes employed in the manufacturing of class S, B and C microcircuits. The differences in failure rate are relative and are indicative of an inherent reliability level. Factors relating to generic type environmental stress and other considerations that impact these failure rates are described in *US MIL-HDBK-217*.

Figure 5.2(c) presents the cost trends associated with the use of discrete components, hybrid microcircuits and monolithic type microcircuits in a complex equipment. It should be noted that the relative reliability levels apply only within the same component technology group (discrete, hybrid or monolithic) and it is not intended to imply that different technologies, using the same class level, will produce an equivalent system reliability. Figure 5.2(d) is indicative of the cost-effectiveness trends associated within the different component technology groups. It indicates that the use of class B microcircuits for many equipments will tend to be most cost-effective.

5.1.2 Semiconductor Selection Guidelines (Fig. 5.3)

Figure 5.3(a) indicates that all JAN, JANTX and JAN TXV semiconductors include inspection tests based on visual and mechanical performance, solderability, temperature cycling, thermal shock, terminal strength, hermetic seal, moisture resistance, vibration, vibration noise, constant acceleration, high temperature life, steady state life, intermittent

US MIL-S-19500
90% confidence level
sampled quality inspections
for all JAN semiconductors

	Destructive test	*Non-destructive test*	*TXV*	*TV*	*JAN*	*COM.*
Group A—Inspection						
Visual and mechanical		●	●	●	●	●
Electrical performance		●	●	●	●	●
Group B—Inspection						
Physical dimensions		●	●	●	●	●
Solderability (heat)	●		●	●	●	
Temperature cycling		●	●	●	●	●
Thermal shock		●	●	●	●	
Thermal strength	●		●	●	●	
Hermetic strength	●		●	●	●	
Moisture resistance	●		●	●	●	
Shock		●	●	●	●	
Vibration fatigue		●	●	●	●	
Vibration (variable frequency)		●	●	●	●	
Vibration noise		●	●	●	●	
Constant acceleration		●	●	●	●	
Thermal strength[a,b]		●	●	●	●	
High temperature life (NONOP)	●		●	●	●	
Steady-state life	●		●	●	●	●
Intermittent life	●		●	●	●	●
Group C—Inspection						
Barometric pressure		●	●	●	●	
Salt atmosphere	●		●	●	●	
TX, TXV screening						
100% process conditioning			●	●		
100% power conditioning			●	●		
100% precap inspection			●			

[a] Method 2036, *US MIL-STD-750*.
[b] Electrical rejects may be used if end points are not required.

Fig. 5.3. Semiconductors. (a) Screening requirements.

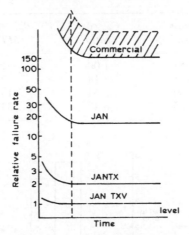

Fig. 5.3.—*contd.* (b) Reliability levels.

life, barometric pressure and salt atmosphere tests. Over and above these, TX semiconductors are further subjected to special screening based on 100% process conditioning and 100% power conditioning. TXV semiconductors further undergo a 100% precap inspection.

Figure 5.3(b) indicates the resulting difference in failure rate that is observed due to the special procedures and processes employed in the manufacture of JAN TXV, TX and JAN semiconductors. The difference in

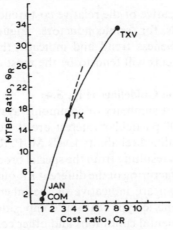

Fig. 5.3.—*contd.* (c) Costs.

	Average cost	C_R, cost ratio	θ_R, relative MTBF	$\dfrac{\theta_R}{C_R}$
COM	30·55	1·00	1·0	1·00
JAN	0·65	1·18	2·0	1·69
TX	2·00	3·64	16·7	4·59
TXV	4·67	8·49	33·3	3·92

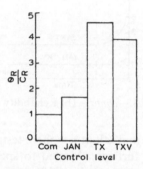

Fig. 5.3.—*contd.*   (d) Cost-effectiveness.

failure rate is relative and is indicative of inherent reliability level. Factors relating to generic type (silicon, NPN, PNP, etc.), environmental stress and other considerations that impact these failure rates are described in *US MIL-HDBK-217*.

Figure 5.3(c) is indicative of the relative cost trends associated with the various control levels for semiconductors. Figuer 5.3(d) shows the associated cost-effectiveness trend and indicates that JANTX semiconductors for many systems will tend to be the most cost-effective.

5.1.3 Resistor Selection Guidelines (Fig. 5.4)

Figure 5.4(a) provides a summary of commonly available resistors as per *US MIL-STD-199*. Of particular interest are the established reliability (ER) resistors with qualified reliability levels M, P, R and S. Figure 5.4(b) depicts the failure rates resulting from the special procedures and processes employed in the manufacturing of the different ER quality assurance levels. The failure rates shown are indicative of the inherent reliability levels. Factors relating to generic type (carbon composition, metal film, etc.), circuit stress, environmental conditions and other considerations that can significantly alter these failure rates are described in *US MIL-HDBK-217*.

Common resistors (US MIL-STD-199)

Type	Specification	Style	ER specification	Style
Fixed				
Composition	US MIL-R-11[a]	RC	US MIL-R-39008	RCR
Film high stability	US MIL-R-10509	RN	US MIL-R-55182	RNR
Film power uninsulated	US MIL-R-11804[a]	RD	—	—
Film power insulated	US MIL-R-22684[a]	RL	US MIL-R-39017	RLR
Wirewound power	US-MIL-R-26	RW	US MIL-R-39007	RWR
Wirewound power chassis	US MIL-R-18546	RE	US MIL-R-39005	RBR
Wirewound precision	US MIL-R-93[a]			
Variable				
Composition	US MIL-R-94	RV	—	—
Film lead screw	US MIL-R-22097	RJ	US MIL-R-39035	RJR
Metal	US MIL-R-23285[b]	RVC	—	—
Non-wirewound precision	US MIL-R-39023[b]	RQ	—	—
Wirewound low temp.	US MIL-R-19	RA	—	—
Wirewound power	US MIL-R-22	RP	—	—
Wirewound precision	US MIL-R-12934	RR	—	—
Wirewound lead screw	US-MIL-R-27208	RT	US MIL-R-39015	—
Wirewound semi-precision	US MIL-R-39002	RK	—	—

[a] Deleted by US MIL-R-1998.
[b] Not in US MIL-STD-1998.

Fig. 5.4. Resistors. (a) Types.

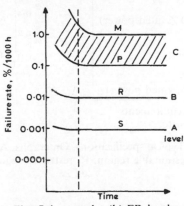

Fig. 5.4—contd. (b) ER levels.

US MIL-R-390088
Sampled quality
inspections for
RCR resistors

High temperature
conditioning

Group I
Visual and mechanical
d.c. resistance

Group II
Resistance–temp. chart
Voltage coefficient
Dielectric voltage test
Insulation resistance

Group III
Low-temp. operation
Temperature cycling
Moisture resistance
Short-time overload

Group IV
Terminal strength
Soldering heat test

Group V
Shock
Vibration, high freq.

Group VI
Life (100 % rated power)

Group VII
Solderability

Group VIII
Life (50 % rated power)

Screening requirements
(RCR resistors)

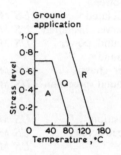

Ground
application

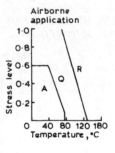

Airborne
application

Fig. 5.4.—*contd.*   (c) Typical specifications. On graphs: A, acceptable region; Q, questionable region; R, restricted region.

Reliability information

The established reliability specification provides for the establishment of a failure rate figure through the single parameter of load life only. This is usually done at 50% of maximum rated power.

Typically, the established reliability specification provides for more frequent

    Moisture resistance
    Burn-in
    Other screening tests

on a 100% basis.

For high reliability, resistors should be derated at 50% or less of maximum rated power, *US MIL-STD-199* presents power derating curves vs high ambient temperatures.

Allen Bradley $\frac{1}{2}$ W, 5%, Qty: 1000–4999	
RC20	RCR 20
0·038	0·038

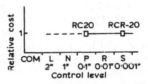

Fig. 5.4.—*contd.* (d) Cost trends.

Figure 5.4(c) indicates the sample screening and inspection tests used for carbon composition resistors in accordance with *US MIL-R-39008*. The screening and inspection tests include high temperature conditioning, visual and mechanical inspections, d.c. resistance measurements, low temperature operation, temperature cycling, moisture resistance, short-time overload, shock, vibration, solderability and rated power life test. Figure 5.4(d) indicates that in certain cases, e.g. carbon resistors, it is cost-effective to use the highest reliability level available.

5.1.4 Capacitor Selection Guidelines (Fig. 5.5)

Figure 5.5(a) is a compilation of common capacitors per *US MIL-STD-198* and includes the following generic types: glass/mica, electrolytic, paper/plastic, ceramic and variable types of capacitors. It is of special interest to note that those capacitors which have established reliability specifications may be available at different reliability quality levels.

Common capacitors (*US MIL-STD-198*)

Type	Specification	Style	Specification	Style
Glass or mica				
Fixed mica dielectric	*US MIL-C-5*	CM	*US MIL-C-39001*	CHR
Fixed mica button style	*US MIL-C-10950*	CB	—	—
Fixed glass dielectric	*US MIL-C-11272^a*	CY	*US MIL-C-23269*	CYR
Electrolytic fixed				
Aluminum, polarized	*US MIL-C-62*	CE	—	—
Non-solid, tantalum foil	*US MIL-C-03965^a*	CL	*US MIL-C-39003*	CSR
Solid electrolyte, tantalum	*US MIL-C-26655^a*	—	*US MIL-C-39003*	CSR
Al_2O_3 electrolyte	*US MIL-C-39018*	CU	—	—
Paper or plastic				
Feed through, RFI, a.c./d.c.	*US MIL-C-11693*	CZ	*US MIL-C-39011^a*	CZR
By-pass, RFI, a.c./d.c.	*US MIL-C-12889*	CA	—	—
Fixed, paper or plastic, a.c./d.c.	—	—	*US MIL-C-14157^a*	CHR
Fixed, plastic	*US MIL-C-19978*	CQ	—	—
Fixed, metalized dielectric, d.c.	—	—	*US MIL-C-55514^a*	—
Fixed, plastic dielectric, d.c.				
Ceramic				
Fixed, temp. compensating	*US MIL-C-20^b*	CC	—	—
Fixed, general purpose	*US MIL-C-11015*	CX	*US MIL-C-39014*	CKR
Variable				
Glass, piston, tubular	*US MIL-C-14409*	DC	—	—
Ceramic dielectric	*US MIL-C-81*	CV	—	—
Air dielectric	*US MIL-C-92*	—	—	—
Vacuum dielectric	*US MIL-C-23183^b*	—	—	—

^a Not in *US MIL-STD-198*.
^b To be added to *US MIL-STD-198*.

Fig. 5.5. Capacitors. (a) Types.

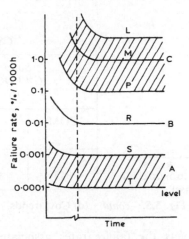

Fig. 5.5.—*contd.* (b) ER levels.

For reliability, capacitors should be derated to 50% or less of maximum rated voltage. Capacitor operating temperature range must be compatible with anticipated equipment operating temperature and hermetically sealed capacitors should be used in environments where anticipated relative humidity can exceed 80%. The designer should also be careful to observe ripple voltage ratings.

Part type	Derating parameter	Derated to % of rated voltage
All	Observe ripple voltage rating	30
Ceramic	Voltage	50
Glass	Voltage	50
Mica	Voltage	60 dipped, 40 molded
Plastic	Voltage	50
Mylar	Voltage	60
Paper	Voltage	50
Tantalum solid	Voltage	50
	Reverse voltage	2
	Circuit impedance	73 Ω/V
Tantalum wet	Voltage	60
	Reverse bias	0
Tantalum foil	Voltage	60

Fig. 5.5.—*contd.* (c) Reliability information.

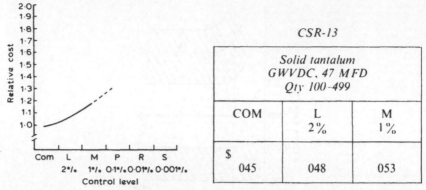

Fig. 5.5.—*contd.*   (d) Cost trends.

Figure 5.5(b) depicts the failure rates associated with the special procedures and processes employed in the manufacturing of components with the different established reliability levels. The failure rates are absolute and are indicative of the inherent reliability levels. Factors relating to generic type, circuit application, environmental stress and other considerations that can impact these failure rates are described in *US MIL-HDBK-217*. Figure 5.5(c) identifies certain derating and circuit application considerations which are generally recommended to circuit designers in order to enhance reliability. Figure 5.5(d) indicates the cost trend associated with the use of a particular established reliability tantalum capacitor. Based on this example, it can be seen that the highest quality level capacitor may not be cost-effective.

5.2   APPLICATION OF SCREENING DURING PRODUCTION (OR MAINTENANCE)

For many new systems especially complex electronic control equipment used in nuclear power plants, the application of proper screening and burn-in procedures will mitigate design and process induced defects and thus provide a higher level of operating reliability. The purpose of reliability screening and burn-in is to compress the early mortality period of a hardware item and reduce its failure rate to an acceptable level as quickly as possible. The rigor of the applied tests and subsequent failure analysis and corrective action efforts determines the degree of reliability improvement. A thorough knowledge of the hardware to be screened and the effectiveness and limitations of the various available screen tests is necessary to plan and implement an effective screening and burn-in program.

Screening generally involves the application of stress on a 100 % basis for the purpose of revealing inherent, as well as workmanship and process induced, defects without weakening or destroying the product. The application of stress serves to reveal defects which ordinarily would not be apparent during normal quality inspection and testing. There are a large number of tests and test sequences that can be applied to remove defects induced at the various levels of fabricated assembly.

For electronic equipment, screen tests can be applied at the part, subassembly and final assembly level. In order to detect and eliminate most of the intrinsic part defects, initial screening is conducted at the part level. Certain part defects, however, are more easily detected as part of an assembly test. This is particularly true of drift measurements and marginal propagation delay problems. Assembly defects, such as cold solder joints, missing solder joints and connector contact defects, can be detected only at the assembly or subsystem level. At high assembly levels, the unit's tolerance for stress is lower and, thus, the stress that can be safely applied is lower. As a general rule, screens for known latent defects should be performed as early in the assembly process as is possible—at the point where higher stress levels and more cost-effective screens can be applied.

The flowchart presented in Fig. 5.6 shows how screening is employed within the design, development and production processes. Screen test planning must start early with the establishment of reliability and cost targets during design. This initial effort is accomplished through an overall reliability analysis and through cost tradeoff studies based on actual failure-rate and failure-mode experience with similar equipment. As the design emerges, detailed reliability predictions and LCC estimates are made for the equipment down to the assembly and component level. During development, an initial screen test plan is prepared that reflects the sequence of screens considered optimum based on anticipated fall-out (or reject) rates, cost-effectiveness and the guidelines presented in this section. Prototype hardware is then built and equipment tests performed. Changes and refinements are made to the screen test plan, as required. At the completion of the demonstration test, a complete (and debugged) production screening test specification is prepared that provides requirements designed to assure a smooth transition from development to production. The screen test specification is then incorporated into the final equipment technical package and the program is transitioned to manufacturing.

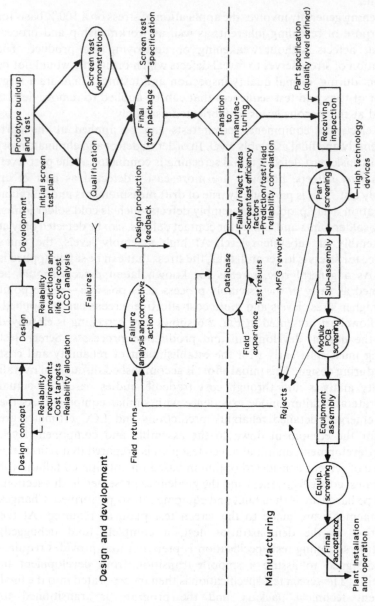

Fig. 5.6. Application of screen testing within the development and manufacturing process.

The idealized manufacturing process, depicted in the lower part of Fig. 5.6, starts with screened parts procured and received to a predetermined level of quality. Screen tests are then applied, as required, at the different levels of assembly. All screen test rejects are analyzed. The results of this analysis are used to identify appropriate design changes and modifications to the manufacturing process, and to reduce, if possible, the overall test burden. All screen test results, including reject rates, failure modes and time-to-failure data are incorporated into a dynamic real-time database from which the effectiveness of the screening test program is continuously assessed and modified as necessary.

In general, screen and burn-in tests can be applied at the three major levels of assembly: part, intermediate (i.e. printed circuit board, module) and equipment. The initial planning and tradeoff studies should take into account the effectiveness and economic choices between part, intermediate and final equipment level screens, and the parameters that must be considered. Table 5.2 provides guidelines to aid in planning and specifying a screening program for electronic control equipment.

Part testing is relatively economical and can be incorporated into supplier specifications. It has the potential for maximum cost avoidance, particularly when applied to complex microcircuits and other high technology devices where reliability is largely dependent on fabrication techniques and process control. Screen stress levels can be matched to requirements which, in general, enable the safe application of higher and more effective stress levels to remove known part defects. Part screens, however, have no impact on the control of defects introduced during subsequent phases of manufacture and assembly or on equipment level problems.

A list of some of the basic screens applicable to various part types with references to standard military part specifications is given in Table 5.3. *US MIL-STD-883* describes testing methods and procedures which are applied to microcircuits (see Table 5.4). Each test covers the apparatus to be used, the procedures and the test conditions. An outline of *US MIL-STD-883*, Method 5004 Screen Tests, is given in Table 5.5.

Part screening is not to be construed as an acceptance test. Rather, it is a process to eliminate weak and potentially unacceptable parts through application of stresses prior to assembly (where the costs associated with defective parts become multiplicative). Screening is a cost-effective procedure to provide qualified parts that meet or exceed reliability and quality targets for assembly into complex equipment. Figure 5.7 illustrates results of the application of a part screen test. It shows how

TABLE 5.2
Screening and Burn-in Guidelines for Electronic Control Equipment

Guideline	Part level	Intermediate level (e.g. PCB)	Equipment (or final assembly) level
Tests should be designed to expose and remove:	Anticipated failure mechanisms for known failure modes, or specific failure mechanisms unique to a given part type (and lot variances) that may not be exposed by standard screens; part types showing excessive failure in test and field experience; mature parts to assure consistent component quality	Parts damaged during assembly; latent defects that escaped part screening; assembly-induced defects	Parts and subassemblies damaged during final assembly; latent defects that escaped part and intermediate screens; harnessing or connector intermittents and other defects induced during final fabrication
Screens should focus on:	New high-technology parts: microprocessors and peripherals, large-scale integrated circuits, special semiconductors	High volume items where majority of fabrication is at intermediate (or PCB) level; high density, high technology assemblies	Low volume items with standard parts and boards, where final fabrication is customized to meet individual application needs

Types of screens to be applied:	Supplier's normal commercial quality screen and burn-in test program should first be evaluated for adequacy. Select screens for effective exposure of anticipated failure mechanisms. Standard screens should be considered (e.g. *MIL-STD-883* for microcircuits, *MIL-STD-750* for semiconductors	Emphasis should be given to temperature cycling with automatic functional testing at ambient, where number of cycles depends on PCB density (with last 1–2 cycles failure free). Number of cycles (or test duration), failure-free criteria and other requirements should be adjusted based on results of subsequent tests and field performance. Screening, generally, should be performed with power off to support production rates and avoid problems with connections, board-to-board switching, cross-talk	Emphasis should be given to an operating burn-in test with power/temperature cycling (with at least 2 complete cycles failure free) where the temperature extremes are set by optimizing: test efficiency versus burn-in facility costs; space requirements; other cost/logistic constraints. Functional testing after screening should be designed to exercise item in self-testing mode
Reliability requirements, criteria	Should not exceed manufacturer's ratings—stress levels must be specified that optimize the process without inducing defects. Tests that are most likely to cause the highest number of rejects should be performed first, to minimize losses and delays	Should not exceed design capabilities of individual parts and material that comprise the assembly	

(continued)

TABLE 5.2—*contd.*

Guideline	Part level	Intermediate level (e.g. PCB)	Equipment (or final assembly) level
Reliability requirements, criteria	Part defects and failures occurring in manufacturing process and in the field should be monitored, and failures analyzed to determine failure mechanisms and conditions, to identify patterns, and to determine if improved screens can be applied at the part level to expose and remove those failure mechanisms The use of plastic encapsulated devices because of their sensitivity to packaging failure mechanisms (i.e. moisture penetration and thermal instability) should be considered carefully and used only where end item environmental variations are minimal or if the assembly/product level packaging system provides adequate moisture protection	Rejected assemblies and PCBs should be analyzed to identify causes of rejects and to institute corrections to the design or process. Reject data should be analyzed to identify trends, failure modes that account for greatest percentage of rejects and to prioritize corrective action efforts. Failure/reject reports should be prepared and maintained showing number of rejects by test, failure mode, cause and corrective action Failure-free criteria should be set based on system requirements and to assure that screening has been effectively completed. Note that rejects during screening indicate the test is effective and potential field failures are being removed from product. During the failure-free period, rejects indicate that the screen test is not complete and rescreening should be performed. Rescreening should be based on a percentage of the original screen and determined from an evaluation of the reject rates and product reliability objectives/risk factors Deliverable spare PCBs and assemblies should be tested to a level equivalent to the system's total screening and burn-in test program	

TABLE 5.3
Part Screens and Burn-in Tests

Screen	Part type				Cost	Effectiveness
	Microcircuits[a]	Semiconductors[b]	Resistors,[c] capacitors,[d] miscellaneous passive parts	Mechanical parts, relays, connectors, fuse, etc.		
Temperature cycling	Extremely useful screen for all part types				Inexpensive	High
High temperature storage	*	*			Inexpensive	Good
Thermal shock	*	*	*	*	Inexpensive	Good
High temperature burn-in	*	*	*		Expensive	Probably the most effective screens for most active parts; in general as part complexity increases, dynamic operating stresses are more effective than static
High temperature reverse bias burn-in	*	* (Transistors)				
Mechanical shock	*	*		*	Moderate	Good
Constant acceleration	*	*			Moderate	Good
Vibration sine variable		*		*	Expensive	Good
Vibration, random				*	Very expensive	High

[a] Per *US MIL-M-38510* and *US MIL-STD-883.*
[b] Per *US MIL-S-19500.*
[c] Per *US MIL-STD-199.*
[d] Per *US MIL-STD-198.*
(Source: Anderson, R. T. and Bass, S., How to control reliability from design through burn-in, *Evaluation Engineering,* March/April, 1981.)

TABLE 5.4
US MIL-STD-883 Test Methods

Method no.	Environmental tests	Method no.	Electrical tests (digital)
1001	Barometric pressure, reduced (altitude operation)	3001·1	Drive source, dynamic
1002	Immersion	3002·1	Load conditions
1003	Insulation resistance	3003·1	Delay measurements
1004·1	Moisture resistance	3004·1	Transition time measurements
1005·1	Steady state life	3005·1	Power supply current
1006	Intermittent life	3006·1	High level output voltage
1007	Agree life	3007·1	Low level output voltage
1008·1	High temperature storage	3008·1	Breakdown voltage, input or output
1009·1	Salt atmosphere (corrosion)	3009·1	Input current, low level
1010·1	Temperature cycling	3010·1	Input current, high level
1011·1	Thermal shock	3011·1	Output short circuit current
1012	Thermal characteristics	3012·1	Terminal capacitance
1013	Dew point	3013·1	Noise margin measurements for digital microelectronic devices
1014·1	Seal	3014	Functional testing
1015·1	Burn-in test		

Mechanical tests

2001·1 Constant acceleration
2002·1 Mechanical shock
2003·1 Solderability
2004·1 Lead integrity
2005 Vibration fatigue
2006 Vibration noise
2007 Vibration, variable frequency
2008·1 Visual and mechanical
2009·1 External visual
2010·2 Internal visual (monolithic)
2011·1 Bond strength
2012·1 Radiography
2013 Internal visual
2014 Internal visual and mechanical
2015 Resistance to solvents
2016 Physical dimensions
2017 Internal visual (hybrid)

Electrical tests (linear)

4001 Input offset voltage and current and bias current
4002 Phase margin and slew rate measurements
4003 Common mode input voltage range
Common mode rejection ratio
Supply voltage rejection ratio
4004 Open loop performance
4005 Output performance
4006 Power gain and noise figure
4007 Automatic gain control range

Test procedures

5001 Parameter mean value control
5002 Parameter distribution control
5003 Failure analysis procedure for microcircuits
5004·2 Screening procedures
5005·2 Qualification and quality conformance procedures
5006 Limit testing

TABLE 5.5
US MIL-STD-883 Method 5004 Screen Test Characteristics

Method 2010—test condition B
Performed normally prior to capping or encapsulation on a 100 % inspection basis
Complex microcircuits may require substitution of alternate screening procedures
(see Method 5004)
Tests include:
  Metallization defects (75 × to 150 × magnification)
    —scratches   —probing
    —voids   —bridging
    —corrosion   —alignment
    —adherence
  Diffusion and passivation layer(s) faults (75 × to 150 × magnification)
  Scribing and die defects (75 × to 150 × magnification)
  Bond inspection (30 × to 60 × magnification)
    —gold ball bonds   —general bonds
    —wedge bonds   —beam lead
    —tailless bonds
  Internal leads (30 × to 60 × magnification)
    —wires
    —beams
  Package conditions (30 × to 60 × magnification)
    —foreign material
    —die mounting
    —die assembly

Method 1008—test condition C
High temperature storage *without* electrical stress applied
Test condition C description:
  150 °C
End point measurements to be performed within 96 h after removal of device from
ambient (150 °C) condition

Method 1010—test condition C
Exposure of parts to extremes of high and low temperatures and alternate exposure
to these extremes (temperature cycling)
Minimum of 10 cycles
Cycle (test condition C) consists of
  1. 10 min at $-65 ^{+0}_{-5}$ °C
  2. 5 min maximum at $25 ^{+10}_{-5}$ °C
  3. 10 min at $150 ^{+5}_{-0}$ °C
  4. 5 min maximum at $25 ^{+10}_{-5}$ °C

Method 1011—test condition A
Thermal shock test
Determine resistance of device to sudden exposures to extreme changes in
temperature

TABLE 5.5—*contd.*

Devices shall be preconditioned by immersion and in intimate contact with a suitable liquid (water) at the temperature specified in step 1 for a minimum of 5 min with immediate part transfer to temperature levels occurring throughout the test (less than 10 s)

A minimum of 15 cycles shall be performed

A cycle consists of (test condition A)

1. 5 min minimum at $100 \, {}^{+5}_{-0} \, °C$
2. 5 min minimum at $-0 \, {}^{+0}_{-5} \, °C$

Method 2001—test condition E
Constant acceleration
Designed to indicate structural and mechanical weaknesses
A constant acceleration is to be applied for a minimum of 1 min in the Y_1 plane
Test condition E specifies a stress level of 30 000 g

Method 1014—fine seal test
Purpose is to determine the effectiveness of the seal of microelectronic devices
Requirements vary depending upon test technique used

Method 1015
Burn-in test
Purpose is to eliminate those devices which would fail as time and stress dependent failures due to manufacturing aberrations (infant mortality or early lifetime)
Per Method 5004, 160 h at 125 °C shall be used as a minimum
The specific test condition has not been identified in Method 5004
Post burn-in test measurements shall be made no more than 96 h after test completion

Method 2009—external visual
Employed at outgoing vendor inspection *or* incoming user inspection
To verify that the materials, design, construction, markings and workmanship of the device are in accordance with the procurement document
Device is to be magnified between 3 × and 10 × with a field of view large enough to contain the entire device
Tests include inspection for:
  Device design, lead identification, markings, materials, construction and workmanship not in specification
  Defects or damage from manufacturing, handling and testing
  Visible evidence of corrosion, contamination or breakage, defective or damaged plating
  Leads which are not intact and aligned in their normal location, free of sharp or unspecified bends and free of twist outside the normal plane
  Leads which are not free of foreign material
  Other defects or features which will interfere with the normal application of the device
  Evidence of any nonconformance with the detailed drawing or applicable procurement document, absence of any required feature or evidence of damage, corrosion or contamination which will interfere with the normal application of the device

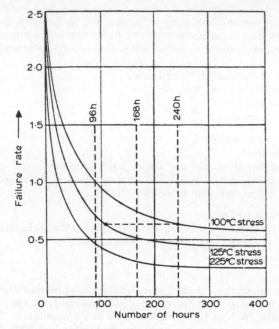

Fig. 5.7.   Part failure rate improvement due to temperature/time stress screening.
(Source: *RDH-376*.)

temperature/time stress screening can improve a part's failure rate. It also shows that by applying a higher temperature stress, i.e. 125 °C instead of 100 °C, comparable failure rate levels can be achieved in less time, i.e. 100h instead of 240 h.

Screening can be inefficient and costly if the screening stresses are not carefully designed to 'attack' the specific defect present in the population. Among the risks associated with screening are:

1. Screen may damage good parts.
2. The nature of defects may change with time.
3. Screens in use may not attack all defects present.
4. Screen is being used for defects no longer in the population.
5. Population of defects may vary for different production lines.
6. Screen is not based on the reliability level that one is attempting to reach.

Intermediate testing is more expensive but can remove defects introduced at the subassembly or module level as well as those intrinsic to the parts.

Because of the several part types incorporated into a board, somewhat lower screen levels must be applied. Generally, special burn-in/temperature cycling facilities are required as well as special automatic test equipment (ATE). In general, some amount of ATE is employed in virtually all large scale screening programs. Automatic testing can not only perform functionality testing rapidly after screening or burn-in of complex boards (or other assemblies) but is also effective in the detection of evasive faults. The latter consists of marginal performance timing problems and other defects arising from part interactions. The extent of the facilities and equipment needed is dependent on the test conditions specified.

Temperature cycling is a highly effective module or subassembly stress test which reveals workmanship and process-induced defects as well as those intrinsic part defects which escaped detection at the part-level screen. Temperature cycling is performed specifically to reveal:

—Assembly defects:
  —delamination
  —fracture
  —insulation cracking
—Part/board bond separating
—Solder problems (cracking opens, etc.)
—Part defects
—Tolerance drift

Figure 5.8 illustrates the environmental conditions and profile under which a typical temperature cycle can be performed. The actual number of cycles

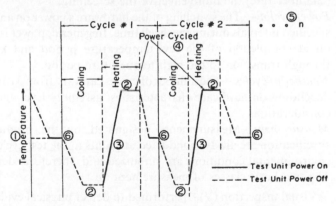

Fig. 5.8. Time/temperature screen test model.

employed is dependent upon board density and part technology and the number initially set represents a baseline for designing the test. Temperature cycling screening, like any quality inspection test, is considered to be a dynamic test where the number of cycles is adjusted, depending on the results of subsequent higher level test or operational performance.

A temperature cycle screen test can be completely specified by the six control parameters (the circled numbers in Fig. 5.8) and the values assigned to them. The significance and impact of each of these parameters is as follows:

1. *Temperature range*—In general, the higher the temperature range, the more defects will be exposed. High and low temperatures are limited by the maximum ratings of the material and the availability of environmental chambers. Low limits of temperature range may be set for -55–$0\,°C$ (typically $-40\,°C$), high limits for 55–$125\,°C$, depending on hardware characteristics, assembly level and facility limitations.

2. *High/low temperature dwell time*—The portion of the cycle that the hardware under test remains at the high or low temperature. Dwell time must include ample time for internal parts to stabilize within $2\,°C$ of the specified temperature extremes and to perform any required measurements.

3. *Transfer time*—The total time to transfer from the specified low temperature to the specified high temperature, or the reverse. In general, the higher the rate of change up to 15 or $20\,°C\,min^{-1}$, (chamber rate) the more effective the screening.

4. *Power cycling*—The switching of the hardware's power on and off at specified intervals during the dwell time. In general, power is turned on at completion of the low temperature period and kept on through transition to the high temperature period.

5. *Number of cycles*—The number of cycles can vary from 3 to over 25 depending on hardware characteristics, assembly level and other considerations.

6. *Measurement*—Measurements designated by the product test specification are always conducted at points in the test cycle where environmental conditions are stabilized and, therefore, definable. The two basic classes of measurement are:

   —Visual inspection (V)—performed to detect physical evidence of defects or damage to the hardware resulting from testing.

—Functional testing (F)—performed to measure a limited number of critical parameters to assure that the test hardware is operating properly. Complete functional testing includes visual inspection and is performed prior to and after screening.

Different time/temperature screen tests can be specified through variations of these parameters. For example, a temperature-cycle screen test may require a transfer time (parameter 3) of 5 °C min $^{-1}$ whereas a thermal shock test would require the transfer time to be less than 10 s with all other parameters the same. A power cycle burn-in can be specified by requiring a constant elevated temperature (parameter 1) for a specified dwell time (parameter 2) and generally power cycled at specified intervals (parameter 4).

Equipment or final assembly level screen testing is expensive but can remove defects introduced at all levels of fabrication. At this point in manufacturing, the permissible stress level may not adequately exercise certain specific parts. However, these high level tests are considered important, even if it is thought that the lower level tests had eliminated all defective parts, board and module defects. The assembly of the remaining components and the boards into the larger assemblies and into the final item cannot be assumed to be free of failure-producing defects. Good parts may be damaged in final assembly, workmanship errors can occur, and system-level design defects may be present. Typical reliability and quality defects found in high-level assemblies or in the final equipment include overstressed parts, improper solder joints and cracked wires due to insufficient stress relief. Despite QC inspections, equipment has even been produced with parts missing. Special burn-in and temperature cycling facilities are required, but little expensive specialized test equipment is necessary because the completed system can be exercised in a nearly self-testing mode.

Typical final assembly screens include:
—Temperature cycling
—Vibration, sine-fixed frequency, random
—High temperature burn-in

As with intermediate level testing a temperature cycle test or a high temperature burn-in at the final assembly level can be effective in revealing workmanship and process induced defects as well as those part defects which escape detection at the part and intermediate level screens.

A low-level vibration test is very effective in revealing solder problems, connector contact problems and loose hardware, and for assuring the mechanical integrity of the overall assembly prior to burn-in, final acceptance testing and release for use.

Recently the use of random vibration, which provides simultaneous excitation of many modes, in contrast to the single frequency sine vibration test, has proven to be an effective screen for isolating workmanship defects found in electronic systems. A screening program developed by the US Navy (*NAVMAT P-9492*) incorporates random vibration as a manufacturing screen. It requires that the equipment under test be hard-mounted to a shake table capable of reproducing random vibration having the power spectral density characteristics shown in Fig. 5.9.

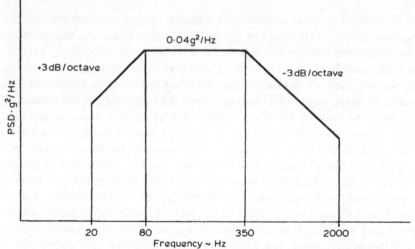

Fig. 5.9.   Random vibration spectrum. (Source: *NAVMAT P-9492*.)

A key step in planning an effective screen test program is the identification of the kinds of failure modes that can occur and the assembly level at which they may be induced. The appropriate screen tests are those which are most effective in accelerating the identified modes, whether they are intrinsic to the part or induced by the manufacturing process. Some of the more common screens are listed below with an indication of their effectiveness.

Temperature cycling—Extremely effective at all levels of assembly; reveals part/PCB defects, solder problems, bond separations, tolerance drifts, mismatches and changes in electrical characteristics.

High temperature burn-in (power cycling)—Effective at all levels of assembly, will reveal time/stress dependent part and process defects.

Vibration, random—Effective primarily at equipment level; reveals solder problems, part/PCB defects, connector problems, intermittents, loose hardware and structural problems.

High temperature storage—Relatively inexpensive screen that can be applied at any level of assembly to reveal time/dormant stress (non-electrical) dependent defects.

Thermal shock—Relatively simple screen that can be applied at the part or module level to reveal cracking, delamination and electrical changes due to moisture or mechanical displacement.

Vibration, sine fixed frequency—Applied at final assembly level to reveal loose hardware, connector contact problems and intermittents.

The planning process involves exercising various options and comparing expected reject or fall-out data to the number of possible defectives that may escape as estimated from an analysis of production inspection reject data and field data on similar systems and equipment.

The rejects that result from screening provide an indication that the test is effective and that potential field failures are being removed from the product. If screening shows very few defects, it is either insufficiently severe, or the item being screened is already reliable. The 'reject rate' is a measure of the number of parts, assemblies or subsystems deviating from specification.

Experience has shown that screen reject rates can vary widely, depending on hardware technology, complexity, process variance (lot by lot) and the nature of the test applied. It should be noted, for example, that if a large number of microcircuits are used in an assembly, removal of even a small number of defects at the part level can have a significant economic impact. It also should be noted that reject rates can vary widely from lot to lot. It is not uncommon for a mature and normally reliable device (where most lots typically show no defects) to occasionally have a lot with an extremely high reject rate. The screen test can prevent the bad lot from causing severe problems later in production.

The effectiveness of the screen test program once implemented in the factory can be evaluated continually and monitored to determine the need to modify individual test criteria and conditions and identify the possibility of earlier detection where the test costs are less and the potential for cost avoidance is higher. It should be emphasized that the initial program represents a baseline for applying the screen tests. A screen test like any quality inspection must be considered as a dynamic test, its duration (or number of cycles) adjusted depending on the results of subsequent higher level tests or field performance. However, the extent and nature of any changes should be determined only through careful review and analysis of the subsequent failures.

5.3 DERATING

Derating can be defined as the operation of a component or electronic part at less severe stresses than those for which it is rated. In practice, derating is accomplished by either reducing stresses or by increasing the strength of the part. Selecting a part of greater strength is usually the most practical approach. Derating is effective because the failure rate of most parts decreases as the applied stress levels are lowered below the rated value. The reverse is also true: the failure rate increases when a part is subjected to higher stresses and temperature. The failure rate model of most parts is stress and temperature dependent.

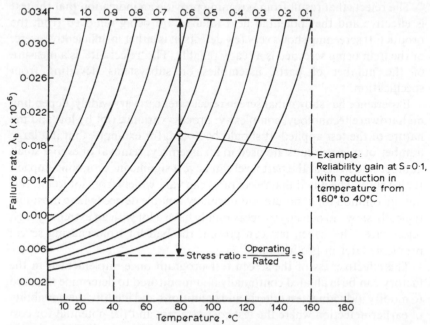

Fig. 5.10.   Failure rate/temperature relationship for group I transistor (silicon, NPN). (Source: *RDH-376*.)

As a general rule, derating should not be conservative to the point where costs rise excessively, e.g. lower than necessary part ratings are selected. Neither should the derating criteria be so loose as to render reliable part application ineffective. Optimum derating occurs at or below the point on the stress–temperature curve where a rapid increase in failure rate is noted for a small increase in temperature or stress. There is, however, a practical minimum to derating. At some minimum stress level, complexity may be necessarily increased to gain performance, thus offsetting the reliability gain accomplished by derating.

For electronic equipment, data on failure rates versus stress are available for a number of parts; *US MIL-HDBK-217* provides failure rate versus stress data for most electronic devices. The curves of Fig. 5.10 are typical of the data that are available. These data can be used to determine the reliability improvement through derating. Guidelines for derating parts used in electronic control equipment are presented in Table 5.6. The table

TABLE 5.6
Derating Criteria at 25 °C (Ambient)

Part type	% Stress for equipment R & M class		
	A	B	C
Microcircuit (125 °C)[a]	50	60	70
Microcircuit (150 °C)	53	62	70
Microcircuit (175 °C)	60	70	80
Semiconductor (100 °C)	50	65	80
Semiconductor (125 °C)	50	60	70
Semiconductor (150 °C)	53	62	70
Semiconductor (175 °C)	60	70	80
Semiconductor (200 °C)	70	75	80
Resistor, F, comp.	50	65	80
Resistor, F, film	60	70	80
Resistor, power, wirewound	45	60	75
Resistor, precision, wirewound	45	60	75
Capacitor, mica	55	65	75
Capacitor, ceramic, temperature compensating	55	67	80
Capacitor, paper/plastic	50	60	70
Capacitor, air trimmer	50	65	80
Capacitor, ceramic	55	67	80
Capacitor, tantalum, wet	48	64	78
Capacitor, tantalum, solid	50	67	78

[a] Maximum temperature.

TABLE 5.7
Derating for Coils, Chokes and Transformers

	Maximum permissible % of manufacturer's rating		
		Voltage	
Inductor type	Current operating	Maximum applied	Transient (maximum)
Coil, inductor saturable reactor	60%	60%	90%
Coil, radio frequency fixed	70%	60%	90%
Inductor general	70%	60%	90%
Transformer, audio	70%	60%	90%
Transformer, pulse, low power	60%	60%	70%
Transformer, power	70%	60%	90%
Transformer, radio frequency	60%	60%	90%
Transformer, saturable core	60%	60%	90%

TABLE 5.8
Relay Derating Chart

Part type	Stress parameter	% stress (allowed)	Remarks
Relay	Rated contact current	50%	The rated contact current for each contact set shall be derated by 50%. Each active contact set must be calculated separately. (a) Active means actually wired to serve a circuit function. (b) Each circuit path through the relay constitutes a contact set.

Notes:
1. Consideration must be given to the type of load to be switched, i.e. inductive, capacitive, lamp resistive, or motor when computing operating current.
2. Rated current means the maximum current for a given type of load which the relay will make, carry and break for its rated life.
3. Adequate contact protection must be provided where applicable.
4. Exercise care in the area of power switching with grounded case relays.

lists maximum stresses relative to three levels of reliability. These levels are defined as follows:

Level A—The most reliable, where the parts are operated well within their ratings.

Level B—The parts are operated within their ratings but not maximized with respect to reliability.

Level C—The parts are operated at or close to their ratings.

Guidelines for derating electromagnetic devices, relays and connectors are presented in Tables 5.7, 5.8 and 5.9 respectively.

TABLE 5.9
Connector Derating Chart

Connector type	Stress parameter	% stress (allowed)
Radio frequency coaxial	Current	50
Multipin	Current	50
Cable	Current	50
All types	Voltage	See below for nonpressured systems

		Working voltage	
Minimum air space	Rated voltage (rms)	d.c.	a.c. (rms)
		Voltage at sea level	
<0·031	600	280	200
0·031	1 000	490	350
0·045	1 500	700	500
0·062	1 800	840	600
0·076	2 250	1 050	750
		Voltage at 50 000 ft altitude	
<0·031	225	100	75
0·031	375	190	125
0·045	525	210	175
0·062	675	315	225
0·076	790	360	360
		Voltage at 70 000 ft altitude	
<0·031	150	70	50
0·031	300	125	90
0·045	375	175	125
0·062	450	210	150
0·076	500	230	165

For mechanical components, failure rate versus stress results from the manufacturer may be applied with stress/strength and reliability theory, i.e. a given part has certain physical strength properties which, if exceeded, will result in failure. Further, this property, as with all properties of non-homogeneous material, varies from part to part. Thus, for a particular part or material an estimate of the mean value and of the dispersion of the strength property is found by testing.

The operating stress imposed on a part also varies. These stresses vary from time to time in a particular part, from part to part in a particular design, and from environment to environment. An estimate of the mean value and the dispersion value of the operating stress must be determined by test, analysis or experiment.

The purpose of stress–strength analysis is to improve the reliability of the design. That is, to find the optimum comparison of stress and strength that will have an acceptable probability of success and compete favorably with other constraints, such as maintainability, cost and availability of material.

The classical approach to mechanical and structural design is to provide enough strength to handle the worst stress a part will encounter. The concept of stress–strength in design recognizes the reality that loads (or stresses) and the strengths of particular items subjected to these stresses cannot be identified as a specific value but have ranges of values with a probability of occurrence associated with each value in the range. The ranges of values (variables) may be described with appropriate statistical distributions for the item in its application environment. Stress–strength design requires knowledge of these distributions. After the strength and stress distributions are determined, a probabilistic approach can be used to determine the reliability measure as a function of stress derating.

5.4 DIAGNOSTICS AND MODULARITY

The maintainability level of an equipment is established by the extent of diagnostics and modularity designed into the equipment. Various options are evaluated during design to determine their impact on maintainability and tradeoff analyses are made with reliability to determine that specific approach which meets or exceeds plant availability objectives at the lowest LCC. These evaluations address the time and cost of locating, replacing and repairing failed parts or components necessary to the operation of an equipment. An effective design for maintainability results in reduced repair time by enhancing the ability of the technician to diagnose a malfunction rapidly, to identify the defective item accurately, and to replace it quickly with a new item without difficulty or delay. This is accomplished through application of built in diagnostics and the use of modular construction. Also, automatic, reliable monitoring of equipment parameters in conjunction with the switching-in of backup or redundant equipment or function will minimize the time for corrective maintenance actions to be taken and reduce the amount of loss of plant function. Consequently, plant availability will be maximized. The maintenance features that must be considered during the design of a complex system or equipment item include: internal, on-line diagnostics that can isolate the failed parts and automatically reconnect to redundant elements; modularity for the easy introduction of redundancy and ease of part (or unit) replacement (for later off-line repair); built-in, plug-in test equipment; remote monitoring control and diagnostic provisions to maximize routine maintenance intervals and tests to detect imminent failures during controlled periods. In addition, effective maintenance requires that sufficient diagnostic and replacement information be prepared and made available for the technician to determine the adequacy of equipment performance and to effect a repair expeditiously.

Designing for ease of maintenance requires providing a means for identifying failure and potential (or marginal) failures and for facilitating fault diagnosis, access to failed units and removal and replacement of failed units. Table 5.10 presents a checklist that can aid in implementing these features.

The term modularization refers to the separation of the components or an equipment into physically and functionally distinct groups or entities, to facilitate identification, removal and replacement. Modular construction provides a means by which the complexity of automated fault isolation techniques may be substantially reduced. By making it necessary to identify a functional group of parts to be interrogated, the logic required to isolate a failure is reduced significantly. In addition, the simplicity, uniformity and improved flexibility of equipment configuration that is characteristic of modularized equipment tends to simplify training of maintenance personnel, reduce the number of skills required, encourage standardization, reduce the number of line items in the supply system and permit simplified upgrading of fielded equipment.

Modularity is achieved through functional design by partitioning the system into physically and functionally self-contained units to facilitate fault isolation, removal and replacement. Partitioning enables equipment units, assemblies and subassemblies to be designed as discrete items or

TABLE 5.10
Maintainability Design Checklist

Use modular design techniques
—Uniform size and shape
—Guide pins and keyed connectors
—Ease of test/checkout
—Quick disconnect
—Minimum number of functions

Use special built-in diagnostics for fault detection (i.e., error warning lights, etc.)

Design for replacement at higher levels

Increase the depth of penetration of fault diagnosis localization features

Use test indications which are less time-consuming and/or less difficult to interpret

Design for minimum diagnostic strategies

Make accessible and obvious both the purpose of the test points and their relationship to the item tested

Improve the quality of technical manuals or maintenance aids

Design access for ease of entry

Reduce the number of access barriers

Reduce the need for isolation access by bringing test point, controls and displays out to accessible locations

Reduce the number of interconnections per replaceable item

Use plug-in elements

Reduce requirements for special tools

---

modules. The application of modular design allows the isolation of faults to a unit which may be removed from the equipment for on-site repair, shipment to a repair depot or throwaway. The equipment may be immediately put back into operation by replacement of a spare module, minimizing on-line maintenance action. Localization of components within modules eliminates long paths and crossovers and simplifies the tracing of signal paths when locating and isolating a failure, as illustrated in Fig. 5.11.

A throwaway maintenance concept for the module should be adopted if the module is relatively inexpensive. This is a logical conclusion of an LCC analysis when it is indicated that repair costs are greater than the cost of a new unit. The logistics of module replacement are directly related to the

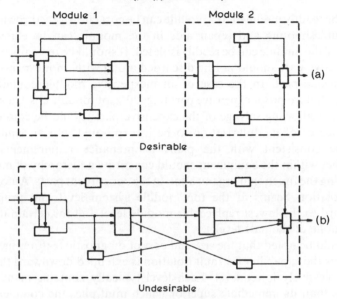

Fig. 5.11. Functional modularization design. (a) Advantages: faults isolated to a unit; on-line maintenance minimized; long paths and crossovers eliminated. (b) Disadvantages: many connections; on-line maintenance not minimized; crossovers and long paths present.

initial design decisions on size and complexity of modules. Repair of equipment can be accomplished by replacement of a module after fault isolation using portable test equipment or built-in test devices. The repair of a module generally requires jigs, fixtures, power supplies and adequate general test equipment.

Based on costs and logistics, a tradeoff must be made during early design to determine whether small inexpensive modules shall be incorporated and designated as throwaway or if larger modules should be used for a possible economy of equipment repair. It is crucial that these decisions be made early during design where changes least affect program costs.

An equipment which implements the throwaway concept of modular design possesses several advantages. Throwaway modules allow savings in repair time, tools, facilities and manpower. They also allow improved standardization and interchangability of modules and assemblies. The use of throwaway modules also imposes several penalties. There is an increased supply burden because modules must always be on hand. Similarly, redesign or retrofit of manufactured units becomes difficult since modules cannot be readily modified. If a module can be cost effectively thrown away, the troubleshooting and repair aids in the module can be eliminated, providing the module can be readily isolated. It should be remembered that an expensive part should not be discarded along with an inexpensive but failure prone part. In the case of an inexpensive module containing a relatively failure proof expensive part (e.g. 70 % of module cost), a plug-in connection allowing salvage of the expensive part should be considered.

The degree of modularization to be incorporated into an equipment must be consistent with the plant maintenance requirements. The assemblies within the equipment should constitute a hierarchy of modules paralleling the functional organization of the constituent parts. Automated fault isolation begins at the top module system level and progresses downward to the lowest replaceable module level and for certain items it can be done at the part level.

It should be noted that the size and cost of diagnostic features increases rapidly as the capability for fault isolation is extended downward through each successively lower level. Each level typically contains many more modules than its immediate superior which multiplies the complexity of sequential switching and sequence selection programming, as well as the number of end-to-end tests that must be performed. But the time expended by the technician in manual fault isolation also jumps sharply as each successively lower module level is manually tested, so that the proportion between acquisition costs and maintenance labor savings must be carefully evaluated.

5.5   OTHER RELIABILITY IMPROVEMENT ATTRIBUTES

In addition to improving the inherent reliability of the design via derating and the incorporation of screened parts, reliability can also be improved through efforts to simplify the design, to improve the equipment's resistance to the environment and by incorporating redundant elements within the equipment.

The functional reliability of the equipment can be greatly increased by redundant design. There are two major classes of redundancy: active redundancy, in which redundant elements are always a functioning part of the equipment, and standby redundancy, where an external element must make a decision to switch to another element that has not been active in the system. Formulae for estimating the reliability of equipment with redundant elements are given in Chapter 2. It should be noted that plant safety is ensured through use of reliable equipment and through application of additional backup systems and equipment as well as through the employment of diversity and alternate modes of operation as established by a plant safety analysis.

The application of redundancy has its costs, since an increased number of elements leads to an increased frequency of maintenance. On the other hand, redundancy means that the equipment is more likely to remain in continuous operation because its functional failure rate is significantly reduced. Many times, because of difficulty in performing maintenance for certain equipment, redundancy may be the only means of achieving adequate equipment reliability.

The achievement of high reliability in equipment can be significantly aided by a dedicated effort to simplify the design, since reliability is, in general, inversely proportional to complexity. This relationship runs counter to the ever increasing demand for more performance, which tends to increase complexity. Thus, one of the steps in designing for high reliability is to simplify the equipment to the maximum extent possible while meeting performance requirements.

Design simplification attempts to meet performance requirements with fewer, more complex and integrated components that reflect higher technology. It should be borne in mind, however, that although reliability is dependent on fewer components, care must be taken to ensure that unusual performance is not required of the newer more complex remaining components and that equipment performance is in fact not degraded. Also, care must be taken not to select parts that may later cause logistic and support problems. Simplification will, in general, yield reliability improvement if the total failure rate of the remaining and replacement components is not higher than the failure rate of the existing equipment and the replacement components do not significantly impact maintenance plans. Cost-benefit analysis must be performed to determine optimum simplification/substitution approaches.

In the case of electronic equipment, design simplification and substitution involves several techniques: (1) the use of proven circuits with known reliability; (2) the substitution of highly reliable digital circuitry for less reliable analog circuitry (where feasible); (3) the use of high reliability integrated circuits to replace discrete lumped constant circuitry; (4) the use of highly reliable components wherever individual discrete components must be used; and (5) the use of designs where a single failure results in reduced output function rather than a complete loss of function.

The equipment, depending on location and site may be subjected to

TABLE 5.11
Environmental Stresses, Effects and Reliability Improvement Techniques in Electronic Equipment

Environmental stress	Effects	Reliability improvement techniques
High temperature	Parameters of resistance, inductance, capacitance, power factor, dielectric constant, etc., will vary; insulation may soften; moving parts may jam due to expansion; finishes may blister; devices suffer thermal aging; oxidation and other chemical reactions are enhanced; viscosity reduction and evaporation of lubricants are problems; structural overloads may occur due to physical expansions	Heat dissipation devices, cooling systems, thermal insulation, heat-withstanding materials
Low temperature	Plastics and rubber lose flexibility and become brittle; electrical constants vary; ice formation occurs when moisture is present; lubricants gel and increase viscosity; high heat losses; finishes may crack; structures may be overloaded due to physical contraction	Heating devices, thermal insulation, cold-withstanding materials
Thermal shock	Materials may be instantaneously overstressed causing cracks and mechanical failure; electrical properties may be permanently altered. Crazing, delamination, ruptured seals	Combination of techniques for high and low temperatures
Shock	Mechanical structures may be overloaded causing weakening or collapse; items may be ripped from their mounts; mechanical functions may be impaired	Strengthened members, reduced inertia and moments, shock absorbing mounts
Vibration	Mechanical strength may deteriorate due to a fatigue or overstress; electrical signals may be mechanically and erroneously modulated; materials and structures may be cracked, displaced or shaken loose from mounts; mechanical functions may be impaired; finishes may be scoured by other surfaces; wear may be increased	Stiffening, control of resonance

Humidity	Penetrates porous substances and causes leakage paths between electrical conductors; causes oxidation which leads to corrosion; moisture causes swelling in materials such as gaskets; excessive loss of humidity causes embrittlement and granulation	Hermetic sealing, moisture-resistant material, dehumidifiers, protective coatings
Salt atmosphere and spray	Salt combined with water is a good conductor which can lower insulation resistance; causes galvanic corrosion of metals; chemical corrosion of metals is accelerated	Non-metal protective covers, reduced use of dissimilar metals in contact, hermetic sealing, dehumidifiers
Electromagnetic radiation	Causes spurious and erroneous signals from electrical and electronic equipment and components; may cause complete disruption of normal electrical and electronic equipment such as communication and measuring systems	Shielding, material selection, part type selection
Nuclear/cosmic radiation	Causes heating and thermal aging; can alter chemical, physical and electrical properties of materials; can produce gases and secondary radiation; can cause oxidation and discoloration of surfaces; damages electrical and electronic components especially semiconductors	Shielding, component selection, nuclear hardening
Sand and dust	Finely finished surfaces are scratched and abraded; friction between surfaces may be increased; lubricants can be contaminated; clogging of orifices, etc.; materials may be worn, cracked, or chipped; abrasion, contaminates insulations, corona paths	Air-filtering, hermetic sealing
Low pressure (high altitude)	Structures such as containers, tanks, etc., are overstressed and can be exploded or fractured; seals may leak; air bubbles in materials may explode causing damage; internal heating may increase due to lack of cooling medium; insulations may suffer arcing and breakdown; ozone may be formed; outgassing is more likely	Increased mechanical strength of containers, pressurization, alternate liquids (low volatility), improved insulation, improved heat transfer methods

(Source: *RDH-376*.)

thermal shock, and rough handling and it may experience shock pulses, vibrations, and various other single and/or combined environmental stresses. Table 5.11, taken from *RDH-376*, lists some of the environmental stresses that can influence equipment reliability. These stresses are identified during design and a determination of their impact on the parts of the equipment can be made. Those parts particularly susceptible to environmental stresses are either eliminated, re-specified to exhibit greater resistance to these identified stresses, or the application stress factors isolated or reduced through thermal or mechanical design to tolerable levels.

Excessive temperature is of special concern since it is often a major cause of reliability degradation. For this reason, each part is evaluated to establish that its thermal design is consistent with the overall system reliability. Following are some of the electronic parts requiring detailed thermal evaluation:

1. Resistors
2. Capacitors
3. Transformers and inductors
4. Semiconductor devices
5. Printed circuit boards

Reducing operating temperature, in general, will increase part reliability. This is accomplished by using thermal designs which reduce heat inputs and which provide for low thermal-resistance paths from heat-producing components to low temperature heat sinks. Following are some general guidelines for reducing temperature:

1. *Resistors*: protect by using leads of minimum length; improve heat transfer to heat sinks by mechanical clamping or by use of encapsulating materials; locate in areas favorable to convection.
2. *Capacitors*: provide thermal insulation when necessary and/or locate remote from heat sources.
3. *Transformers and inductors*: improve heat dissipation by incorporating cooling fins; provide conductive paths for better heat transfer and locate to utilize convection cooling when possible.
4. *Semiconductor devices*: maximize heat transfer between the device and its mounting by using large intimate contact areas and by specifying thermal gaskets or compounds when required; locate remote from high temperature parts and use heat sinks with fins positioned to make best advantage of air or coolant flow; improve

radiation characteristics with the use of printed or coated surface treatment.

5. *Printed circuit boards*: specify larger area conductors where practical; segregate heat producing elements from heat sensitive components; use intermediate metal core layers in multilayer systems, and provide good conduction paths from these layers to support members and intermediate heat sinks; use protective coatings and encapsulants for improving heat transfer to lower temperature supports and heat sinks.

Chapter 6

Reliability Engineering and Control

This chapter provides specific methods and procedures for engineering reliability into nuclear equipment and components during acquisition and for maintaining high reliability levels during plant operation. The methods and procedures cover the engineering program elements, i.e. prediction, FMECA, degradation control, assessment and feedback and others described in Chapter 3, that would make up an overall life cycle reliability assessment–evaluation–improvement methodology as outlined in Chapter 2. These procedures, when applied during development (and operation) as part of a life cycle reliability methodology, are performed to assess the reliability of an equipment as it progresses through acquisition and operation and to flag and rank items impacting reliability which can be fed back to the design and development process. They support the determination of the cost-effectiveness of proposed equipment changes. Results of equipment reliability analysis including MTBFs, failure modes, criticalities (as well as effective changes) and their LCC factors can also be incorporated into a dynamic real-time database from which the effectiveness of the reliability methodology (as well as the overall reliability program) can be continually assessed. The database would also represent a primary experience pool for designing new equipment with respect to reliability.

In order to assure that specified reliability levels are achieved, not only must reliability be planned for and designed into the equipment but must also be properly engineered during production and sustained during plant integration and operation. The evaluation and subsequent improvement of reliability is shown conceptually in Fig. 2.1. The figure shows the development of an equipment as it progresses from design, development and ultimately to plant operation. It indicates that the inherent reliability level (or maximum achievable) is dependent on the components selected,

254

stress–strength factors and other reliability design attributes. As the system progresses from design to development, a drop in reliability level is incurred. This drop can be minimized and reliability can grow toward its inherent level, through application of a well executed engineering program. The extent and rigor with which the reliability engineering program elements are implemented dictates the maximum level of reliability to be obtained. Initial reliability estimates provide a benchmark for subsequent reliability assessment and improvement efforts that are designed to mitigate degradation during production and plant use. Included in these efforts are tasks to assure that reliability is not compromised due to poor or improper manufacturing (or maintenance) processes, and to assure that the operating environment is properly accounted for in design considerations. Other reliability efforts are performed to measure and track reliability throughout acquisition and plant operation.

Some of the techniques described in this chapter are used to obtain a quantitative measure of reliability. Other techniques serve to aid in determining where a design may be improved. Quantitative evaluations are applied throughout the acquisition–operation cycle in order to assure that

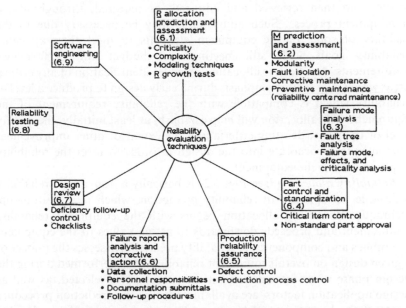

Fig. 6.1. Classification of reliability engineering procedures. (Numbers in parentheses are the sections which deal with that class of procedure.)

the equipment will function reliably under actual plant operational, maintenance and environmental conditions. Figure 6.1 depicts the categories into which these methods and procedures can be classified. Procedures pertaining to each of these categories are presented in the following sections.

6.1   RELIABILITY ALLOCATION, PREDICTION AND ASSESSMENT

The reliability of a complex equipment is a function of the reliabilities of the assemblies and components which it comprises. The equipment reliability requirement is allocated among its elements based on specific complexity factors or other considerations to establish basic target goals for subsequent lower level design as well as for reliability prediction and assessment efforts. The reliability relationship between the various elements of an equipment is defined and each element's contribution to the overall system reliability is evaluated. The allocation is initially performed at the beginning of the development cycle for input to early design tradeoff studies and then reviewed and adjusted, as required, throughout the development process. Such adjustments may be necessary ·due to the inability of some of the equipment elements to meet their allocated reliability goals while still meeting cost budget and performance requirements. An effective allocation provides identification of any critical areas which may require emphasis during early design to produce a level of reliability which is compatible with the reliability requirement of the equipment. The allocation will help establish, at least initially, the quality level of the parts, derating criteria and the extent of other improvement attributes to incorporate into the design in order to meet the reliability requirement for the equipment.

Reliability allocation (see Fig. 6.2) is basically a 'top–down' reliability technique, as opposed to reliability prediction, which is a 'bottom–up' technique. Reliability allocation begins with the equipment reliability requirement and proceeds downwards to assign values to the equipment assemblies and components. A reliability prediction assesses the impact of a given design on overall equipment reliability. It is performed during the design phase when information on the actual parts selected, as well as design application factors, are available. Ideally, the prediction procedure starts at the lowest defined hardware level to determine, from detailed diagrams, the reliability of the individual parts, and then proceeds upwards

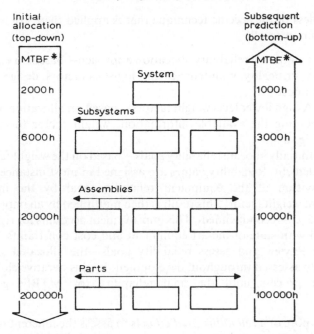

	Allocation		*Prediction*
Step 1.	Set overall system reliability requirement.	Step 1.	Determine reliability of lowest level items.
Step 2.	Apportion reliability goals among various elements.	Step 2.	Integrate upward and arrive at estimate of overall equipment reliability.

Fig. 6.2. Reliability allocation/prediction approach. (*Sample data.)

from part to higher assemblies, using appropriate reliability models, until a result is arrived at for the overall equipment reliability.

There are many approaches to allocating reliability. Reliability can be apportioned on the basis of (a) complexity, (b) criticality, (c) maintenance and logistic considerations, (d) design effort, (e) acquisition costs, and so on. Simple models such as those based on criticality or expected complexity, are generally used because of the lack of detailed information during the early design stage.

Regardless of the specific technique that is applied there are four basic steps common to all:

Step 1: Establish a reliability allocation approach—based on equipment complexity, criticality, maintenance and logistics factors, design effort or acquisition cost.

Step 2: Assign lower level weight factors—based on allocation approach and considering the available information; weight factor formulae are given in Fig. 6.3.

Step 3: Initially allocate reliability goals—based on the weight factors for each subelement. Reliability values are assigned in most instances, by the simple division of the equipment reliability goal by the individual subelement weight factor. This result is then multiplied by an improvement factor (125 % is recommended). This improvement accounts for risk factors associated with state-of-the-art equipment and cost constraints.

Step 4: Review and assess reliability goals—the allocated goals are continually assessed throughout development by this iterative closed-loop allocation process until the final reliability (or MTBF) goals are established.

The purpose of a *reliability prediction* is to assess the inherent reliability of a design during development prior to the build-up of hardware. Prediction establishes a quantitative value for reliability, permits reliability to be traded as a design parameter and provides a quantitative means with which to assess any discrepancy between the specified equipment reliability requirement and lower level allocations and that of the design. Prediction also provides inputs for: (1) reliability growth; (2) reliability demonstration; (3) design reviews; (4) acquisition cost estimates; (5) logistics and support cost estimates; and (6) equipment improvement programs.

As the design is initiated, and its parts complement and configuration details are identified, various techniques are used to estimate and predict the inherent reliability level of an equipment. Early in development, the failure rate of the parts are computed and integrated into the appropriate reliability model and the reliability of the equipment is estimated. A reliability prediction can be made at any time during development but its accuracy depends on the detail and quality of the design information available. As the design is detailed and configured the reliability prediction techniques and models applied to evaluate the reliability of the design are also more detailed to reflect a greater level of hardware definition. During early design when the equipment is defined only at the top level, and only rough estimates of part counts by part type are available, gross models are applied

Techniques	Weighting formula	Allocated subsystem reliability	Definition of terms
Equally critical subsystem in series configuration	None	$R = R_s^{1/n}$	n—number of subsystems R_s—systems reliability
Non-equivalent subsystems in series configuration	$W_i = \dfrac{C_i}{C_1 + C_2 \ldots C_n}$	$R_i = (R_s)^{W_i}$	C_i—complexity of ith subsystems
Consideration of subsystem importance and complexity	$m_j = \text{MTBF}_i$ $= \dfrac{k_i \cdot t_i}{\left(\dfrac{n_i}{N}\right)(1 - nR_s)}$	$R_i = 1 - k_i(1 - \exp(-t_i/m_i))$	k_i—probability system fails if i fails n_i—number of components in ith subsystem

Fig. 6.3. Weight factors for reliability allocation techniques. (Source: *RADC-TR-67-108 (Vol. 2)*.)

to predict reliability. As the design progresses and the configuration of the major assemblies and components are configured, more detailed stress–strength prediction techniques are used to estimate equipment reliability with much higher levels of accuracy.

Part count predictions are performed during early design using average stress levels. As design progresses, detailed stress analysis predictions are performed based on the selected parts and their individual stress levels. The stress analysis predictions are refined throughout the detailed design phase, until the design is finalized. Descriptions of the part count method and the stress analysis technique follow.

6.1.1 Part Count Prediction

The part count method provides an estimate of reliability based on the number of parts by type. The technique uses information derived from available engineering information and does not require part-by-part stress and design data. Input failure rates are derived from industry wide data sources, plant specific data, as well as from published handbooks such as *US MIL-HDBK-217* for electronic parts, *RADC-TR-75-22* handbook for non-electronic components and *IEEE STD-500-1977*. The following steps describe the part count prediction technique.

Step 1: Define the equipment.

Step 2: Define what constitutes a failure.

Step 3: Develop a reliability block diagram and math model using series/parallel reliability models (see Chapter 2).

Step 4: Determine the number of parts of each type in all blocks of the block diagram.

Step 5: Compute the reliability of each block by multiplying the total number of parts of each part type by its generic failure rate. A part count prediction worksheet is presented in Fig. 6.4. The worksheet can serve as an aid in the computation of block failure rates.

Step 6: Compute the reliability for the equipment, taking into account the failure rate of each block and its probability relationship to the other blocks as described by a reliability block diagram and math model.

Generic failure rates for various electronic parts are presented in Table 6.1. The failure rates are derived from *US MIL-HDBK-217* based on generalized part information when considering 'high rel' part processing (e.g. B level or equivalent microcircuits, JANTX semiconductors, etc.) and considering a controlled application environment. When more exact

Project _____ Sheet ____ of ____

System _____ Date _____

Reliability block description _____

Environment: _____ Temperature: 35°C ☐   55°C ☐

Other _____

Serial number	Part description	Qty	Generic failure rate, λ $(\times 10^{-6}\,h)$	Total failure rate, λ_T $(\times 10^{-6}\,h)$
			Total reliability	

Fig. 6.4. Part count reliability prediction.

TABLE 6.1
Electronic Part Failure Rates

Part description	Failure rate ($\times 10^{-6}$)
Microelectronic devices	
Bipolar	0·02–3·3
	(depending on number of gates)
MOS	0·02–3·0
	(depending on number of gates)
Discrete semiconductors	
Transistors, S1, NPN	0·000 5
Transistors, S1, PNP	0·000 7
Diode, general purpose	0·000 1
Diode, Zener	0·000 5
Diode, LED	0·001
Resistors	
Composition	0·005
Film	0·001
Wirewound	0·008
Power	0·01
Variable wirewound	0·01
Variable non-wirewound	0·02
Capacitors	
Paper	0·01
Plastic	0·004
Mica	0·000 5
Glass	0·000 3
Ceramic	0·004
Tantalum electrolytic solid	0·006
Aluminum electrolytic	0·07
Variable ceramic	0·03
Variable, piston type	0·1
Motors	1·6
Relays	
General purpose	0·14
Reed	0·11
Connectors	
Plug	0·006
PCB	0·003
Printed wiring boards (two-sided)	0·001
Inductive devices	
Pulse transformers	0·003
Audio transformers	0·006
Power transformers	0·02
Radio frequency coils, fixed	0·002
Switches	
Toggle	0·001
Sensitive	0·15
Rotary	0·33

information is not available, the failure rates presented in Table 6.1 can be used as a first approximation.

6.1.2 Stress–Strength Prediction

The stress analysis prediction technique requires the use of detailed part models plus calculation of stress values for each part prior to determining its failure rate. Each part is evaluated in its system application based on mechanical, thermal and electrical stress analyses. The stress–strength reliability prediction technique follows the same procedure (except for step 5) as the part count technique. This procedure is as follows:

Step 1: Define the equipment.

Step 2: Define what constitutes a failure.

Step 3: Develop a reliability block diagram and math model using series/ parallel reliability models (see Chapter 2).

Step 4: Determine the number and types of all parts in each block of the block diagram.

Step 5: Compute the reliability of each block, taking into account the basic failure rate of each part and its application stress factors; *US MIL-HDBK-217* provides failure rate vs stress models for electronic parts.

Step 6: Compute the reliability of the overall equipment, taking into account the reliability of each block and its probability relationship to the other blocks as described by the reliability block diagram and math model.

The procedure is facilitated by using a worksheet similar to the one depicted in Fig. 6.5 for electronic equipment. The worksheet given in Fig. 6.5 is designed to be used in conjunction with part models (λ_b) and adjustment factors ($\P_i$) given in *US MIL-HDBK-217*. The sheet provides for tabulation of: part description, electrical stress factors, thermal stress factors, basic failure rate and various environmental and quality adjustment factors as well as the final part failure rate. The worksheet provides a means to establish the inherent reliability of the design by evaluation and control of stress factors (i.e. derating, temperature, etc.), based on part failure models and related data.

US MIL-HDBK-217 provides failure rate data for a large number of electronic and electromechanical parts and components. The failure rate data implies an exponential probability density function (reliability $= \exp(-\lambda t)$, where λ is constant with time). The general form of the *US MIL-HDBK-217* model is:

$$\lambda_p = (\lambda_b)\,\P_E\,\P_A\,\P_Q \cdots \P_n$$

STRESS ANALYSIS – RELIABILITY PREDICTION WORKSHEET																										
ENGINEERING DATA														RELIABILITY ANALYSIS												
1	2	3	4			5			6	7	8	9	10	11	12	13	14	15	16	17	18	19	20	21	22	23
LINE	REF DES	PART DESCRIPTION	LOAD, VOLTAGE, DISSIPATION						Stress Ratio Or Tj	Appl Or Hot Spot Value Or Cap Freq & Wave F'M	Const Or Wire Or Cmplxy (Note 1)	Cycles Or Service Grade	Contact Qty Or Form Or Z_{CKT}	λ_b $\times 10^{-6}$ Hrs.	Π_V Π_{S2} Π_L	Π_R Π_{RW} Π_{SR} Π_{CV}	Π_E Π_{EG}	Σ_E	Π_{ER}	Π_Q OR Π_A	Π_{CS} Π_F Π_G Π_P Π_f	Π_C	Π_N Π_{TAP} Π_P (PIN)	N Σ CYC	Π CYC	λ_{PART} $\times 10^{-6}$
			ACTUAL			RATED																				
			I	V	W	I	V	W																		
1																										
2																										
3																										
4																										
5																										
6																										
7																										
8																										
9																										
10																										

ENG ________ NAME ________ DATE ________ PROJECT ________ ASSEMBLY ________ ENVIRONMENT

REL ________ NAME ________ DATE ________ UNIT ________ ASSEMBLY PART NO ________ ASSEMBLY FPMH

NOTES LIST CONNECTOR BODY MATERIAL HERE

SUM

Fig. 6.5. Stress analysis—reliability prediction worksheet. (Source: *RDH-376*.)

where: λ_p is the total part failure rate; λ_b is the basic part failure rate (data is provided in terms of failure rate versus operating temperature for each generic type: specific values are computed as a function of different stress levels, e.g. power, voltage or current); $\P_E$ is the environmental adjustment factor which accounts for the influence of environment other than temperature (it is related to operating conditions such as vibration, humidity, etc); $\P_A$ is the application adjustment factor (it depends on how the part is used, and takes into account secondary stress and application factors that are considered to be reliability significant); $\P_Q$ is the quality adjustment factor (it is used to account for the degree of manufacturing control under which the part was fabricated and tested prior to its shipment); $\P_n$ represents adjustment factors which are used to account for cycling effects, construction type, and other design and application characteristics.

As can be seen from the above, basic stress dependent failure rates, λ_b, are modified by suitable adjustment factors dependent upon the environment, quality and use expected for that part. The quality of a part, for example, has a direct effect on the part failure rate and appears in the models as the factor $\P_Q$. Many parts are covered by specifications that have several quality levels; hence the part models have values of $\P_Q$ that are keyed to these quality levels. Such parts with their quality designators are shown in Table 6.2. The detailed requirements for these levels are clearly defined in their applicable specifications. Some parts are covered by older specifications, usually referred to as non-ER, that do not have multi-levels of quality. These part models generally have two quality levels designated 'Mil. Spec.' and 'Lower'. If the part is procured in complete accordance with the applicable specification, the $\P_E$ value for 'Mil. Spec.' is used. If any requirements are waived, or if a commercial part is procured, the $\P_Q$ value

TABLE 6.2
Parts with Multi-level Quality Specifications

Part	Quality designators
Microelectronics	S, B, B1, B2, C, C1, D, D1
Discrete semiconductors	JAN TXV, JANTX, JAN
Capacitors, established reliability	L, M, P, R, S
Resistors, established reliability	M, P, R, S
Coils, molded, radio frequency, established reliability	M, P, R, S
Relays, established reliability	L, M, P, R

for 'Lower' is used. The $\P_E$ and $\P_A$ factors are used in all models and the other factors apply only to specific part models. The full range of values for the various modeling factors are provided in *US MIL-HDBK-217*. It must be emphasized that the part quality levels can be degraded in design and manufacturing. The achievement of high quality as measured by the *US MIL-HDBK-217* quality factors requires that the parts are adequately derated and that they are built into the hardware with a manufacturing and quality control process commensurate with the high part quality.

Reliability assessment is the process of determining the reliability of an equipment when using test and/or other empirical data in lieu of handbook data. It is performed during the later stages of the development/production cycle and during plant operation providing a more accurate (as opposed to reliability prediction) measurement of reliability and feedback to the R & D process. Reliability assessment is used to determine if the equipment is meeting its specified or allocated reliability requirements, to aid in the measurement of reliability growth, and thus plays a prominent role in the final design decision and production acceptance. Using the system reliability model (developed for reliability prediction) along with data stemming from reliability tests or plant operation one may establish whether the required reliability level is being met. The assessments can be compared with one another over a period of time to establish trends, extrapolate results to some future times and to assess the operating situation on an on-going basis.

A key objective of reliability assessments performed during plant operation is to establish a basis for assigning equipment improvement priorities. The reliability assessments can be used as the basis to perform sensitivity analysis to develop criticality numerics for each of the assemblies, components and supporting elements that make up the equipment. The criticality numerics provide a method of ranking (i.e. prioritizing) each hardware element according to changing operational experiences. The hardware elements can be listed in order of decreasing criticalities (truncated at some selected low number). For example, such a list would contain the top twenty components or material problem areas ranked in decreasing order of concern. The list could be updated monthly and provide a running account of the most notable operating problems. This list would allow managers to focus on efforts to formulate improvement recommendations for those problem areas which have the highest priority in terms of plant safety and performance, reliability or economic impact.

Bayesian statistics are often used in the performance of reliability assessments. This approach allows prior information (e.g. predictive) obtained for a random process to be applied to the analysis of more recent information, such as test plant operational and failure data. Thus, not only can reliability prediction estimates, or the latest test information, be used exclusively and independently of one another to assess reliability, but a final assessment can also be made that reflects both information sources.

6.2 MAINTAINABILITY PREDICTION/MAINTENANCE PLANNING (RELIABILITY CENTERED MAINTENANCE)

Maintainability is defined as the probability that a hardware item will be retained in or restored to a specified operating condition within allowable time limits, using available test equipment, facilities, personnel and spare parts, and that it will be performed in accordance with prescribed procedures. Maintainability prediction is the process of estimating the parameters that describe this probability. As with reliability prediction, it is an analytical process, based on design characteristics and maintenance features (i.e. test points, self-check features, accessibility, modularization, adjustments, etc.), to determine the ease and speed with which maintenance operations can be performed and failures be diagnosed and corrected. In general, maintainability is composed of two parts:

1. Corrective maintenance—the action performed, as a result of failure, to restore an item to a specified condition.
2. Preventive maintenance—the action performed in an attempt to retain an item in a specified condition by providing systematic inspection to detect and prevent incipient failure.

Maintainability predictions are performed after the basic engineering data are available in order to provide a meaningful quantitative evaluation of design characteristics. Maintainability predictions, like reliability predictions, become more detailed and accurate as the system design becomes more thoroughly defined. These predictions provide an indication of compliance with specified maintainability requirements and provide confidence that subsequent maintainability demonstration tests will yield satisfactory results.

Predicting maintainability involves the determination of MTTR using a failure rate (λ) obtained from reliability prediction studies and maintenance time factors (R_p) derived from a review of system or component design characteristics. Conceptually, the repair of hardware items after the occurrence of a failure necessitates the initiation of a corrective maintenance task which ultimately results in the interchange of a replaceable part or assembly. In order to achieve a complete 'repair', various activities both before and after the actual interchange are necessary. These activities can be subdivided into the following elements:

1. *Localization time*—The objective of localization is to eliminate as many as possible of the unfailed functions from further consideration by performing rapid tests (frequently involving only operating controls, displays and/or monitoring devices) before proceeding with the more difficult diagnostic techniques of fault isolation.

2. *Isolation time*—The time associated with tracing a failure down to a replaceable item through the use of test equipment.

3. *Disassembly time*—That time associated with gaining access to the replaceable part, up to the point of interchange.

4. *Interchange time*—The time associated with the physical removal of a failed item and its replacement with a new item.

5. *Reassembly time*—The time associated with disassembly, except that the steps are performed in reverse order.

6. *Alignment time*—The time associated with the manipulation of operating and maintenance controls and mechanical parts so as to bring the equipment within its specified operating ranges.

7. *Checkout time*—The time associated with verification that the repair has restored the equipment's normal performance.

The composite time for all the above activities is called the repair time, R_p. In order to provide weight factors for the expected number of corrective maintenance actions, the failure rate of each replaceable component/part/assembly is used. The failure rate and repair time are combined to arrive at a corrective maintenance action rate. This process is repeated for each replaceable part/assembly in the system. From the maintenance action rates $(R_p \lambda_p)$ derived for each replaceable item, the MTTR can be determined using the following expression:

$$\text{MTTR} = \frac{\sum R_p \lambda_p}{\sum \lambda_p}$$

The prediction process involves preparing a functional level diagram and determining the repair time for each replaceable item.

A sample functional level diagram is shown in Fig. 6.6. Each branch of

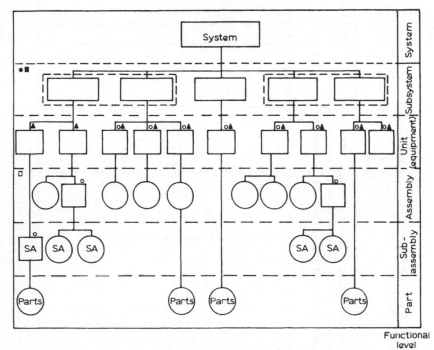

Fig. 6.6. Functional level diagram (modified). Levels of (●) localization, (▲) isolation, (○) access, (□) alignment, (■) checkout.

the diagram is terminated with a circle which indicates the item(s) that will be replaced to correct failures in that branch. The diagram, once completely structured for the system, will reflect the overall maintenance concept and the complete replacement breakdown for all items that the system comprises. The levels at which localization, isolation, access, alignment and checkout occur will be determined and noted on the diagram in accordance with the following description:

Level of localization—the functional level to which a failure can be located without employing accessory support equipment.

Level of isolation—the functional level to which a failure can be located using support test equipment at designated points.

Level of access—the access level for a replaceable item is that level to which disassembly must be accomplished in order to gain access to the item that is to be replaced and from which reassembly must be accomplished after item replacement. This can be determined directly from the functional

CORRECTIVE MAINTENANCE WORKSHEET

System ——————
Model number ——————
Date ——————
Sheet —— of ——

Replaceable item	Qty (N)	Total failure rate ($N\lambda$)	Average time to perform corrective maintenance tasks							R_p	$N\lambda R_p$
			Localization	Isolation	Disassembly	Interchange	Reassembly	Alignment	Checkout		

Fig. 6.7. Corrective maintenance worksheet.

level diagram as the level of the first rectangular block above the replacement item.

Level of alignment—the functional level at which alignment must be done following replacement of a variable item or unit containing sensitive mechanisms requiring alignment or adjustment.

Level of checkout—the functional level at which system/component operation can be verified using self-test or other testing facilities.

After the function level diagram is structured, the next step in the analysis process is to estimate the repair time for each replaceable item depicted in the diagram. Figure 6.7 presents an analysis worksheet that can be used to aid in this process. The worksheet provides a tabulation of all maintenance time elements defined within the body of the procedure. Time elements for each replaceable item are estimated by using the functional level diagram in conjunction with actual maintenance time experience factors or average maintenance time data as presented in *US MIL-HDBK-472*. These time estimates are to take into account the maintenance characteristics of the system as reflected in the functional level diagram.

Preventive maintenance activities include inspection and calibration, detection and correction of incipient failures, periodic replacement of short life parts, adjustments, lubrication and routine checkouts. Figure 6.8

Description of preventive maintenance task	(F)/year	Time req'd (M_p)	FM_p

Fig. 6.8. Preventive maintenance worksheet.

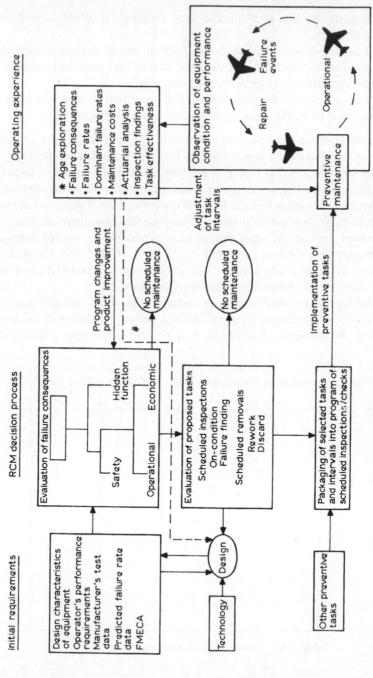

Fig. 6.9. The life cycle RCM process in the airline industry. (Source: *AD-A066979.*)

presents a worksheet that can be used in the analysis process. Use of the worksheet in the analysis procedure is described as follows:

1. Identify and record on the worksheet all applicable preventive maintenance tasks—this is accomplished through a review of operating and maintenance manuals, reliability studies and other available engineering data and information that define the performance, reliability, safety and maintenance characteristics of the system.
2. Determine and record the frequency of occurrence of each task.
3. Determine and record the time required to perform each task.
4. Finally, compute and record the total down time for preventive maintenance.

One of the key elements of an effective reliability engineering program is the establishment of effective preventive maintenance tasks and requirements. A relatively new program first initiated by the airlines (see Fig. 6.9) and now applied by the DoD for deriving optimum maintenance plans is *reliability centered maintenance* (RCM). RCM is a concept which uses an analytical methodology, or logic, for establishing specific preventive maintenance tasks for complex systems or equipment. Intrinsic to RCM is the identification of critical failure modes and deterioration mechanisms through engineering analyses and/or field experience to determine the consequences and the most effective apportionment of maintenance activities.

RCM is based on the premise that maintenance cannot improve upon the reliability of an item established during its design. Proper maintenance can only preserve reliability. The RCM concept recognizes this concept in the establishment of a maintenance program geared to the preservation of inherent reliability and safety. The concept uses decision logic to evaluate the failure modes of maintenance significant items (MSI) which directly relate to equipment functions for the purpose of constructing sets of maintenance tasks. The logic process forces all scheduled maintenance into one of the following three categories:

1. On-condition maintenance requirements—scheduled inspections or tests designed to measure aging or deterioration of a component or structure; based on the deterioration found, the hardware item either undergoes maintenance or remains in service.
2. Hardtime maintenance requirements—scheduled removal tasks at predetermined fixed intervals of age or usage.

3.   Condition monitoring maintenance requirements—unscheduled tasks consisting of routine monitoring during normal operation whereby components are allowed to fail or where impending failure can be detected prior to occurrence.

Recent concepts by the airlines involve separating the maintenance requirements into the following five tasks:

1.   Lubrication/servicing.
2.   Operational checks.
3.   Inspection/functional check.
4.   Restoration.
5.   Discard.

Figure 6.10 shows the manner in which RCM decision logic is applied to a system or equipment to establish the maintenance requirements. The RCM decision logic consists of nine 'trigger' questions. Five of these questions (1, 2, 3, 4 and 6) establish the criticality of the component under evaluation. An inspection or service is required to anticipate failure before someone is injured or costly damage occurs. An inspection may also be required to detect hidden failures which can affect safety, performance or compliance with administrative or legal procedures. At this point, there are still options as to how each maintenance task can be effectively performed. The remaining four questions (7, 9, 10 and 12) lead the analyst to the maintenance category that will perform the mission at minimum cost: condition monitoring, on-condition, hard time or some combination of the three. Each equipment must be thoroughly analyzed by knowledgeable technical personnel to determine appropriate answers to the item under evaluation.

It must be emphasized that the RCM logic is applied to each MSI in order to determine the maintenance task requirements; however, individual failure modes must be addressed during the application of the logic. Thus, for a given repairable item or component, different scheduled tasks could be arrived at due to the different failure modes and their characteristics. As an example, a given component might undergo condition monitoring during normal operations to detect the majority of predicted failure modes for the component while still having an on-condition or hardtime requirement due to a failure mode that is not detectable during routine operator monitoring.

RCM can be directly applied to nuclear components and structures to determine which components require maintenance and what the requirements are. For on-condition maintenance actions, detection prior to failure

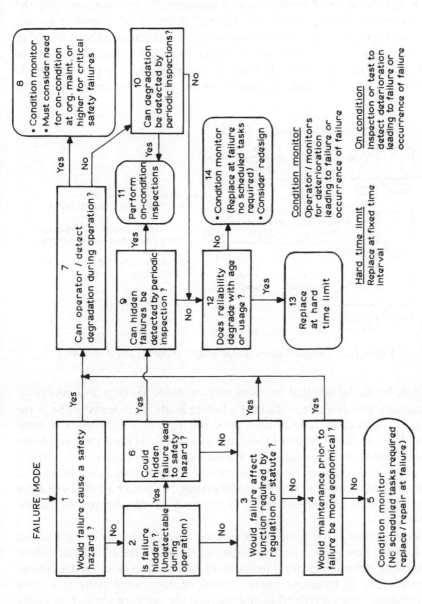

Fig. 6.10. Reliability centered maintenance logic diagram. (Source: *DA-P-750-40.*)

can be accomplished only if the inspection occurs during the period between the onset of noticeable and unacceptable deterioration and the occurrence of failure. The term 'Time of onset' (Tos) is applied to this period. Figure 6.11 depicts the Tos concept. Tos is statistically estimated in order to select a least cost option that provides high calculated values of reliability and availability. It must be estimated for every failure mode exhibiting detectable deterioration prior to failure. The interval selected

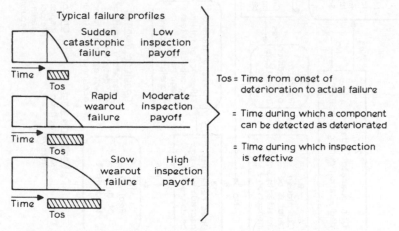

Fig. 6.11.   Failure theory, Tos concept. (Source: *DA-P-750-40*.)

should be the largest that provides an acceptable likelihood of successful detection. For example, if Tos for a failure mode is estimated to be in the range 2–4 months, a monthly check will provide essentially the same likelihood of detection prior to failure as a daily check with a much lower expenditure of resources. For reliability failures (safety not directly affected), it is suggested that the interval be set near the shortest interval of the range estimated for Tos. For safety failure, a smaller interval would be appropriate—perhaps one-third to one-half of the shortest estimated interval.

Replacement at hardtime limits must be supported by statistically sound test data or perhaps statistical analysis of field data. The failure distribution should be determined in addition to the mean time or usage to failure. Replacement at hardtime limits is most effective when there is a close correlation between reliability and age, e.g. the variance about mean time to failure is narrow. Although 'wearout' is a classic concept, the existence of well-behaved component wearout mechanisms are not all that prevalent in the real world. In a comprehensive study (AD-A066979) of age-reliability patterns, it was found that only about 6% of all components involved in modern aircraft exhibited pronounced wearout characteristics of the type which are suited to 'hardtime' maintenance concepts. Items in this class are generally simple (so-called 'single-cell') components such as parts which are subject to metal fatigue or other kinds of mechanical wear which are well correlated with time, or time in use. Detailed statistical analyses of component experience data must be applied to determine the relationships between reliability and component aging and used as the basis to establish hardtime limits.

Recognition of the limitations of hardtime maintenance approaches has led to the introduction of 'on-condition' maintenance concepts based on inspection. This approach is effective whenever potential failure (reduced resistance to failure) can be ascertained reliably and inexpensively. Tire wear is a classic example. The most sophisticated maintenance program is based on concepts of 'condition monitoring' in which failure potential is under constant surveillance. This is a technology limited approach which previously appeared only in very high value (safety or cost sensitive) items. With the emergence of new classes of effective and inexpensive sensors and microprocessors, 'condition monitoring' is becoming much more prevalent.

RCM logic is applied to the individual failure modes of each repairable item in the equipment identified by FMECA, through a progressive determination of how impending failures can be detected and corrected in order to preserve, to the degree possible, the inherent levels of reliability and safety in the system. RCM identifies specific preventive maintenance tasks and requirements for:

—detecting and correcting incipient failures, either before they occur or before they develop into major defects;
—reducing the probability of failure;
—detecting hidden failures that have occurred;
—increasing the cost effectiveness of the system's maintenance program.

A properly structured maintenance program, which includes RCM, integrates in an efficient way many of the relevant assurance programs (i.e. reliability, maintainability and safety) and other special studies which also serve the common objective of orienting the design and development phase toward a practical, serviceable and affordable product. It outputs data to other tasks such as logistics, level of repair and maintenance/design tradeoff analysis. Essential to the overall process is the availability of detailed engineering reliability and maintainability (R & M) data.

The organization of maintenance activities according to either the three categories or the five tasks establishes the basic form of the eventual maintenance program and thereby impacts operating and support costs and the manner in which reliability is assured. The driving force in RCM analyses is to reduce the scheduled maintenance burden and support cost incurred by the system while maintaining required levels of system availability.

The RCM logic process is based upon the following criteria:

1.   Scheduled maintenance tasks should be performed on non-critical components only when performance of the scheduled task will reduce the LCC of ownership of the system/equipment.

2.   Scheduled maintenance tasks should be performed on critical components only when such tasks will prevent a decrease in reliability and/or deterioration of safety to unacceptable levels, or when the tasks will reduce the LCC of ownership of the system/equipment.

As with other reliability and logistic analyses and tasks, the logic process will be reapplied as available data moves from a predicted state to actual operational experience values with a higher degree of certainty. In developing the initial maintenance program using the RCM process, it is frequently necessary to make action decisions without adequate information. The RCM decision hierarchy is structured to yield sure-safe practices in these situations through the selection, 'by default' of the most conservative course. This practice of employing default logic is not only the safest course, it is also the most expensive.

Consequently, one of the most urgent steps to be accomplished once an equipment is operational and an experience base begins to accumulate, is to reassess all RCM default decisions. The objective is to eliminate excessive maintenance costs while retaining established and required levels of reliability and safety.

Once all components have been initially subjected to the RCM logic process, a system analysis is performed to arrive at an overall maintenance plan. This system analysis merges individual component requirements into a maintenance plan by optimizing the frequency of schedule maintenance requirements and the sequence for performing individual scheduled tasks.

The maintenance plan is a description of the requirements and tasks to be accomplished in achieving, restoring or maintaining the operational capability of the complete system. The plan describes how the hardware will be maintained and its relationship to the overall maintenance concept. It is generally obtained as a product of the early design phase updated as necessary during the course of the acquisition program and is reassessed in the light of plant experience data. The plan provides a definition as to what constitutes a repair action and what should be the scope of the maintenance activities planned for execution at the various levels of repair. It specifically identifies logistic support requirements including:

(a) Maintenance tasks, including requirements to support the equipment at each level of repair.
(b) Spare provisioning.
(c) Tool and test equipment, including calibration equipment and calibration requirements.
(d) Manpower training and skill levels.
(e) Maintenance manuals and data.
(f) Training manuals.
(g) Support equipment/facilities.
(h) Shipping and transportation.
(i) Quality control.
(j) Configuration management.

The plan also establishes requirements and/or interfaces for reliability program tasks, maintainability program tasks, LCC analysis and other related program tasks or elements.

In addition to the scheduled maintenance task requirements identified during application of the RCM logic, any scheduled tasks that were assumed in establishing the reliability characteristics of the system/equipment under the reliability program must either be included in the maintenance plan or identified as being omitted from the maintenance plan. Inherent failure rates and failure modes and effects might need adjusting if an assumed scheduled maintenance action is omitted from the maintenance plan after application of the RCM logic.

Even after a properly structured maintenance program for a given equipment is put in place it remains a dynamic element within the on-going RCM process. From a RCM point-of-view, it is mandatory that a system be maintained for monitoring and compiling the reliability characteristics of the hardware under actual operating conditions.

Within the RCM life cycle process, the in-service period not only constitutes the end-item environment, it is also the time frame in which some of the most vital RCM data and information are generated and collected. The maintenance infrastructure in-place for an operating equipment is, in fact, an information management system with the following principal components:

—A system for reporting failures and their consequences.
—A system for continuous assessment of age condition (aging) behavior of significant items.
—A system for rigorously controlling the introduction of new maintenance tasks (mandating the use of the RCM process).
—A system for periodic reassessment of on-going tasks and for purging those which no longer provide an RCM supportable value.
—A system for expeditiously dealing with unanticipated failures at all levels of criticality.

The information collected within this system is critical to a number of functions. It constitutes, first of all, the primary database for essential corrective feedback to the RCM decision process. It is also extremely important, especially in the early operational period, that all RCM default decisions be reassessed in the light of any new available data in order to eliminate, on a sound basis, any unnecessary or excessive maintenance tasks. It is equally important that any newly encountered (unanticipated) failures be fed back into the system and subjected to the RCM decision process. The experience base is also used for a continuing logical reassessment of task requirements from a cost reduction point of view. The reliability experience database is also used as an input to provisioning and logistic plans and as a reliability corporate memory for the design of new systems and components.

In order to establish and maintain an effective RCM program, a utility must routinely utilize an extensive array of reliability and failure information sources. The manner in which this data is developed, organized and used, is governed by priorities which favor safety-dictated necessity over economic desirability. The ranking of information, starting with the most urgent, is as follows:

—Any information relative to failures which have a direct impact on plant safety.
—Any information relative to failures which have a direct impact on operational capability and availability.
—Information, derived from (any level of) inspection, which indicates potential failure conditions.

—Failure mode information for all units removed from service because of malfunction.

—General condition information on all parts in units removed from service because of malfunction.

The flow of information into and within the RCM program is a continuing process, but the prior-to-service functions draw on characteristically different data sources.

The primary determinant of equipment reliability is the design itself and, in the maturing discipline of nuclear equipment design, a strong sensitivity to plant safety and equipment reliability has already become institutionalized. The realization of the reliability inherent in the equipment (or component) design is dependent on the selection and use of properly rated and screened components and the use of proper reliability management and control techniques during manufacture and assembly. Data on component reliability is obtained from the suppliers, and from such centralized industry wide sources as the NPRDS which collects experience data from the plants. The equipment manufacturer reliability control plan and especially the documented results of various screening and burn-in processes also constitute a rather basic input to the early phases of the RCM process in terms of establishing initial reliability projections.

Critical component lists which are established for a specific design, identify functionally critical elements and thereby set certain minimum reliability standards for them. The design configuration (and all of its associated reliability data) and the equipment function description (and its associated stress environment) constitute the basis on which an FMECA is performed. The resultant FMECA provides a strong foundation for the RCM decision logic and is therefore, one of the most important data sources during the prior-to-service period.

In addition to using all of these well-defined sources and procedures, it is often necessary to fill certain gaps by relying on the seasoned judgement of those who are exercising the RCM decision logic and to rely heavily on their collective experience with similar items.

Much of the more useful in-service RCM data is derived from information systems which were put in place for other operational and logistic purposes. The extraction of relevant data is done through a carefully established process based on the RCM information priorities previously discussed.

The plant log is one of the most important of these existing information systems. Aside from dealing with routine operational and performance parameters, these logs also document any unsatisfactory conditions observed or reported during operation. These observations are organized to alert maintenance personnel who, in turn, annotate the repairs made as a result of these reports. After initial use and review, the plant logs are kept on file for an extended period. Another important routine source is the plant maintenance information system which records all scheduled maintenance tasks performed. Many plants computerize this data and then arrange it so that it can be fed directly from the plant log. It provides useful information on maintenance frequency and cost for specific items.

6.3   FAILURE MODE ANALYSIS CONCEPTS

A key task within a well structured reliability engineering program is failure mode analysis. In view of the complexity of nuclear equipment and components the proper application of failure mode analysis is necessary to assure attainment of operational reliability and plant safety objectives. It involves identifying potential modes of failure (both hardware and software) and determining the effect of each failure mode on system or equipment operation. Failure mode analysis provides a means to identify critical areas for corrective action (e.g. redesign, more reliable parts, etc.) during development prior to the build-up of hardware and the performance of costly equipment tests at a time when changes can be implemented easily.

A complete failure mode analysis is especially required in complex hardware/software systems where a great degree of interaction is involved. The analysis covers:

1.   All parts.
2.   All possible component failure modes.
3.   The probability of failure for each failure mode.
4.   The effects on the system or subsystem caused by each failure mode.
5.   Each failure mode cause.
6.   Possible means of correction or prevention for each failure mode.

Failure mode analysis is a disciplined technique for analyzing a design, evaluating the potential manner in which failures can occur and tabulating results in such a manner that responsible design personnel can determine if it is desirable to allocate resources to compensate for a potential failure mode or to incorporate fault compensating or tolerant features into the equipment. Fault compensating methods include top–down modular design, requirements verification/validation, design (computer code) walkthroughs and structured programming. Fault tolerance features include incorporation of self-checking and redundant software.

The systematic identification of potential failure modes and their effects, using formal analysis techniques, is directly applicable to nuclear systems and equipment. The more complex the system the greater the interaction between its constituent components and the greater the need for a formal and systematic process to identify and classify effects.

Specifically, failure mode analysis can be used to:

1. rank the most serious failure modes;
2. determine needs for redundancy (or diversity) fail-safe design features, derating and/or design simplification;
3. identify untestable redundant areas;
4. determine the need to select more reliable materials, part, equipment and/or components;
5. identify single failure points;
6. identify critical items for design review, configuration control, and traceability;
7. provide the logic model required to quantitatively predict the probability of anomalous conditions of the equipment;
8. identify safety hazard areas;
9. assure that the test program is responsive to identified potential failure modes and safety hazards;
10. aid in identifying untestable functions;
11. establish allowable use time or cycles with respect to components where wearout is dominant;
12. pinpoint key areas for concentrating quality, inspection and manufacturing process controls;
13. establish data recording requirements including recording frequency or monitoring requirements during testing, checkout and plant operation;
14. support maintenance and logistics planning and maintainability analysis by providing input data for RCM decision logic analysis as well as information for selection of preventative and corrective maintenance points and development of troubleshooting guides; and
15. support plant maintenance by providing fault isolation sequences and defining alternate-mode-of-operations.

Two techniques are generally used in failure mode analysis. These are listed below and further described in the following paragraphs:

1. *Failure mode, effects and criticality analysis (FMECA)* represents a 'bottom-up' analytical approach to failure analysis.
2. *Fault tree analysis (FTA)* represents a 'top-down' analytical approach to failure analysis.

The FMECA approach to failure mode analysis involves systematically identifying and tabulating failure modes at the lowest level of assembly and determining their effects at higher levels of assembly and ultimately the effect on the equipment or system. An initial FMECA can be performed early in development and, because only limited design definition may be available, only the more obvious failure modes may be identified. As design definitions are developed, the analyses can be expanded to successively more detailed levels.

FMECA can be characterized as a systematic method of cataloging failure modes and assessing their consequences on the system. FMECA can be performed utilizing either actual failure modes from field data or hypothesized failure modes derived from design analyses, reliability prediction activities and experiences relative to the manner in which hardware/software fails. In its most complete form, failure modes are identified at the lowest level of direct concern to the system designer. In addition to providing insight into failure cause and effect relationships, the failure mode and effects analysis provides a disciplined method to proceed module-by-module through the system to assess failure consequences. Failure modes are analytically induced into each component and failure effects are evaluated and noted, including severity and frequency (or probability) of occurrence.

The first step in the failure mode and effects analysis is to list all failure modes at the lowest practical level of assembly or equipment structure. In the case of hardware failures are due to design, manufacturing or maintenance defects, overstressing and wearout. With software, failures are due to human errors committed during design including the specification and coding process and maintenance. These errors include:

1. Incomplete, ambiguous or untestable requirements.
2. Deficient design representation.
3. Poor coding practices (lack of standards, inspection).
4. Typographical errors.

For each failure mode listed, the corresponding effect on performance at the next higher level is determined. The resulting failure effect becomes, in essence, the failure mode that impacts the next higher level. Iteration of this process results in establishing the ultimate effect at the system level. Once the analysis has been performed for all failure modes, it is usually the case that each effect or symptom at the equipment level is caused by several different failure modes at the lowest level. This relationship to the end effect provides the basis for grouping the lower level failure modes for effective corrective action.

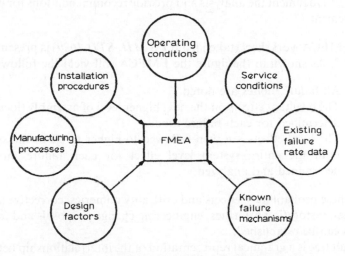

Fig. 6.12. FMECA information support sources.

Probabilities for the occurrence of the system effect can be calculated, using this approach, based on the probability of occurrence of the lower level failure modes (i.e. modal failure rate times time). Based on these probabilities and a severity factor assigned to the various system effects, a criticality number can be calculated. Criticality numerics provide a method of ranking the system level effects derived previously. Criticality numerics also provide the basis for corrective action priorities, engineering change proposals or field retrofit actions.

Figure 6.12 depicts the information sources that affect the ultimate performance of the system or equipment being analyzed and which must be considered in the FMECA.

In its most complete form an FMECA is performed to the part level.

Included are six steps as follows:

Step 1: Define the hardware system and its requirements.

Step 2: Establish a logic block diagram (the reliability prediction block diagram can be used as a starting point for this analysis).

Step 3: Set assumptions and ground rules for performing the analysis.

Step 4: Identify worksheet requirements, including failure modes, effects, failure detection methods, etc.

Step 5: Evaluate criticality of the failure modes.

Step 6: Document the analysis and provide recommendations for design improvement.

An FMECA worksheet (taken from *US MIL-STD-1629*) is presented in Fig. 6.13. As shown in the figure the FMECA will yield the following:

(a)   All failure modes are noted.

(b)   The resultant effects on the next higher level of assembly (local) are determined for each failure mode.

(c)   The local effects are then followed to higher levels of the system until the entire system level effect for each failure mode is determined and analyzed.

From these probabilities, effects and criticality numerics, corrective action priorities, restoration priorities, engineering change proposals and retrofit actions can be established.

A fault tree is a graphical representation of the interrelationship between a specific event occurring (a failure) and the ultimate effect it has upon the system. It also is an iterative documented process which can be utilized to identify basic system faults, establish their probabilities of occurrence and ultimately establish their cause and effects. In contrast to the FMECA process, an FTA is a 'top–down' approach to failure study. During preliminary design the analysis is performed to identify failure modes and formulate corrective action suggestions. After final design the analysis is based on detailed design drawings and is performed to show that the system is acceptable with respect to reliability and, as necessary, suggests modifications to the final design.

The performance of the FTA methodology, involves: first, the structuring of a detailed logic diagram that depicts the basic faults that can lead to system failure; next, the use of computational techniques to analyze the basic faults and determine failure mode probabilities; and finally, a detailed fault matrix which includes all system failure modes, their

FAILURE MODE, EFFECTS, AND CRITICALITY ANALYSIS

Assembly level
ref. drawing

Engineering:
Date:
Sheet:

Item identification and function	System operational mode	Failure mode and cause	Failure effect		Failure detection method	Compensating provisions	Criticality analysis					Remarks
			Local	System			Severity level	Part, λ_p	Failure mode contribution α	Failure mode probability β	Loss frequency, $\lambda = \alpha\beta\lambda_p$	

Fig. 6.13. Example of FMECA worksheet format. (Source: *US MIL-STD-1629*.)

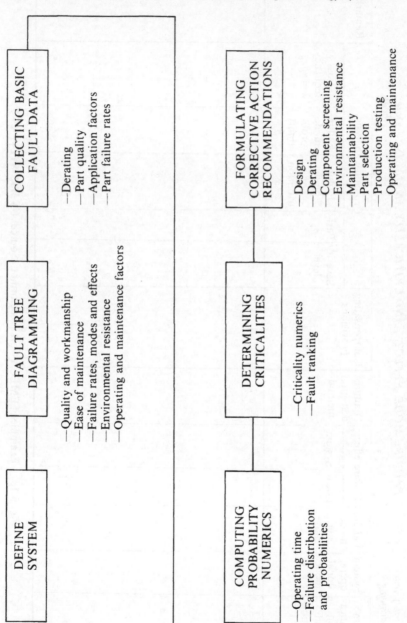

Fig. 6.14. Steps and factors involved in the application of fault tree analysis.

probabilities of occurrence, and corrective action suggestions that when implemented would eliminate (or minimize) those faults considered critical.

The procedure with which to implement an FTA process includes the following steps. (Figure 6.14 presents some of the factors associated with each step.)

Step 1: Define the top event(s) to be studied. This involves determining the equipment failure (or degradation of operation performance) to be studied. Since an FTA is 'single event' oriented a diagram must be constructed for each event to be analyzed. All events, whether they be malfunctions or failures of lower assemblies, are analyzed with respect to the top single event under study.

Step 2: Construct a fault tree diagram. This requires the development of a detailed logic diagram which depicts all possible sequences of events in lower subsystem levels leading to the event under study. This requires a thorough knowledge of the equipment design, functional operation, plant maintenance procedures and environmental application factors. The fault tree is constructed using basic logic elements, as described in Fig. 6.15. When the fault tree structure is complete, the undesired event is completely defined in terms of:

(a)   basic faults (hardware and human) whose occurrence alone or in combination can result in the defined hazard regardless of their apparent frequency of occurrence;

(b)   independent input events; and

(c)   basic faults (e.g. component failure modes) for which failure rate data are available or may be estimated.

Through the use of this methodology, a graphical portrayal which depicts the paths leading to each successively lower fault level for the specific event is achieved.

Step 3: Collecting basic fault data. Failure rate data are necessary inputs for determining occurrence probabilities and assessing criticality. The data consist of two general classes:

(a)   component failure rate data; and

(b)   human error rate data.

In general, the component failure rates are determined through a review of component items identified as faults on the fault tree. This involves reviewing the failure modes of each basic element which comprises the identified fault and establishing a model failure rate, based on historical

Definitions of symbols

Event—usually a fault, resulting from the combination of more basic faults and/or conditions and which can be developed further.

Basic fault—usually a specific circuit, component or human error which can be assigned a probability of occurrence.

Undeveloped fault—a fault not developed further as to its causes because of lack of information, time, or value in doing so.

And gate—the output event occurs only when all of the input events are present.

Or gate—the output event occurs when one or more of the input events are present.

Inhibit gate—similar to an 'and' gate, however, used to include application of a conditional event.

Normal event—an event expected to occur in normal operation.

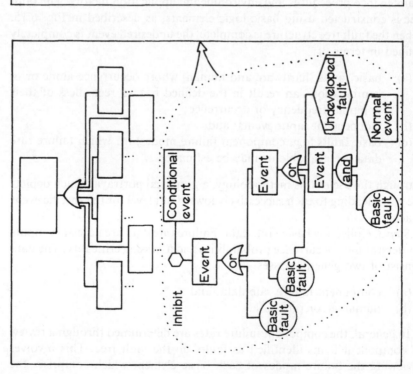

Fig. 6.15. Sample fault tree diagram.

generic part data and available design application information. Standard reliability prediction techniques can be used to estimate these failure rate numerics.

Human error rates mean the expected rate at which a failure caused by operating or maintenance personnel takes place, whether intentionally or unintentionally. It is very difficult to obtain an error rate since very little data exist regarding this area. Since a large scale database is lacking, human error rates can be developed through subjective techniques based on discussions with personnel familiar with the system operation and maintenance environment. These techniques involve detailing each human error depicted on the fault tree into basic task elements. The intent is to define small segments of human performance—where an error rate can be more easily assessed. Assessing the error rate for these individual elements would involve a literature survey, including a review of currently available human error data and/or prior estimation information from personnel familiar with the operational elements. The final error rate numerics must account for the nature of human performance and its sensitivity to learning, fatigue and other behavioural factors.

Step 4: Computing probability numerics. This involves computing the occurrence probabilities for all basic faults, events and conditions (top faults) based on the combinatorial properties of the logic elements in the fault tree. The analysis involves repeated applications of basic probability expressions for the fault tree logic gates. Given a fault tree diagram whose basic faults and output event probabilities are computed, starting with the lowest levels and continuing to the higher levels in the tree:

And Gate

$$P(A) = \prod_{i=1}^{n} P(x_i)$$

Or Gate

$$P(A) = 1 - \prod_{i=1}^{n} 1 - P(x_i)$$

where $P(A) =$ output probability, $P(x_i) =$ probability of the ith input and $n =$ number of inputs.

Step 5: Determining criticalities. Criticality is a measure of the relative seriousness of the effects of each fault. It involves both qualitative engineering evaluation and quantitative analysis, and serves to provide a basis for ranking the faults for corrective action priorities. The object is to assign a criticality numeric to each fault based on its occurrence probability and its contribution to the overall probability for the fault condition under study.

Criticality can be defined quantitatively by the following expression:

$$CR = P(x_i)P(H/x_i)$$

where $P(H/x_i)$ is the conditional probability of the overall hazardous condition given that the basic fault (x_i) has occurred.

Computerized techniques can be used to determine criticality numerics.

Step 6: Formulating corrective action recommendations. This is the final step of the FTA—reviewing and evaluating all data in order to formulate general corrective suggestions. These suggestions can be related quantitatively to the fault elements and failure modes identified by the fault tree analysis. These suggestions, in general, would involve:

—Areas of redesign.
—Component part selection.
—Design and procurement criteria.
—Maintenance procedure.
—Inspection procedures.
—Quality controls.
—Special safety instruction.

The scope and extent of the suggested corrective measures would depend on the faults identified and their criticality and should be considered in relation to their effectiveness, practicality and cost.

A fault matrix can also be prepared to aid in the analysis. Included in the matrix would be elements such as:

1. Basic fault identification number and description.
2. Failure mode.
3. Probability of occurrence.
4. Effect on next higher assembly.
5. Effect on system.
6. Criticality numeric.
7. Corrective action suggestions.

A sample fault matrix is presented in Fig. 6.16 that depicts the above elements.

Fault identification number	Basic fault description	Failure mode	Probability of occurrence ($\times 10^{-4}$)	Effect on next higher assembly	Effect on system	Criticality numeric	Corrective action suggestions

Fig. 6.16. Sample fault matrix.

6.4   PART CONTROL AND STANDARDIZATION

A cost-effective parts program, consisting of the use of properly screened and qualified parts that are adequately derated, is the essence of an effective reliability engineering program and the best assurance of reliable hardware. The task of selecting, specifying, assuring proper design application and, in general, controlling the quality and reliability of the parts used in nuclear systems, particularly electronic control equipment, requires extensive engineering effort during design and procurement. It is a multidisciplinary undertaking involving the best efforts of component engineers, reliability engineers, design engineers and procurement personnel. The total effort includes tasks to: determine part criticality and reliability; establish approval, qualification and standardization procedures; prepare part specifications and ultimately to procure parts which meet the performance reliability and total cost requirements.

A general rule for part control is that standard parts are used wherever possible. Standard parts are defined as those which by virtue of systematic testing programs and a history of successful use in equipment, have demonstrated their ability to consistently function within certain specific electrical, mechanical and environmental limits and, as a result, have become qualified to US military specifications (*MIL-STDs*). US military specifications, which thoroughly delineate a part's substance, form and operating characteristics, exist (or are in preparation) for practically every known type of electronic part. For example:

1.   *US MIL-STD-1562*—microcircuits.
2.   *US MIL-STD-701*—semiconductors.
3.   *US MIL-STD-199*—resistors.
4.   *US MIL-STD-198*—capacitors.

MIL-STDs also exist which cover the subject of testing methods applicable to MIL-specified components. For example:

1.   *US MIL-STD-202*—for electronic parts.
2.   *US MIL-STD-750*—for semiconductor devices.
3.   *US MIL-STD-883*—for microelectronic devices.

If a standard part is not available, special attention is then given to selection of the best non-standard part. This involves evaluation of the part, its reliability history, construction and its potential failure modes as well as a determination of cost-effectiveness.

The selection and control effort also includes the preparation of

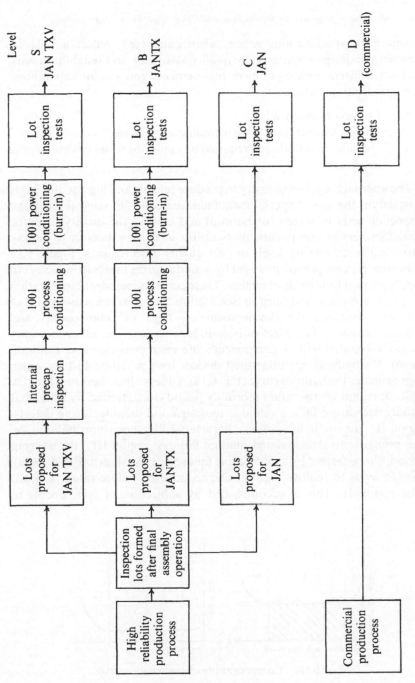

Fig. 6.17. Illustration of manufacturing processes for semiconductors.

procurement specifications which, when completed, reflect a balance, between the design requirements, quality assurance and reliability needs, that is consistent with equipment requirements and vendor capabilities. The specifications cover:

1. Lot acceptance testing.
2. QA provisions (including incoming inspection).
3. Qualification testing as required by application and environmental conditions.

The approach used by industry to produce reliable and high quality parts is based on the use of special manufacturing controls, conditioning and inspection tests to screen (or burn-in) and control the quality of parts. Manufacturers of component parts utilize separate processes to supply parts that meet varying levels of part quality requirements. Figure 6.17 illustrates process procedures used by manufacturers of semiconductors to supply parts at four levels of quality. These controls include the application of special screening and burn-in (see Section 5.2). This is a major point of difference between the development of 'high rel' components and commercial parts. The effect of burn-in is to remove the infant mortality period associated with a component's life characteristic curve (bathtub curve). Virtually all manufactured devices have a failure history which approximates the bathtub curve (Fig. 6.18). Observe that there are three life regions: region 1—the infant mortality period characterized by a high but rapidly decreasing failure rate due to design and manufacturing defects; region II—the useful life period, characterized by a constant failure rate, due primarily to stress-related random failures; region III—the wearout period, characterized by an increasing failure rate. As shown (Fig. 6.18(b)), burn-in tends to remove region I and reduces the failure rate associated with region II. This is accomplished by subjection of components to

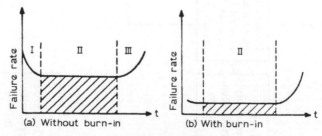

(a) Without burn-in   (b) With burn-in

Fig. 6.18. Component life characteristic curves.

predetermined stress levels, with consequent removal of weak parts, prior to installation in equipment.

6.5   PRODUCTION RELIABILITY ASSURANCE

Degradation in reliability can occur as a result of defects induced by the manufacturing (or maintenance) process and the fact that the quality inspections and tests that are developed to detect and eliminate these defects are not perfect. Reliability degradation is an important factor to be considered with all types of hardware but it is especially important in new hardware where there is a large degree of manufacturing learning. It is through the analysis of the cause of such defects, and the subsequent corrective action, that reliability degradation can be controlled.

Reliability degradation is the result of many causal factors (see Chapter 2). Included among these are built-in flaws due to faulty workmanship, deviations from initial design intent, handling damage and other process stresses. Examples of the types of early failures which may occur due to manufacturing deficiencies are:

1. Poor connections.
2. Poor welds or seals.
3. Improper positioning of parts.
4. Chemical impurities.
5. Contamination of surfaces or materials.
6. Poor soldering of parts.
7. Improper securing of component elements.
8. Bending or deformation of materials.
9. Warping in materials.
10. Electronic part damage due to improper handling (e.g. electronic static discharge).

Reliability predictive efforts that are based on, or computed through, conventional stress–strength prediction concepts, as described in Section 6.1, reflect the reliability potential of a system as it is expected to perform during its useful life period. These estimates depict the inherent reliability of the design as defined by its engineering documentation, basic stress–strength design factors and gross application, manufacturing and quality factors. Stress–strength prediction estimates do not represent the actual operating reliability unless the equipment has reached complete maturity, where design failures and early manufacturing and quality defects have been eliminated.

Manufacturing induced defects can be classified in terms of quality and latent defects. The quality defects are those which are generally observable and detectable through standard inspection procedures, or other tests which provide an indication of equipment condition and capability of operation. Latent (or reliability) defects are detectable only through the application of various screen tests or after some period of normal operation. Latent defects, if not controlled through properly devised and applied inspections, screens and test procedures, contribute significantly to production reliability degradation and early field (infant mortality) failures. Section 5.2 describes the application of screens to eliminate latent defects.

Manufacturing defects can be further subdivided into two classes, intrinsic and induced. The intrinsic defects are based upon the inherent weakness (failure rate) of an equipment's component parts and are controlled during design through part selection and derating. The induced defects are those which are a direct result of manufacturing process stresses and/or operator and inspector error. These defects are controlled by the reliability engineering program, first during design through the specification of materials that allow for manufacturing variations, then through improved manufacturing by control of the fabrication process to lower induced defects—both quality and latent—and finally through inspections and tests that are specifically designed to detect and eliminate defects induced by the process.

Standard inspections and tests are designed to, and actually will, remove the more obvious quality defects; they are not designed to remove latent defects. Stress screen tests designed to convert latent defects (or unobservable defects) into defects which may be removed by conventional testing methods must also be applied. Included among these tests are temperature, temperature cycling, vibration, on/off cycling, input voltage, input voltage cycling and non-destructive tests (NDT), including: (1) thermal; (2) electromagnetic–electronic; (3) sonic–ultrasonic; (4) visual; (5) functional; and (6) optical. However, no test or inspection station is perfectly efficient. Thus, some defects will escape to later production stages and then be removed or, more likely, pass through to operation and thus lower the achieved level of reliability.

A process and inspection analysis can be performed to identify areas where production process reliability improvements are required. The analysis involves defining the initial production or reconditioning process in terms of a detailed production flowchart and estimating the rate of defects (both quality and latent) introduced and removed at various steps within the process. Defect rates are followed through the flow allowing calculation of the number of defects remaining at the end of the process. A comparison of defect rates leaving a process with defect rate entering the process allows estimate of reliability degradation.

The procedure involves performing a process and inspection analysis to compute an average MTBF that accounts for the introduction and removal of defects. Such an analysis, in its simplest form, will show where large degrees of unreliability (defects) are introduced and thus provide a basis to formulate and implement corrective action in response to the overall reliability growth process.

This initial manufacturing MTBF (outgoing from production MTBF) is computed from the following expression:

$$\text{MTBF}_a = \text{MTBF}_i \pi_p$$

where: MTBF_a = initial manufacturing MTBF, MTBF_i = the inherent MTBF and is computed from part failure rate models (i.e. *US MIL-HDBK-217D*), π_p = overall production degradation adjustment factor, i.e.

$$\pi_p = \frac{D_i}{D_i + D_o}$$

where D_i = the inherent defect rate that is computed from MTBF, i.e.

$$D_i = 1 - \exp(-t/\text{MTBF}_i)$$

and

$$\text{MTBF}_i = 1/\lambda_i$$

D_o = outgoing production defect rate (includes both quality and reliability defects) and is computed from a production reliability and inspection analysis.

Figure 6.19 depicts the steps involved in performing a production reliability analysis. It shows a complete reliability analysis leading to an average MTBF that accounts for stress-related failures during the useful life period as well as early failures due to process-induced latent defects. The analysis involves first evaluating the equipment design to determine the inherent (designed-based) MTBF. Once the design analysis is completed, a

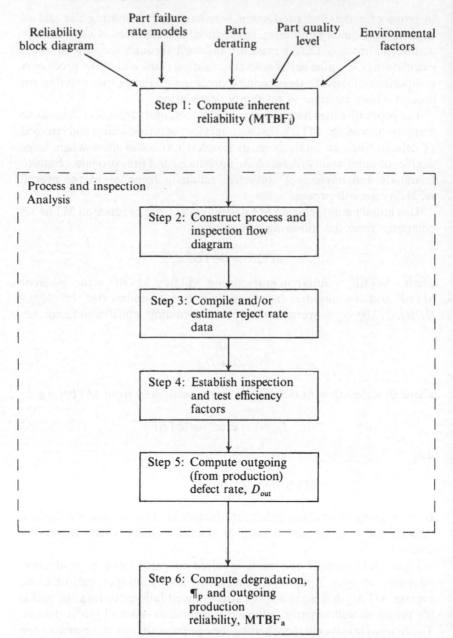

Fig. 6.19.   MTBF (outgoing from production) estimating process.

process and inspection analysis is performed to determine the outgoing defect rate, D_o, and ultimately the factor, $\P_p$, that accounts for degradation in reliability. The analysis accounts for the efficiency of the quality inspections and the strength of applied screens to detect defects. 'Screening strength' is defined as the probability that a stress screen will transform a latent defect into a hard failure (given that there is a latent defect present) and that the failure will be detected by the screen. The output of the two analyses are then combined to yield an outgoing from production MTBF estimate.

The analysis, as depicted in Fig. 6.19, involves the following steps:

Step 1—Compute the reliability of the system as it enters the manufacturing process. The initial estimate of reliability is derived from design-based MTBF prediction techniques such as that given in *US MIL-HDBK-217*.

Step 2—Construct a process and inspection flow diagram. The construction of such a flow chart involves, first, the identification of the various processes, inspections and tests which take place during manufacturing and, second, a pictorial presentation describing how each process flows into the next process or inspection point. Figure 6.20 presents a basic and highly simplified process flow diagram to illustrate the technique. The size of the diagram will vary depending on the complexity of the process. A very simple item may require a flow diagram which contains only a few processes, whereas a complex item may involve several subprocesses combined to form the overall process.

Step 3—Establish reject rate data associated with each inspection and test. For analysis performed on planned processes, experience factors are used to estimate the reject rates. The estimated reject rates must take into account historical part/assembly failure modes in the light of the characteristics of the test used to detect that failure mode. Some of the tests that are used to detect and screen process-induced defects and which aid in this evaluation are discussed in Section 5.2. For analysis performed on current processes, actual inspection reject rate data can be collected and utilized.

Step 4—Establish inspection and test efficiency factors. Efficiency is defined as the ratio of defects removed (or rejects) to the total defects in the fabricated items. Efficiency factors are based on past experience for the same or a similar process when such data exist. For newly instituted or proposed inspection and screen tests having little or no prior history as to how many defects are found, estimates of inspection and test efficiency can

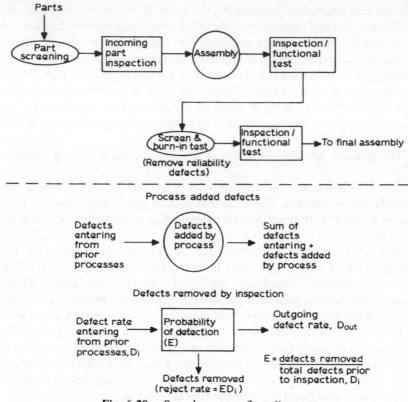

Fig. 6.20.   Sample process flow diagram.

be made using a submodel that takes into account pertinent factors and attributes involved in or related to inspection, such as:

—Complexity of the part/assembly under test (e.g. simple part, easy access to measurement).
—Test design adequacy (e.g. designed to detect specific failure modes? does it reflect accept/reject criteria for significant parameters of unreliability as defined by FMEA?).
—Test procedure adequacy.
—Test equipment complexity.
—Inspector experience (e.g. highly qualified, several years in quality control).
—Time allocated for inspection.
—Sampling plan.

Weighting factors can be applied to each of the inspection attributes and used to estimate efficiency. The efficiency of an inspection can also be expressed in terms of a defect detection probability. A perfect or error-free inspection would have an associated numeric value of 1. The individual factors (such as inspector experience) which influence inspection efficiency can be expressed as probabilities (e.g. probability of inspector error) and combined to determine the overall inspection efficiency probability. As an example, assume there are four independent factors which influence a particular inspection. Further assume that the probability of each factor is 0·9. Then the probability of inspection (i.e. inspecting efficiency) is $0·9^4$ or about 0·66. Thus, even though the probability of each factor is relatively high, the collective probability or the inspection efficiency is relatively low. This illustrates the difficulty of obtaining a perfect inspection.

Step 5—Compute the outgoing defect rate based on the reject rates (from step 3) and the efficiency factors (from step 4) using the process flow diagram developed during step 2. Note that for a given inspection with a predefined efficiency factor, E, that the number of defects of a fabricated item prior to its inspection can be estimated from the measured or estimated rejects, i.e.

$$E = \frac{\text{rejects}}{\text{total defects (prior to inspection)}}$$

The number of outgoing defects is simply the difference between the number prior to inspection and that removed by the inspection.

Step 6—Compute reliability degradation based on the ratio of the inherent design-based reliability (step 1) and the outgoing from manufacturing defect rates (step 5).

Hardware reliability can be improved through successive application of the above analysis. Those processes where large numbers of defects are being introduced can be isolated and then corrected or changed. This is accomplished by introducing an improved process or by applying a screen test (or sequence) to remove the defects. The inclusions of a screening test will increase the initial cost of the system, but the cost avoidance due to increased factory productivity (i.e. lower rework, scrap rate, etc.) and, more importantly, lower support cost should more than offset the initial cost. To be most cost-effective, particularly for large complex hardware, the application of the reliability and inspection analysis should be first applied to subsystems and equipment designated as critical by methods such as FMECA, described in Section 6.3.

6.6 FAILURE REPORTING, ANALYSIS AND CORRECTIVE ACTION (FRACA)

A uniform and coordinated mechanism for reporting, analyzing and initiating corrective actions for failures that occur during acquisition and plant operation must be established. FRACA is a key element in 'failure recurrence control' for both newly developed and operational equipment. A well-structured FRACA program includes provisions to assure that failures are accurately reported and thoroughly analyzed and that corrective actions are taken on a timely basis to reduce or prevent recurrence. It includes documenting the sequence of events that occurs upon detection of a failure, including the analysis, corrective action, personnel responsibilities, scheduling and the applicability of the analysis to reliability growth, demonstration and operational acceptance.

The program requires that engineering (and statistical) investigation be performed to determine the cause and cure of failures. The analysis of failures is important throughout the hardware acquisition operation improvement process but is most cost-effective during development when most reliability growth should occur. In production (as well as during operation) failure analysis is performed to correct deficiencies which jeopardize the achieved (or demonstrated) reliability.

FRACA is an essential element of a reliability control program. A well-designed FRACA system will provide a uniform mechanism for reporting failures, to focus the talents of qualified experts in the analysis of failure, to determine causes and remedies and to make these findings known to the appropriate engineers and designers to enable them to formulate and implement corrective action.

The FRACA system applies to failures which occur at the plant, the manufacturer's facilities and at other locations such as subcontracted test laboratories. It is applicable to:

1. Design verification tests.
2. Preproduction tests, including formal reliability growth, development, qualification and reliability demonstration.
3. Production tests, including receiving inspection, in-process inspection and acceptance.
4. Plant surveillance test.
5. Plant operation.

Failure is defined as the cessation of the equipment or any of its components to perform its specified function(s). This implies that the system (or component) had performed its intended function prior to transitioning to a failure state. Failures may manifest themselves in two separate and distinct ways, namely:

Catastrophic failure: Characterized by an abrupt and complete cessation of the specified function.

Degradation failure: A failure which manifests itself as a relatively slow change in performance beyond pre-established limits.

The reliability engineering organization is generally responsible for instituting and managing the FRACA system. It establishes policy, provides direction and monitors the status of FRACA investigations. Its specific responsibilities include:

—Assigning identification numbers to reports received from the responsible inspection and testing organizations, completing the reports and determining the need for further analysis and corrective action.
—Conducting failure analysis and corrective action investigations and preparing analysis and corrective action reports.
—Submitting report copies to program management and to other internal organizations.
—Maintaining the FRACA experience base.

The cognizant plant operating or manufacturer's testing organizations are responsible for initiating failure reports promptly as they are observed.

An abbreviated flowchart showing the sequence of events that would take place when a failure occurs is given in Fig. 6.21. Note that a failure is defined as any deviation from acceptable limits called out in the equipment specification. Also, any operating discrepancy that requires an unscheduled adjustment or calibration to be made (except normal operating adjustment or scheduled maintenance procedure), after initial satisfactory operation of the affected equipment, is defined as an equipment failure for reporting purposes.

Upon discovery of a failure during a formal test, the operator stops the test, initiates a failure report, logs the discrepancy and notifies the reliability organization. The operator then proceeds with action in accordance with the applicable test plan. The report form shown in Fig. 6.22 can be used for reporting failures.

The reliability organization would verify the accuracy of symptomatic and causal failure information, especially for failures which occur when the equipment is located at outside testing facilities. A preliminary failure

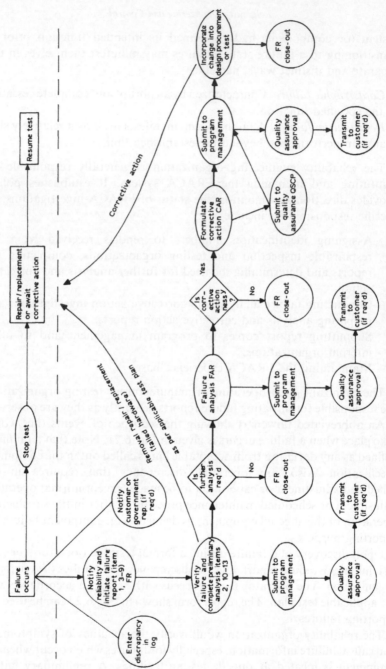

Fig. 6.21. Flowchart of FRACA activities.

FAILURE REPORT

① System...
   Project no. .... Product order no. ....

③ Assembly.................. Subassembly .... Component ....

   .... No. .... No. .... No. ....

② Report no...............................
   Date..
④ Date of occurrence....................
   Time ....

   Total operating time.............. h

⑤ Failure discovered during
   ☐ Design verification test (explain)
   ....
   ....

   ☐ Pre-production system/component test
      Reliability growth ☐   Reliability demo ☐
      Development   ☐   Other   ☐
      Qualification   ☐   Describe ....
   ☐ Production test
      Acceptance ☐
      In process ☐
      Receiving ☐

⑥ Test procedure no. ....
   paragraph no. ....
   Test equipment...... Serial no. ....
   ....
   ....

⑦ Disposition (failed test hardware)
   Authorized by ....
   Date...
   To production control for repair ☐
   Order no...........................
   Scrap ☐
   Hold for MRB review ☐
   Hold pending failure analysis/correction
   action ☐

⑧ Failure description (where appropriate describe test and environmental conditions)
   ....
   ....

   Effect on ....
   System operation...

⑨ Test operator signature .... Date ....

   Below for reliability engineering use   Preliminary investigation description

⑩ ....
   ....

   Further analysis required   Yes ☐   (FAR req'd)   No ☐   Failure considered closed
   Maintenance action taken to restore system/component to operation
   ....

   Replacement item(s) .... Adjustment(s) ....

   Downtime   Date/time system/component returned to test...................

   Diagnostic ....
   Remove/replace/checkout.............................
   Delay.......................................
   Total ....

   ⑪ Failure classification
      Relevant   ☐   Non-relevant   ☐
      Chargeable ☐   Non-chargeable ☐

   ⑫
      Initiated
      Failure ☐

⑬ Reliability engineering signature.. Date ....

Fig. 6.22. Failure report form.

investigation is performed to determine the cause of the failure or if further analysis is necessary. Based on directions from program management, repairs or replacements are made and the test continued or the test held up pending failure analysis and the incorporation of fully verified corrective changes.

Failure analysis is performed, if necessary, on failed equipment. The services of design engineers, materials analysts, outside part vendors or other experts are employed to assist in the analysis and to formulate the corrective action necessary to prevent recurrence. In order to remain cognizant of in-process activities and aware of significant trends or discrepancies which may affect the inherent reliability of the equipment, screening board reports, summary sheets, in-process discrepancy reports, etc., are monitored. Based on any observed trends, an investigation may be initiated to determine the cause and a failure report completed and processed for the area under investigation. In the case of an electronic part failure, an autopsy on the failed part may be conducted to determine and/or verify the actual cause of failure. The component analysis methods used, including test, X-ray, dissection, SEM, chemical analysis, etc., to determine failure causes are described. The analysis form given in Fig. 6.23 can be used for reporting the results of all failure analyses.

Where analysis shows corrective action is necessary, a corrective action request (see Fig. 6.24) is initiated and submitted for appropriate action. Corrective action is requested from the cognizant organization. Corrective measures are recommended that may eliminate the failure mechanism and are described in the appropriate space. These measures could involve:

1.   System/equipment redesign.
2.   Part selection criteria.
3.   Part derating criteria.
4.   Reliability growth and demonstration.
5.   The application of tests to weed out specific failure mechanisms.
6.   Special in-process fabrication inspections and tests.
7.   Special reliability assurance provisions.

When quality assurance receives notice of action taken, the corrective action request form (see Fig. 6.24) is completed and returned to the reliability organization indicating the corrective action that was taken.

Failure reporting analyses and corrective action forms are designed to meet the needs of the individual system development and production program as well as the organizational responsibilities, requirements and constraints of the manufacturer's organization. The forms shown in Figs.

Failure analysis report				① Reference failure report no.		
③ Repair record, replacement parts				② Date		
Part no.	Mfg.	Mfg. part no.	Ckt. symbol	Part serial no.	Part lot code	Prod. ord. no.

Adjustments

④ Description of analysis (use additional sheets, if necessary)

⑤ Corrective action to be requested
Yes ☐   No ☐

⑥ Signature   Date

Fig. 6.23. Failure analysis report form.

	Reference failure report
Corrective action request	Report no. Date

To_____ Department_____

Description of problem

Recommendation

Signatures: Reliability manager _____ Date _____
            Engineering manager _____ Date _____
            Program manager_____ Date _____

Action taken

Signature _____ Date _____
Return completed form to reliability by _____ (Date)

Engineering change order no.	Signature_____ Date_____

Fig. 6.24. Corrective action request form.

6.23 and 6.24 are considered applicable to most hardware programs. They can be adjusted or modified to meet a specific requirement as specified in the applicable program and/or test plan. They include entries for part identification data, conditions under which failures occurred, operating parameters indicating degradation, replacement part(s), repair times, references to applicable plans or procedures and complete details leading up to or surrounding the failure incident.

The following paragraphs describe the individual entries on each form and how they are to be completed. The numbers refer to the circled numbers on the forms.

6.6.1 Failure Report (FR) (see Fig. 6.22)

Upon discovery of a failure the responsible test operator shall complete items 1 and items 3–9 of the failure report form as follows:

Item 1—Record equipment name (including model and serial number).

Item 2—The initiator of the FR will leave this space blank—reliability will insert the FR number and date after receipt of report (with item 1 and items 3–9 completed).

Item 3—Identify the system hierarchy (including the component, subassembly and assembly, as applicable) in which the failure occurred and their associated drawings. When a component or subassembly is tested separately or not yet installed in its higher assembly, identify the assembly for which it is intended.

Item 4—Record date and time in which failure occurred; record total system operating time in hours (note: elapse time indicators shall be incorporated in test hardware, or an equivalent technique as described in the applicable test plan, and used to report accumulated operating time).

Item 5—Check the appropriate test where failure occurred.

Item 6—Record the appropriate test procedure and paragraph number and identify the test equipment (including serial numbers) used to perform the test.

Item 7—Check the disposition of the failed test hardware. The test supervisor shall sign and date the form in the space provided.

Item 8—Describe the failure symptoms, including the visual characteristics of the hardware, the effect on operation and, where appropriate, the test and environmental conditions. Reference to the test procedures and specific paragraph numbers shall be made as applicable. If discernible, describe what appears to have caused the failure to occur.

Item 9—Sign and date the report in the space provided. The test supervisor shall initial the report next to the signature of the test operator.

Upon completion of the above items, the test supervisor shall forward the report to the reliability organization for further processing. The reliability organization shall enter the following information:

Item 2—Record FR number—All failure reports which are prepared as a result of the provisions of this procedure shall be given a unique number. The reports shall be consecutively numbered without regard to the project. Numbers shall be assigned and entered in the space provided upon verification of the failure.

Item 10—Describe the results of the preliminary investigation to determine the nature and cause of the failure. Indicate the extent of maintenance action necessary to restore operation and whether further analyses are necessary to determine root causes. If parts are replaced, and/or further analysis is necessary, an FAR form must be completed (see Fig. 6.23). If not the FR shall be considered closed. If applicable, describe the maintenance actions taken to restore the system/component to operation. Identify item(s) replaced and adjustments made (note: specific information concerning replacement parts and special adjustments shall be recorded on the FAR). Record approximate downtime including time to diagnose failure, remove and replace failed hardware and to checkout, align and verify that failure has been corrected, and any delay times to acquire test equipment, replacement parts, etc. Record the date and time the system/component was restored to test.

Item 11—Record whether the failure is considered relevant or non-relevant. Classification for demonstration and production acceptance tests must be based on specific criteria provided in the applicable test plan.
Relevant failures include:
(a) Intermittent failures.
(b) Unverified. Failures which cannot be duplicated or are still under investigation or for which no cause could be determined.
(c) Verified. Failures not otherwise excluded as non-relevant.

Non-relevant failures include:
(a) Installation damage.
(b) Accident or mishandling.
(c) Failures of the test facility or test-peculiar instrumentation.
(d) Equipment failures caused by an externally applied overstress condition in excess of the approved test requirements.
(e) Normal operating adjustments (non-failures) as prescribed in the approved equipment operating instructions.

For relevant failures record whether the failure is chargeable or non-chargeable according to criteria provided in the applicable test plan.

Item 12—Check this box if there is definite knowledge that the failure was caused by human error. For example, voltage improperly applied by the test operator.

Item 13—Sign and date the FR in the spaces provided.

6.6.2 Failure Analysis Report (FAR) Form (see Fig. 6.23)

The reliability organization shall complete all items on this form, inserting the required information in the space provided.

1. *Reference failure report number:* The original failure report number shall be entered in this space.
2. *Date:* The date that analysis activities are begun shall be inserted in this block.
3. *Repair record:* Appropriate information on the replacement part (where applicable) shall be entered in this space. Include description of any adjustments beyond normal operating adjustments made to restore equipment to operation.
4. *Analysis:* The results of the analysis shall be described in this portion of the form. Where the analysis is performed on an electrical part, pertinent part identification data (serial number, date code, etc.) shall be entered in this space. The analysis shall list the cause of failure insofar as is possible. Consideration shall be given to applicable methods of failure analysis including test, x-rays, dissection, chemical analysis, microphotography, etc. In addition, the analysis shall show any secondary failures caused by the primary failure.
5. *Corrective action to be requested:* Depending upon the results of analysis, this block shall be checked either yes or no.

6.   *Signature and date:* The failure analyst shall sign and date the form at the completion of the analysis.

6.6.3 Corrective Action Request (CAR) Form (see Fig. 6.24)

The reliability organization shall complete the first half of the form giving the reference failure report number, the date, a description of the problem and recommendations for corrective action. The lower portion of the form shall be completed showing the action taken to correct the problem (and the need for follow-up action). The organization completing this form shall sign and date it in the space provided. Where engineering change orders are generated as a result of this request they shall be entered in the space provided. The reliability organization shall acknowledge receipt of the completed form by means of a signature by an authorized person at the bottom of the form.

Maintaining accurate and up-to-date records through implementation of the data reporting, analysis and corrective action system described in this section provides a dynamic, continually expanding experience base. This experience base, consisting of failures occurring during test and corrective responses, can be applied to the development of subsequent hardware development programs. Furthermore these experience data can be used to:

—assess and track reliability;
—perform comparative analysis and assessments;
—determine the effectiveness of reliability and quality activities;
—identify critical components and problem areas;
—compute historical component failure rates (for new design reliability prediction in lieu of generic failures such as *US MIL-HDBK-217*).

6.7   DESIGN REVIEW

Design reviews are performed to assess the reliability LCC, performance and various other characteristics of the equipment at major design and testing milestones. An overall design review program is established that includes both internal and formal reviews with utility participation consistent with the requirements of the procurement specifications and the program plan. An effective design review program provides for reviews of all equipment elements down to the component level including subcontractor's design efforts. Plans are prepared that fully define the nature and scope of the review, the participants and their responsibilities, and the deficiency follow-up control procedures. In addition, the design review includes the preparation and application of a detailed and comprehensive checklist and criteria against which the design can be evaluated. The checklist is prepared relative to the design phase under review.

Informal reviews of the design are performed, in addition to the formal reviews, on a continuing basis by the reliability engineering organization with system, design and component engineers participating, as required. These informal reviews cover:

1. R & M allocation and prediction.
2. Part selection criteria.
3. Design application factors and derating criteria.
4. FMECA data.
5. Applicable test data.

These informal reliability reviews are conducted frequently during early design when changes can be implemented most effectively. Design changes occurring during later design stages, which many times involve changes to drawings, part selection, procurement and approval or replacement of existing hardware, will be considerably more costly. The continuous review of the design with emphasis during the early phases of the development program facilitates detection and correction of actual or potential problems prior to finalization of the design.

Formal design reviews provide the means for assessment and monitoring of the equipment designer/manufacturer effort and generally coincide with major program development milestones. The prime purpose of a formal design review is to provide assurance to development management that significant factors and considerations, particularly those affecting reliability, maintainability and safety, have been reflected into the design effort. The design review provides:

1. A means of solving interface problems.
2. Confidence that experienced personnel are involved in the design detail.
3. A record of why decisions were made.
4. A knowledge that equipment will tie together and be compatible.
5. Support to the utility, the designer and the manufacturer in making tradeoff decisions.
6. A greater probability of achieving a fully mature design that meets all requirements.

The formal design review program first considers the overall system design approach, the techniques and disciplines to be applied, etc., and then as the design progresses, broadens to include a detailed review of such factors as part selection and application criteria, derating, failure rates, thermal analysis, environmental application stresses and subsystem integration and interface problems. Design review activity is controlled and evaluated by assessing checklists, internal design review functions, and the techniques by which discrepancies uncovered are followed up and corrected. For most new procurements, the planning and conduct of preliminary design reviews (PDR) and critical design reviews (CDR) are required. These are discussed in the following paragraphs.

The PDR is performed after completion of the paper design, prior to detailed design and buildup of hardware for development and reliability growth testing. The PDR is performed at the equipment assembly and part level to assure that the preliminary design approach will satisfy specified requirements, that the applied technologies and processes that are within the state-of-the-art, that design, manufacturing and test facilities are available, that the equipment can be designed, tested and built in a timely fashion, and that the manufacturer's personnel (and their subcontractors) are technically qualified. Pending results of the PDR, the program will proceed to the detailed design phase.

Specified information that is reviewed during the PDR includes:

1. Personnel, delineating their experience and organizational assignments.
2. Design review checklists, prepared for the PDR as well as the subsequent CDR.
3. Design review plan (hardware milestone identification, etc.).
4. PDR data package including:
    (a) Program plans.
    (b) Critical parts list.
    (c) Circuit analyses, including worst case, tolerance and transient studies.
    (d) Part selection and application data.
    (e) R & M allocations and predictions (part count).
    (f) Maintenance concept.
    (g) Preliminary integrated test plans (i.e. reliability growth, R & M demonstration, reliability screening and acceptance).
    (h) Special studies (e.g. R & M and cost tradeoffs).
5. Deficiency follow-up action, including problems to be studied, target dates for completion and methods of follow-up to assure completed actions.

The CDR is held after completion of the detailed design and development phase, after completion of development or reliability growth testing, and prior to the conduct of qualification testing and the subsequent release of engineering drawings to manufacturing. This review provides the greatest potential for identification and correction of detailed design problem areas because it is conducted down to the part level at the time when the design is considered to be complete, when all development tests have been completed, and the output data from the PDR and the contractors' informal reviews are available. The CDR encompasses much of the same objectives and scope as the PDR; however, since this review represents the final opportunity to evaluate the design prior to qualification testing and manufacturing, it may prevent the start of costly testing and manufacturing of hardware with design related problems.

A specific data package to support the conduct of the CDR is prepared by the manufacturer. Examples of the types of data necessary to support a CDR includes:

1. System R & M predictions (stress analysis).
2. Total parts list, identifying non-standard parts, if any, with appropriate justification.
3. Past derating application data.
4. Part failure rate data and sources.
5. FMECA.
6. FTA.
7. Circuit analysis.
8. Development and reliability growth test results.
9. R & M demonstration test plans and procedures.
10. Production screening and acceptance test plans and procedures.
11. Technical configuration data, including block diagrams, schematics, detailed drawings, etc.
12. Approved design changes.

6.8 RELIABILITY TESTING

Achievement of reliable equipment requires the application of well-planned and properly executed tests during development and production. It requires a test program that emphasizes detailed planning early in the system life cycle to structure the most cost-effective tests consistent with specified requirements, equipment characteristics and budgeting constraints. An integrated test plan, which is part of the overall methodology to specify optimum reliability levels (Chapter 3), must be prepared to ensure reliability achievement in the plant.

An overall reliability engineering program must be structured, emphasizing the performance of cost-effective reliability tests at key points during the development and production cycles. These tests must be designed to insure reliability growth and to demonstrate compliance with specified requirements.

An effective reliability test program involves designing and applying growth and demonstration tests during development and screening and acceptance tests during production. These tests are defined as follows:

1. *Reliability growth* tests are designed to identify problem areas, expose latent defects or underscore deficiencies, such that the implemented corrective action causes incremental reliability growth as the test progresses.
2. *Reliability demonstration* tests are designed to apply statistical control and confidence factors to periods of equipment operation for the purpose of showing compliance with reliability or MTBF requirements.
3. *Screening* tests are designed to force out latent defects non-destructively during manufacturing.
4. *Reliability acceptance* tests are designed to show that the hardware item under test meets its functional or performance requirements on an item by item basis within the production situation.

Reliability growth is defined as the improvement process during which hardware reliability increases to an acceptable level. Due to defects induced in the fabrication process, the measured reliability of newly fabricated hardware is much less than the potential reliability estimated during design (see Fig. 6.25). The purpose of a growth process, especially a reliability growth test, is to achieve acceptable reliability during plant use. Achievement of acceptable reliability is dependent on the extent to which testing and other improvement techniques have been used during development to 'screen out' design and fabrication flaws and on the rigor with which these flaws are analyzed and corrected.

A primary objective of growth testing is to provide methods by which hardware reliability development can be dimensioned, disciplined and managed as an integral part of overall development. Reliability growth testing also provides a technique for extrapolating the current reliability status (at any point during the test) to some future result. In addition, it

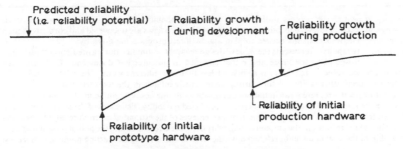

Fig. 6.25. Reliability growth process.

provides methods to assess the magnitude of the test, analyze and fix (TAAF) effort prior to the start of the development, thus allowing for tradeoff decisions.

For electronic systems, the model used most often for reliability growth processes, and in particular growth testing, is one originally published by J. T. Duane. Essentially, this model provides a deterministic approach to reliability growth in which the system MTBF versus operating hours relationship falls along a straight line when plotted on log-log paper. That is, the change in MTBF during development is proportional to T^α where T is the cumulative operating time and α is the rate of growth corresponding to the rapidity with which faults are found and changes made to permanently eliminate the basic causes of the fault observed.

Details pertaining to the planning and specification of a growth test program (based on the Duane model) for a newly designed system are shown in Fig. 6.26. The value of the parameter α can vary between a minimum of 0·1 (which can be expected in a program where no specific consideration is given to reliability) and a maximum of 0·5 (where an aggressive reliability program with management support is implemented). In the cases of minimum growth rate, growth is largely due to a solution of problems impacting production and from corrective action taken as a result of user experience. Maximum growth rate results from a formal stress oriented dedicated test program, designed to aggravate and force defects and vigorous corrective action. For example, Fig. 6.26 shows an item with an MTBF potential (inherent reliability) of 1000 h and an initial MTBF (starting point) of 100 h. Thus, at the minimum growth rate (α) of 0·1, the achievement of an MTBF of 200 h (double the initial MTBF) requires 100 000 h of cumulative operating time. This is the case when no specific attention is given to reliability growth. However, if the growth rate can be accelerated to the maximum value of 0·5 (by growth testing and

Reliability growth is the process during which hardware reliability increases to an acceptable level; achieving this acceptable level depends upon the rigor with which flaws are uncovered and corrective action implemented. The four elements needed to plan a growth test program are given by:

① Inherent reliability—represents the value of design reliability estimated during prediction studies, and which may correspond to the value above that specified in procurement documents. Ordinarily, the contract specified value of reliability is somewhat less than the inherent value. The relationship of the inherent (or specified) reliability to the starting point greatly influences the total test time.

② Starting point—represents an initial value of reliability for the newly manufactured hardware; usually falls within the range of 10–40% of the inherent or predicted reliability. Estimates of the starting point can be derived from prior experience or are based on percentages of the estimated inherent reliability. Starting points must take into account the amount of reliability control exercised during the design program and the relationship of the system under development to the state-of-the-art. Higher starting points minimize test time.

③ Rate of growth—depicted by the slope of the growth curve which is, in turn, governed by the amount of control, rigor and efficiency by which failures are discovered, analyzed and corrected through design and quality action. Rigorous test programs which foster the discovery of failures, coupled with management supported analysis and timely corrective action, will result in a faster growth rate and consequently less total test time.

④ Calendar time/test time—represents the efficiency factors associated with the growth test program. Efficiency factors include repair time, operating/non-operating time as they relate to calendar time. Lengthy delays for failure analysis, subsequent design changes, implementation of corrective action or short operating periods will extend the growth test period.

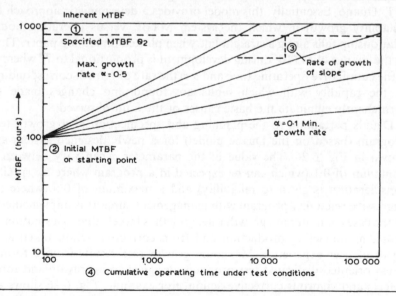

Procedure:
1. Specify total test time (item 4) for the growth test and number of units to be subjected to growth testing.
2. Define growth rate.

Fig. 6.26.   Reliability growth test program. (The growth plot shown is from Anon., *Research Study of Radar Reliability and its Impact on Life Cycle Costs for the APQ-113, -114, -120, and -144 Radar Systems,* 1972.)

formal failure analysis activities), then only about 400 h cumulative operating time is required to achieve an MTBF of 200 h.

Reliability demonstration can also be specified which involves formal testing conducted per *US MIL-STD-781*. This standard defines test plans, environment exposure levels, cycle times and documentation required to demonstrate formally that the specified MTBF requirements of the equipment have been achieved. Demonstration tests are normally conducted after growth tests in the development cycle using initial production hardware. Reliability demonstration testing conducted per *US MIL-STD-781* carries with it a certain statistical confidence level. The more demonstration testing that is performed, the higher the resulting statistical confidence, whereas the more reliability growth testing that is performed, the higher the actual reliability. Depending on program requirements, funding and other constraints, equipment testing may maximize growth testing and minimize demonstration testing (resulting in a high MTBF at a low confidence) or it may minimize growth and maximize demonstration (resulting in a lower MTBF at a high confidence). These concepts are shown graphically in Fig. 6.27.

Screening tests are employed to eliminate incipient failures from the manufacturing process. Screening is the application to a component or assembly of a stress test, or tests, to reveal inherent weaknesses (and thus

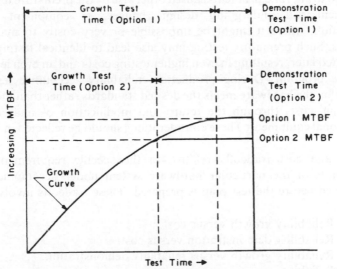

Fig. 6.27. Reliability growth and demonstration testing tradeoff. (*Source: RDH-376.*)

incipient failures) without destroying the integrity of the assembly. Methods for screening are discussed in Chapter 5.

Reliability acceptance testing is a periodic series of tests performed to indicate continuing 'acceptable' production by determining individual unit compliance to reliability criteria. The tests are designed to simulate in-service stresses and are intended to provide unit verification of the demonstrated reliability level. *US MIL-STD-781* provides procedures and requirements for reliability acceptance testing for equipment that experiences an exponential failure distribution.

A well-structured test plan involves the implementation of an integrated test program instead of separately managed testing programs. This prevents both duplications and omissions in testing and also provides a single test baseline in parallel with a closely integrated program on reliability assessment and tracking. The approach emphasizes the intimate tie-in of the reliability assessment and tracking effort with the specified hardware requirements and underscores the role of reliability growth as an input to the various decision points during development and production.

The test plan is prepared early in the hardware development program. It should describe all test activities with clearly identified planning and implementation milestones. Note that reliability testing performed late in the development phase as opposed to other times in the life cycle, is designed to alleviate problems caused otherwise by the premature testing of equipment still undergoing design changes, or equipment during production, where it might be impossible or very costly to make any changes. Such premature testing may also lead to identical testing being conducted later, resulting in even higher testing costs and an even less cost-effective program. During production, reliability acceptance tests can assure that the hardware meets the desired standards rather than incurring additional costs through the return and modification of unacceptable operational equipment. These considerations should be reflected in the test plan.

Test and cost tradeoffs relative to the specific requirements and constraints of the particular hardware system under development are performed before the test plan is prepared. These tradeoffs involve:

1. Reliability growth versus cost.
2. Reliability demonstration versus cost.
3. Reliability growth versus reliability demonstration.
4. Reliability screening test versus cost.
5. Reliability acceptance testing versus cost.

As previously discussed and shown in Fig. 6.26, reliability growth testing will improve an equipment's reliability level as a function of the amount of testing hours involved (the more reliability growth testing performed, the higher the actual reliability). Since the cost of growth tests is proportional to the number of testing hours, growth testing and its associated reliability improvement can be traded off against cost. The following criteria should be taken into consideration when allowing for growth testing in the overall integrated test plan:

1. The contractor's plan for reliability growth testing should demonstrate a vigorous TAAF program which emphasizes comprehensive and detailed failure analysis activity, and identifies relationships between various time factors, growth rates and starting/end points.
2. Specific criteria should be formulated as part of the integrated test program and should identify, for example:
    (a) Specified MTBF.
    (b) Predicted MTBF (inherent)
    (c) Starting point.
    (d) Growth rate.

Methods for specifying and predicting MTBF have been discussed. The starting point is, in most instances, approximately 10% of the inherent predicted MTBF. The exact starting point for a given system is dependent on the extent of the analytical tasks performed during early design, prior to the build-up of prototype hardware. A full development program, which emphasizes reliability prediction, FMECA and design review, would result in a high starting point. The growth rate is a function of the intensity of FRACA effort.

Growth plans should include the cumulative test time required to grow to the specified MTBF, the number of test units subjected to growth tests and the anticipated test time per unit. In addition the contractor's growth plans shall indicate realistic time factors which recognize that, in order to grow under a constant level of corrective action, sufficient downtime must be allowed for adequate implementation of corrective action before restarting the growth tests. The plans should include:

1. Calendar time/month available.
2. Test time/calendar time.
3. Description of test cycle (environment on/off time).

The progress of growth testing should be tracked. Logs and data forms that record the number of units on test, test time accumulated, failures, corrective actions and level of reliability, of MTBF, achieved during a given time period should be maintained.

Reliability demonstration testing, as conducted per *US MIL-STD-781*, carries with it a certain statistical confidence level. Since the cost of demonstration testing is proportional to the total number of testing hours, statistical confidence can be traded off against cost. It must be emphasized that reliability demonstration and acceptance testing is most applicable in situations involving items produced and deployed in substantial quantities. Growth testing, on the other hand, emphasizes stress and environmental testing of a small number of equipments to identify design and fabrication deficiencies and thus is most applicable to nuclear equipment and components. As the amount of effort involved in each testing procedure increases, so its associated cost increases. Thus, a tradeoff decision must be made for each test type.

The key to determining the make up (i.e. test type, extent and rigor) of the test program considered most cost-effective for a given equipment is to first determine the maximum level of reliability achievable within the state-of-the-art and then to determine the level of reliability to specify, which optimizes LCC and meets plant safety objectives (see Fig. 3.1).

If the specified reliability (or MTBF), for example, is close to the maximum that can be achieved within the state-of-the-art (i.e. if there is little room for reliability improvement) then a very vigorous and intensive test program should be structured and implemented. The program, in that case, would then include full reliability growth tests, demonstration tests, screening tests and production acceptance tests to assure compliance to specified requirements and with high confidence. However, if the specified value is not stringent and there is ample room for reliability improvement, then the test program would not have to be extensive (see Table 6.3).

Once the extent of testing has been determined through a review of the reliability specified and its relationship to the state-of-the-art, then evaluations and tradeoffs should be made to determine what tests to include or exclude.

The nature of the procurement (i.e. new development, commercial) will dictate to a large extent the type of tests. If the hardware to be procured is a commercial product, no growth or demonstration testing should be included. However, depending upon the required reliability level, acceptance and screening testing may be required. Modified commercial procurements may, however, call for some growth and demonstration testing (along with the acceptance and screening test requirements) while new developments will be dependent solely upon the required R & M levels.

Other factors which should be reviewed are the number of units to be procured along with the equipment's complexity and criticality. The complexity of an equipment plays a significant role in the determination of cost-effective testing in that the reliability of a more complex equipment is more difficult to improve (and thus achieve) than a simpler equipment with the same specified reliability level. Therefore, a higher level testing program will be required. Equipment criticality must also be considered. A highly critical system will always demand high levels of testing focusing on growth and screening tests.

TABLE 6.3
Reliability Test Levels as a Function of Reliability

State-of-the-art level	*Equipment characteristics*[a]	*Test program*
Highest, $R_2 \sim R_3$	Equipment(s) offers the least room for reliability improvement Most of the improvement techniques are included	Rigorous
Moderate, $R_2 < R_3$	This equipment(s) offers moderate room for reliability improvement Some of the improvement techniques are incorporated	Moderate
Lowest, $R_2 \ll R_3$	This equipment(s) offers the greatest room for reliability improvement Few of the reliability improvement techniques are incorporated	Low

[a] Apply for a given equipment design.

Once the tradeoff decisions have been made, the test plan should be formulated early in the program to allow adequate time for the preparation and subsequent approval of the test procedures prior to actual testing. The test plan and its procedures should cover sample size, measurement techniques and associated error, environmental factors, statistical techniques and variables involved in data analysis. As part of the plan, system requirements should be identified in sufficient detail to direct concentration of the test efforts to those parameters and functions most critical to meeting the system requirements. The plan should establish a test program baseline defining required test parameters and the methodology for their measurement. It should include:

1. The test objectives and requirements.
2. A test flowchart that defines all tests and subtests.
3. The test methodology to satisfy evaluation criteria.
4. A matrix of test resources required (i.e. identification of the test units, the number| to be tested, support personnel, etc.).
5. The test conditions (i.e. environmental, operational and performance profiles and duty cycle).
6. The test schedules expressed in calendar time, including test milestones, preventive maintenance interval and test program review schedule.
7. The required precision/confidence of measurements and measurement standards and criteria.
8. The test ground rules, chargeability criteria and interface boundaries.
9. The test facility and equipment descriptions and requirements.
10. Blocks of time and resources designated for the incorporation of design corrections.
11. Data collection and recording requirements.
12. Closed loop failure reporting and analysis requirements including timing for corrective actions.
13. Description of preventive maintenance to be accomplished during test.
14. Status reviews and assessment milestones.
15. Final disposition of test items.

6.9 SOFTWARE RELIABILITY ENGINEERING AND CONTROL

An effective software reliability engineering program requires careful planning and close management from the very beginning of the equipment acquisition cycle. It also demands the application of highly disciplined engineering practice from conceptualization through plant operation. Figure 6.28 shows the relationship of some of the major hardware/software engineering tasks during each phase of the acquisition cycle and at each milestone. The figure indicates that the hardware and software proceed through development concurrently. A system must not be approved for

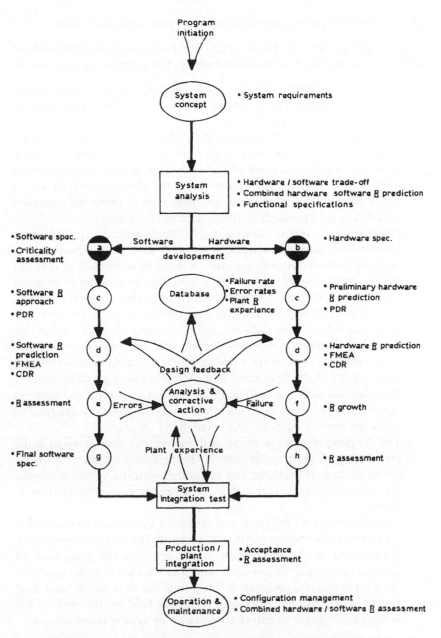

a Software req'n   b Hardware req'n   c Preliminary design   d Detailed design
e Coding / debug   f Fabrication and test   g Software system test   h Hardware qual.

Fig. 6.28. System hardware/software development.

advancement to the next phase until both hardware and software have satisfied all requirements of the earlier phase. The engineering tasks are to:

—Produce a statement of requirements during the conceptual design phase covering system functions as well as the constraints (design, cost, etc.) that the system must meet.

—Translate the system requirements into well-defined functional specifications. This includes performing detailed hardware/software tradeoffs to select the most effective design approach among various alternatives. Also, a computer program design specification is prepared and a design review is generally held to assess the adequacy of the selected approach to the system requirements.

—Develop and verify, during early development, the software specification interface requirements and data requirements. This includes preparing functional and detailed flowcharts. The detailed flowcharts are used to define the information processing in terms of logical flow and operations to be performed by the computer program. Also the relationship between the computer program and the interfaces between the software and the hardware are defined at this time. A preliminary computer program specification is prepared and, prior to coding and testing, a design review is usually held to establish the integrity of the flowcharts and the preliminary specification.

—Translate the software design into actual program code and perform initial testing of the code. This initial testing is normally designed to check for correct outputs using predefined inputs.

—Test the computer programs against the requirements as stated in the preliminary program specifications. Once tested, a software documentation package is prepared and integrated with the system hardware documentation and a complete program product specification is finalized.

—Load the computer program and conduct a system test to insure that the system performance meets requirements. The system is completely documented at the completion of the development phase and all changes resulting from previous tasks and tests are incorporated into the supporting documentation including the flow charts and final system specifications. A design review is held at the end of the development phase to assure compliance to system requirements.

Considerable emphasis must be placed on fault avoidance techniques and on checking for correctness at every stage of the design and development process due to the fact that it is impractical to exhaustively test software and computer systems. System faults, both hardware and software, are caused by human error, defects introduced during manufacture and wearout/overstressing. System reliability can be considerably improved through systematic application of fault avoidance, fault removal and fault tolerance techniques. In addition, reliability can be assured through rigorous formulation of a system specification and application of FMECA.

Fault avoidance and fault removal techniques are specifically aimed at hardware and software faults caused by human error. Fault tolerance methods are aimed to minimize consequences of faults due to human error and faults due to wearout/overstressing. Rigorous formulation of a system specification affects reliability in two different ways: it facilitates checking for and proving of the correctness of algorithms and it also prevents misunderstandings of what constitutes a system failure. FMECA, if carried out during the formulation of a system specification, provides a basis for specifying higher reliability requirements for those subsystems whose failures may have particularly critical effects. Furthermore a most effective way of avoiding the occurrence of human errors and, as a consequence, avoiding hardware and software faults, is to keep the complexity of hardware and software under strict control.

There is much interaction between hardware and software development. The interplay of analysis, test, verification and validation is essential throughout the development cycle in order to ensure software quality and total system reliability. Each of the tasks not only contributes to the total program, but also provide timely inputs to other tasks in relation to system and software milestones. Quality assessments and reliability analyses, for example, are performed as part of early system studies (tradeoffs) to establish the optimum levels of reliability to be achieved in both hardware and software design. These analyses extend through design and development to further define reliability, to establish a basis for integration tests and finally through assessments performed during system test. The test program includes program/system testing during development to force out design errors and system integration and acceptance testing prior to delivery to assure with confidence that the requirements are met.

Some of the essential reliability program tasks and activities necessary to support the acquisition and operation of computerized systems are:

1.   Provide support during the early concept phase by performing appropriate technical and system reliability analysis. The system . analysis should identify control functions and parameters and

associated criticality of failure, safety related issues, special conditions of use (and possible misuse), environmental and human factors which have the potential for introducing error and malfunction, and the mechanics of the man/machine interface.

2. Establish reliability requirements for the software by defining system requirements that describe the goals and objectives to be achieved by the software routines and how they will be integrated with the hardware requirements.

3. Select computer and peripheral devices through detail consideration and tradeoffs of component availability, cost, figure needs, flexibility, application requirements, vendor support and reliability.

4. Perform system engineering (prediction, FMECA, etc.) during hardware/software design and development to assure that reliability is achieved and that critical hardware failure modes and software errors cannot contribute to unsafe operation of the system.

5. Establish reliability requirements for microprocessor-based devices as well as the peripherals from a review of manufacturer's data, experience and other sources, such as *US MIL-HDBK-217*.

6. Prepare an integrated test plan that provides a full description of the tests to be performed to demonstrate the acceptability of the hardware and software designs.

7. Implement a well-documented configuration management system that provides traceability of the system configuration as well as all changes to the hardware and associated software after final acceptance.

The contractor's reliability and quality assurance program is essential to system development and acquisition. Military systems generally require a contractor software quality program prepared in accordance with *US MIL-S-52779*. The contractor, in accordance with *US MIL-S-52779*, must design and maintain an effective and economical software quality program that includes procedures which make data available to the US Government adequate for use in establishing software acceptance criteria. *US MIL-S-52779* delineates requirements covering:

—Tools, techniques and methodologies.
—Computer program design.
—Work certification.

—Documentation.
—Computer program library controls.
—Reviews and audits.
—Configuration management (CM).
—Testing.
—Corrective action.

The purpose of the program is to assure that software developed, acquired or otherwise provided under the contract complies with the requirements of the contract. It is intended that the program be effectively tailored and economically planned and developed in consonance with, or as an extension of, the contractor's other quality assurance, administrative and technical programs. The term 'program', as used herein, identifies the collective requirements of the specification. The program shall require periodic assessment and, where necessary, realignment of the program to conform to changes in the acquisition program.

There are many factors that affect software reliability planning. These include system application, development and maintenance life cycle considerations, organizational factors and funding constraints. The reliability engineering program is to assure that software developed or acquired complies with requirements.

Various software quality assurance programs could be structured to reflect system criticality categories. For example, criticality can be categorized in terms of critical, essential and non-essential as defined in Chapter 3.

A software program matrix is presented in Table 6.4. This matrix identifies some of the program elements relative to each system criticality category. This matrix concept provides a basis for planning and formulating the make up and provisions of a complete and effective software quality/reliability assurance program. For example, the critical category can be defined, as shown in the table, as requiring that a full compliment of QA elements be planned and implemented. For the essential category, the requirements for some of the elements may be eliminated (or relaxed). The non-essential category would require only a limited amount of QA activity. It should be noted that each individual software development program must be structured and tailored to coincide with its specific criticality category, system requirements and meet its specified QA requirements.

The outputs of each of the program tasks can be defined in terms of a deliverable document or data item. This documentation is necessary to

TABLE 6.4
Software Program Matrix

Tasks	System criticality		
	Critical	Essential	Non-essential
Establishment of system requirements	Required	Required	Required
QA planning	Required	Required	Recommended
Software specification	Required	Required	Required
Reliability analysis and assessment	Required	Recommended	Not applicable
FMEA (combined hardware/software)	Required	Recommended	Not applicable
Subcontractor controls	Required	Required	Required
Design reviews and audits	Required	Required	Recommended
Testing	Required	Required	Required
Error analysis	Required	Required	Recommended
Corrective action	Required	Required	Recommended
Configuration management	Required	Required	Required
Data collection analysis and feedback	Required	Recommended	Not applicable

provide a basis for completely specifying and planning development/ production programs and to demonstrate that the quality tasks are adequately implemented. The purpose of the documentation is to disseminate information, record data, document design/production decisions (and the underlined logic) and to report program status.

A configuration management system must be planned and implemented to ensure that the identity and integrity of the complete software system (i.e. hardware design and program package) is maintained. Configuration management is of particular importance for computer controlled systems because of the abstract nature of software, the volume of documentation and the fact that incorrect program changes can be made relatively easily. Configuration management is a formal process of controlling the operational software status and reviewing proposed improvements, to best direct maintenance effort and to minimize the impact of maintenance and testing on operational service. The software components (i.e. the computer programs and their associated documentation) must be treated as configuration items equivalent to hardware components.

Specifically, configuration management, as it is related to software quality control, applies administrative direction and surveillance to: (1) identify and document the functional and physical characteristics of a configuration item; (2) control changes to those characteristics; and (3) record and report change processing and implementation status. Configuration management is required in varying degrees throughout the entire system acquisition life cycle, generally beginning at a low level and increasing in detail as the system progresses through design and test. The sequence of analysis, design, coding, checkout and test—and some of the related events and support activities—occur at least once for every configuration item during system acquisition. In addition, the basic elements of this sequence may occur on a smaller scale in the conceptual or validation phase or in both of these if there is a need for simulations or other exploratory computer activity. Also, this sequence is repeated for every software problem correction or modification that is made. During testing, this sequence may be repeated hundreds of times within a short period. Loops within the computer program life cycle also occur when it is necessary to change results of an earlier stage and then incorporate the consequences of this change in all intervening stages.

Configuration management is achieved by imposing a procedural framework around the configuration items and applying systematic administrative techniques in order to identify and document the characteristics of the item, control changes to these characteristics and record/report on the implementation of approved changes throughout the system life cycle. The configuration of the software is identified and established at discrete points (baselines) during the life cycle so that subsequent changes to the configuration can be evaluated and systematically controlled. Thus, effective configuration management maintains the integrity and traceability of the software throughout the life cycle.

The primary focus of configuration management is that of change management with a major objective of attaining maximum efficiency in processing changes. The change control function governs the changes to all components of the software system to ensure that they modify the configuration only in ways desired and approved. This is accomplished by the use of procedures for the evaluation, approval, implementation and documentation of all changes. Change control provides an orderly approach to software change. It provides a mechanism for:

—changing software components
—recording and tracking change requests (including present status of the request)
—recording actual changes made
—typing change records to the actual software change
—regression testing after changes

A change control function provides for informal changes, such as those made to components under development, as well as more controlled, formal changes to stabilized and/or released software.

In addition, a well-functioning configuration control system requires that the operating software is traceable back to a set of documents. This ensures that controlled changes are carried out with maximum benefit. A status accounting function is established to provide a mechanism for maintaining a record of how the system has evolved and where the system is at any time relative to what appears in this documentation or to what has been agreed upon. Status accounting involves tracking and reporting the status of all components of the software system. It is the means whereby the other configuration control activities are recorded. Because of the large amounts of information involved, this function is usually, at least partially, automated. The specific configuration management procedures and functions are developed based on the scale and complexity of the software system.

During the system concept phase the objectives and general definition of the requirements for the software are established. Feasibility studies, cost–benefit analyses and the documentation prepared within this phase are determined by the procedures and practices of the procuring activity. During development, the requirements for the software are determined and the software is then defined, specified, programmed and tested. Documentation is prepared within this phase to provide an adequate record of the technical information developed. During operation the software is maintained, evaluated and changed as additional requirements are identified.

Within development the requirements for the software and documentation are first defined, including functional requirements and data requirements. Then, various design alternatives, specific requirements and functions to be performed are analyzed and a design is specified. Documents which may be prepared include the system/subsystem specification, program specification, database specification and test plan. During coding and debugging, documents which may be prepared include the user's manual, operating manual, program maintenance manual and test plan. After the software is tested, the software and documentation are evaluated in terms of readiness for implementation and a test analysis report is prepared. The documentation associated with the software development phase are defined as follows:

Functional requirements document—provides a basis for the mutual understanding between users and designers of the initial definition of the software, including the requirements, operating environment and development plan.

Data requirements document—provides, during the definition stage of software development, data description and technical information about data collection requirements.

System/subsystem specification—specifies, for analysts and programmers, the requirements, operating environment, design characteristics and program specifications (if desired) for a system or subsystem.

Program specification—specifies, for programmers, the requirements, operating environment and design characteristics of a computer program.

Database specification—specifies the identification, logical characteristics and physical characterstics of a particular database.

User's manual—describes the functions performed by the software in non-ADP terminology, such that the user organization can determine its applicability, and when and how to use it; it can serve as a reference document for preparation of input data and parameters and for interpretation of results.

Operating manual—provides computer operating personnel with a description of the software and of the operational environment so that the software can be run.

Program maintenance manual—provides the maintenance programmers with the information necessary to understand the programs, their operating environment and their maintenance procedures.

Test plan—provides a plan for the testing of software, detailed specifications, descriptions, and procedures for all tests, and test data reduction and evaluation criteria.

Test analysis report—documents the test analysis results and findings, presents the demonstrated capabilities and deficiencies for review, and provides a basis for preparing a statement of software readiness for implementation.

Development and implementation of a formal and well-documented error data collection and feedback system, supported by error analysis and corrective action, is an important part of the overall program documentation and a key element in an effective software quality program. The data recovery and feedback program should be designed to be compatible with, and incorporate data from, other system data collection efforts during acquisition and operation. An effective software data system provides

---

Program:

Module:

Error conditions:
    Input conditions:
  Description of failure:

Effect importance:

Execution time since last failure:	Total run time:
Date: Time:	Signed:

Program statements involved:
  Line: Statement:

Error source:
    Code: Design: Specification:

Correction recommended:
  Code:
  Design:
  Specification:

Date: Signed:	Approved:
Correction made/tested: Date: Time: Signed:	
Program master amended: Date: Time: Signed:	

Fig. 6.29. Sample software error/problem report. (Source: P. D. T. O'Connor, *Practical Reliability Engineering*, Heyden, London, 1981.)

output information that can be used for:

—Software reliability assessments.
—Comparative analysis and assessments.
—Compilation of error rates.
—Error sources.
—Determination of the effectiveness of software tools, error detection methods and quality management concepts.

The data system should provide traceability and identity to the software items with errors, indicate the accumulated operational (or use) times to discovery of the error and provide descriptions of the errors and their sources or causes. Cumulative plots of error events versus operating time should be prepared and maintained. Also, periodic summary reports should be prepared providing:

1.   Error rates by software item.
2.   Source/cause classification, e.g.: type; arithmetic, logic,

input/output, severity; cause or source; detection method; time or stage of detection; parents, siblings or offspring.

3. Recommended and accomplished corrective action.
4. General software reliability analysis that correlates designs, predictions and field experience.

Plans should be prepared that describe the specific mechanisms for the collecting, and for subsequent feedback, of operational and maintenance data during deployment as well as during system integration testing. This includes preparation of detailed instructions, including data collection forms, and the delineation of responsibilities for implementation.

All errors, whether discovered during checking, validation or use, must be reported with full details of program operating conditions at the time. The corrective action report should state the source of the error (specification, design, coding) and describe the changes made. A software error reporting and corrective action procedure is just as important as a failure reporting system for hardware. Error reports and corrective action details should be reported using a form such as that shown in Fig. 6.29 and retained with the module or program folder as part of the development record.

Appendix A: Definitions

Accelerated test—A test in which the applied stress level is chosen to exceed the level stated in the reference conditions to shorten the time required to observe the stress response of the item or magnify the response in a given time. To be valid, an accelerated test must not alter the basic modes and/or mechanisms of failure or their relative prevalence.

Acceleration factor—The ratio between the times necessary to obtain a stated proportion of failures for two different sets of stress conditions involving the same failure modes and/or mechanisms.

Acceptable Quality Level (AQL)—The quality standard associated with a given producer's risk which is prescribed by the customer or quality engineer for the products on order. It is usually expressed in terms of percent defective per hundred units.

Acceptance sampling—A procedure in which decisions to accept or reject are based on the examination of samples.

Acceptance tests—Tests to determine conformance to design or specifications as a basis for acceptance. They may apply to components, equipments, subsystems or systems.

Acceptance Test Procedure (ATP)—The configuration controlled, explicitly defined sequence of tests to which the system is subjected for purposes of establishing its acceptability for entry into operational service.

Active element—A part that converts or controls energy; e.g. transistor, diode, electron tube, relay, valve, motor, hydraulic pump.

Active repair time—That portion of downtime during which one or more technicians are working on the system to effect a repair.

Adequacy—Sufficient generating capability to meet the aggregate peak electric loads (MW) and energy requirements (MWh h^{-1}) of all customers at all times.

Algorithm—A step-by-step process for the solution of a problem. Usually developed in outline or as a flowchart before coding, i.e. setting it into computer language.

Analog—Refers to data in the form of continuously changing physical quantities—waves—or devices that operate on it.

Applications program—A program that actually solves a specific problem, as opposed to the programs that perform system functions, such as assemblers, etc.

Assembler—Software that translates source code statements in a symbolic language (assembly language) into object code. The assembler assigns location in storage to successive instructions and replaces symbolic addresses by machine language equivalents.

Assembly—A number of parts or subassemblies joined together to perform a specific function.

Assembly language—An English-like programming language which saves the programmer the trouble of remembering the bit patterns in each instruction; also relieves him of the necessity to keep track of locations of data and instructions in his program. The assembler operates on a 'one-for-one' basis in that each phrase of the language translates directly into a specific machine-language word, as contrasted with high level language.

Assurance—The relative confidence or certainty that specific program objectives will be achieved.

Availability—The probability that a material, component, equipment, system or process is in its intended functional condition at a given time and therefore is either in use or capable of being used under a stated environment.

Availability, inherent (A_i)—The availability potential of a given design configuration under ideal support conditions (i.e. no logistics waiting time). Inherent availability, which includes only corrective maintenance time, mean time to repair (MTTR), is given by:

$$A_i = \frac{\text{MTBF}}{\text{MTBF} + \text{MTTR}}$$

Availability, intrinsic—The probability that the system is operating satisfactorily at any point in time when used under stated conditions, where the times considered are operating time and active repair time.

Availability (operating)—The percentage of time in a period that the system, process or plant is operating or is available to operate (ready status). This measure ignores partial outages, i.e. if the system is producing any product at all, it is considered to be 'available'.

Baseline—The documented, approved description of the system at any point in time. Baseline items are distinguished from exercises, which represent non-approved or not-yet approved trade study items, candidate change times or recommended items. Depending upon the complexity of the system being developed, the baseline may be described in several distinct steps. Upon approval of each step, which may be marked by a development program milestone, the baseline may be replaced under formal change control.

BAUD—The rate at which data are transmitted over a serial link, e.g. a telephone line, in bits per second. The format for data transmission is 10 or 11 bits per character, so 300 BAUD is about 30 characters per second.

Bug—Something which causes the computer program to malfunction. Usually, but not always, due to an error in the logic of the program.

Burn-in—A common form of reliability screen where items (parts, assemblies or products) are operated prior to their ultimate application to stabilize their characteristics and to identify early failures.

Burn-in dynamic—A high temperature test conducted with the items under test subjected to simulated actual operating conditions. The test simulation is commonly termed exercising, as input data signals and/or operating voltages are continuously varied during the test.

Burn-in static—A high temperature test conducted with the active elements of the item under test connected to each other or to external voltages which do not vary throughout the test.

Byte—A group of adjacent bits. A specific portion of a binary word. The most common size byte is 8 bits, although 16 is also used, as well as 4 and 12.

Capacity—The net power output for which a generating unit or station is rated.

Capacity factor—A percentage calculated from the ratio of product actually produced in a period to the product that would be produced if the process system or facility operated at full rated capacity for the period.

Capacity, gross available—The gross dependable capacity modified for equipment limitation at any time.

Capacity, gross dependable—The gross maximum capacity modified for ambient limitations for a specified period of time, such as a month or a season.

Capacity, gross maximum—The maximum capacity that a unit can produce over a specified period of time.

Central Processor Unit (CPU)—Part of a computer system which contains the main storage, arithmetic unit and special register groups. It performs arithmetic operations, controls instruction processing and provides timing signals.

Certification—The process of obtaining regulatory agency approval for a function, equipment or system, by establishing that it complies with all applicable government regulations.

Change control—The process of evaluating, approving and documenting changes to the system.

Characteristic, operating—The curve which describes the probability of acceptance of a lot of various values of process average.

Checkout—Tests or observations of an item to determine its condition or status.

Combined environmental test—A test performed during which more than one environmental stress is imposed.

Compatible hardware/software prediction models—Suitable interpretation of hardware and software mathematical relationships for combined computation so as to make feasible prediction of the system reliability.

Compiler—Software that translates source code statements in a high order language, such as FORTRAN or PASCAL, into assembly language or object code. A compiler translates the complete program, which is then executed. Every change in the program requires a complete recompilation.

Complexity level—A measure of the number of active elements required to perform a specific system function.

Computer—Electronic machinery which, by means of stored instructions and data, performs rapid complex calculations or compiles, correlates and selects data. Examples are analog and digital processors, data processors, information processors, real-time control processors, electronic calculators, hybrid computers, communications processors and microprocessors.

Computer data—A representation of facts, concepts or instructions in a structured form suitable for acceptance, interpretation or processing by communication between computer equipment. Such data can be external to (in computer-readable form) or resident within the computer equipment and can be in the form of analog or digital signals.

Computer equipment/computer hardware—Devices capable of accepting and storing computer data, executing a systematic sequence of operations on computer data or producing computer outputs. Such devices can perform substantial interpretation, computation, communication, control or other logical functions. Examples are central processing units, terminals, printers, analog/digital converters, tape drives, disks, drums, microprocessors and automatic test equipment.

Computer program—A series of instructions or statements in a form acceptable to computer equipment, designed to cause the computer equipment to execute an operation or operations. Computer programs include operating systems, assemblers, compilers, interpreters, data management systems, utility programs, sort-merge programs, and maintenance/diagnostic programs, as well as applications programs such as payroll, inventory control, operational flight, satellite navigation, automatic test, crew simulator and engineering analysis programs. Computer programs may be general-purpose in nature or be designed to satisfy the requirements of a specialized process or a particular user. Programs are essentially written at three levels: (1) binary (can be directly executed by the MPU); (2) assembly language (symbolic representation of the binary); (3) high-level language (such as BASIC), requiring a compiler or interpreter.

Computer resources—The totality of computer equipment, computer programs, computer data, associated computer documentation, contractual services, personnel and computer supplies.

Computer software—A combination of associated computer programs and computer data required to command the computer equipment to perform computational or control functions.

Computer system—An interacting assembly consisting of computer equipment, computer programs and computer data.

Computer system documentation—Information that describes the technical details of the computer system over its life cycle. Documentation includes, but is not limited to, equipment design specifications, engineering drawings, operators' manuals, technical orders, computer software documentation, systems specifications, run diagrams and interface specifications.

Confidence level—Statistical boundaries limiting an estimate with a specified risk.

Confidence limits—Extremes of a confidence interval within which the true value has a designated chance (confidence level) of being included.

Configuration management—A technical and administrative process used to identify, control and account for engineering documents describing the functional and physical characteristics of components, equipment, systems or a process. It is also used to track and control hardware to conform to the documentation.

Corrective maintenance—All unscheduled inspection, testing or repair activities performed on equipment, following its failure, for the purpose of restoring the equipment to satisfactory operating condition.

Critical Design Review (CDR)—A review to verify the adequacy of the design to satisfy the requirements in the system requirements document and the software requirements document, to establish a firm design baseline, to assess risk areas and to approve commencement of qualification, verification and validation activities. Documentation requirements may include: design description document(s), engineering analyses and plans and procedures.

Critical item—A procedure, material, component, or item of equipment whose failure could significantly affect safety, performance, environment, schedule or cost.

Criticality—A measure of the severity of a failure in relation to required performance, hazards to material or personnel, and maintenance cost.

Damaging overstress—An induced or natural overstress which exceeds the required capability of the item under test and causes partial or catastrophic failure.

Debug—The development process to locate, identify and correct programming mistakes, including omissions from software.

Debugging (burn-in)—A process of shaking down each item of finished equipment that is performed prior to placing the item in use. During this debugging period, weak system elements are expected to fail and be replaced by elements of normal quality (statistically) that are not subject to similar early failure. The debugging process may involve exposure to all field operational stresses. The debugging process is not, however, intended to detect inherent weaknesses in system design, which should have been eliminated in the preproduction stages by appropriate techniques. The debugging process eliminates the parts subject to infant mortality.

Defect—A characteristic which does not conform to applicable specification requirements and which adversely affects or potentially affects the quality of a device.

Degradation—A gradual deterioration in performance as a function of time.

Delay time—The component of downtime during which no maintenance is being accomplished on the item because of technician alert and response time, supply delay, or administrative reasons.

Demonstrated—That which has been proven by the use of concrete evidence gathered under specified conditions.

Derating—The intentional reduction of the stress–strength ratio in the application of an item, usually for the purpose of reducing the occurrence of stress-related failures.

Derating, seasonal—The difference between gross maximum capacity and gross dependable capacity.

Derating, unit—The difference between gross dependable capacity and gross available capacity.

Design life—The expected time or cycles, based on the design of the item, during which the item remains operationally effective and economically useful before wearout.

Design margin—The self-imposed restriction on a design more severe than either specified or operational use requirements.

Design review—Meeting held during the design process to critically examine the product design, configuration, design documentation, test program planning and test data.

Destructive testing—Testing of any nature that may materially affect the life expectancy of the item tested, whether or not failures occur during the test.

Diagnostic—A program or routine used to diagnose system malfunctions.

Digital—Refers to data in the form of discrete units—'on/off' or 'high/low' states—and to devices operating on such data.

Downtime—The period of time during which an item is not in a condition to perform its intended function.

Durability—An element of reliability, defined as the probability that an item will successfully survive to its projected service life or rebuild point (whichever is the more appropriate durability measure for the item) without experiencing a durability failure. A durability failure is considered to be a malfunction that precludes further operation of the item under consideration and is of such consequence (in terms of cost and/or time to restore) that the item must be replaced or rebuilt.

Effectiveness—The capability of the system or device to perform its function.

Element—One of the constituent parts of anything. An element, in fact, may be a part, a subassembly, an assembly, a unit, a set, etc.

Embedded computer system—A computer system that is integral to a larger system whose primary function is not data processing. Embedded computer systems are considered different from automatic data processing systems primarily in the context of how they are developed, acquired, and operated in a using system.

Engineering, human—The science of studying the man–machine relationship in order to minimize the effects of human error and fatigue and thereby provide a more reliable operating system.

Engineering quality—The science of establishing quality acceptance and evaluation criteria such as acceptance sampling plans, control charts, classification of defects and tests.

Environment—The aggregate of all the external conditions and influences affecting the life and development of the product.

Environmental stress screening—The process or method whereby a group of like items are subjected to the application of physical climatic stresses or forces (or combinations thereof) to identify and eliminate defective, abnormal or marginal parts and manufacturing defects.

Equipment—One or more units and necessary assemblies, subassemblies and parts, connected or associated together and including all necessary interconnecting cabling, hydraulic lines, accessories, etc., to perform an operational function (e.g. radio receiving set, missile, radar set). An equipment is not normally a replaceable item.

Error—Any discrepancy between a computed, observed or measured quantity and the true, specified or theoretically correct value or condition. A conceptual, syntactic or clerical discrepancy which causes one or more faults in the software.

Failure—The cessation of the ability of a system or any of its elements to perform a specified function or functions.

Failure analysis—The study of a specific failure to determine the failure mode, mechanisms and/or the circumstances that caused the failure.

Failure, catastrophic—Failure that is both sudden and complete.

Failure, chance—That failure which occurs at random within the operational time of an equipment, after all efforts have been made to eliminate design defects and unsound components and before wearout becomes dominant.

Failure, degradation—Failure which results from gradual deviation in characteristics beyond specified limits but does not cause complete loss of the required function. (Note: Such failures can many times be anticipated by prior examination.)

Failure, dependent—Failure which is caused by the failure of an associated item(s).

Failure effect—A description of the consequence of the failure in terms of operating or performance characteristics; e.g. shutdown, loss in efficiency, safety hazard.

Failure free criteria—An acceptance requirement that is imposed during the later part of a screen test that requires no failure for a specified period or number of cycles. Successful completion of the failure free period (or number of cycles) provides an indication that the screening has been effectively completed depending on the expected failure rate and length of the failure-free period.

Failure, independent—Failure which occurs without being related to the failure of associated items; not dependent.

Failure law, exponential—The exponential failure law states that the probability of survival P_S of an equipment operating for a time, T, is a function of the mean life, m, or of failure rate, λ, as expressed by the following:

$$P_S = \exp(-T/m) \qquad P_S = \exp(-\lambda T)$$

Failure mechanism—The physical process or occurrence that caused a failure (e.g. stress corrosion cracking, operator error, equipment malfunction, relay contacts welded by overload and bearings frozen by contamination with foreign material).

Failure mode—A particular way in which failures occur, independent of the reason for failure; the condition or state which is the end result of a particular failure mechanism.

Failure Mode and Effects Analysis (FMEA)—Identification and documentation of each significant failure mode of each item and the impact of the occurrence of that mode of failure on the component, other components, and the overall operation of the system.

Failure (noncurtailing)—A component failure that occurs with no effect on the output of the plant.

Failure, random—Any failure whose cause and/or mechanism make its time of occurrence unpredictable, but which is predictable only in a probabilistic or statistical sense.

Failure rate (λ)—The number of failures of an item per unit measure of life (cycles, time, etc.). During the useful life period, the failure rate, λ, is considered constant.

Failure Reporting And Corrective Action (FRACA)—A systematic and comprehensive method of reporting failures and a means for implementing the corrective maintenance indicated by these failures.

Failure, secondary—Failure of an item caused either directly or indirectly by the failure of another item.

Failure, soft—Failure or error that occurs under stress conditions but disappears at ambient, e.g. occurs at elevated temperatures but disappears as the temperature is lowered.

Failure, wearout—Failure that occurs as a result of deterioration or mechanical wear and whose probability of occurrence increases with time.

Fallout rate (reject rate)—The percent of defective, abnormal or marginal items that are rejected or fallout from the application of a screen test.

Fault—An attribute which adversely affects the reliability of a device.

Fault detection time—Time between the occurrence of a fault and the point at which it is recognized that the system or equipment does not respond to operational demand during the mission sequence.

Fault localization—A man–machine task to determine which particular major unit of equipment is at fault, by making use of malfunction symptoms, test equipment and features built into the equipment.

Fault Tree Analysis (FTA)—A method for relating a process of system failure to equipment, component or materials failure modes using fault trees. A fault tree is a model that graphically and logically represents the various combinations of possible events, fault and normal, occurring in a process or system that leads to the top event. Process or system elements may include hardware, software and human and environmental factors.

File—A logical block of information, designated by name, and considered as a unit by a user. A file may be physically divided into smaller records.

Firmware—A program permanently fixed onto a memory chip (ROM), i.e. software in a hardware support.

Fixed-instruction computer (stored-instruction computer)—The instruction set of a computer is fixed by the manufacturer. The users will design application programs using this instruction set (in contrast to the micro-programmable computer for which the users must design their own instruction set and thus customize the computer for their needs).

Fixed sinusoidal vibration—Vibration excitation with a constant level and frequency, with a sinusoidal waveform.

Flowchart—A graphical representation for definition, analysis or solution of a problem, in which symbols are used to represent operations, data, flow, equipment, etc.

Forced Outage Index (FOI)—The percentage of forced outage hours to available hours plus forced outage hours.

$$FOI = \frac{FH}{AH + FH} \times 100$$

Forced outage rate—The ratio of forced outage hours to operating hours, plus forced outage hours.

Function—Each special purpose performed by a system, subsystem, unit or part.

Functional configuration audit—A formal examination of test data, prior to acceptance, to verify compliance of measured performance with specification requirements.

Functional test—A test that directly or indirectly measures a specific function of equipment or a component.

Hazard—Any real or potential condition that can cause injury or death to personnel or damage to or loss of equipment or property.

Hazard rate, $H_{(z)}$ (instantaneous failure rate)—At a particular time, the rate of change of the number of items that have failed divided by the number of items surviving.

High-level language—Programming language which generates machine codes from problem- or function-oriented statements. FORTRAN, COBAL and BASIC are three commonly used high-level languages. A single functional statement may translate into a series of instructions or subroutines in machine language, in contrast to a low-level (assembly) language in which statements translate on a one-for-one basis.

High order language (high level language)—A type of source language which is problem- or function-oriented, that enables code to be written in a more readily understandable form than object code and can be automatically translated into object code. Most HOLs, such as FORTRAN or PASCAL, are not restricted to application on only one type of computer.

Host computer—Any computer used to develop software for another (target) computer.

Human factors—A body of scientific facts about human characteristics. The term covers biomedical and psychosocial considerations in the areas of human engineering, personnel selection, training, life support, job performance aid and human performance evaluation.

Infant failure—Failures which occur early in the operating life of a part, assembly or product.

Input device—Any machine that allows you to enter commands or information into the computer's main (RAM) memory. An input device could be a typewriter keyboard, an organ keyboard, a tape drive, a disk drive, a microphone, light pen, a digitizer or electronic sensors.

Instruction—A set of bits that defines a computer operation, and is a basic command understood by the CPU.

Integrated Circuit (IC)—A complex, microscopic circuit on a chip of silicon. Large scale integration is usually equivalent to 100 to 1000 gates. Small scale integration is usually equivalent to < 10 logic gates.

Interchangeability—The ability to interchange, without restriction, like equipments or portions thereof in manufacture, maintenance or operation.

Interface—A circuit that allows one type of electronic unit to communicate with another electronic or mechanical device. Chiefly used as a buffer between computers and mechanical devices. One-chip interfaces now exist for most peripherals.

Internal system control—Any device(s), automatic or manual, that controls the operation of a system without external stimulus.

Interpreter—A computer program that converts high-level language statements to 'machine' code for direct computer operation. Unlike a compiler, however, it executes these statements immediately, instead of later. BASIC is an interpreter language.

Isolation level—The functional level to which a failure can be isolated using accessory test equipment at designated test points.

Item Item denotes any level of hardware assembly; i.e. system, subsystem, equipment, components, part, etc.

Kilobyte—A kilobyte is precisely, 1024 bytes.

Language—In relation to computers, any unified, related set of commands or instructions that the computer can accept. Low-level languages are difficult to use but closely resemble the fundamental operations of the computer. High-level languages resemble English.

Life Cycle Cost (LCC)—The total cost of acquisition, operation, maintenance and support of an item throughout its useful life.

Life cycle cost analysis—A function whose objective is to optimize the economics resulting from costs expended for design, construction, operation and maintenance of equipment, a component, a system or a process. LCC analyses are significant for:

—Assisting engineering in design tradeoffs by providing a baseline of total life cycle costs for all major design alternatives.
—Providing a basis for determining the least cost involved in other major project alternatives (e.g. maintenance concept development, planning system operation and support activities, and maintenance planning).

Life test—a test, performed on a group of items, continued until a specified percentage of the group fails or some predetermined minimum operating time period has elapsed.

Load factor—The ratio of the actual energy supplied during a designated period to the energy that would have been supplied if the peak load were to exist throughout the designated period.

Logic—The term used to designate that part of the computer's circuitry that makes logical decisions.

Logical decision—The capability of the computer to decide if one quantity is greater than, equal to or less than another quantity and then the use of the outcome of that decision as a cue to proceed in a given way with a program.

Logical operator—Symbols used in programming to represent the operating of logic including AND, OR, NOT. Expressions containing these symbols are often called Boolean expressions.

Loss-Of-Load Probability (LOLP)—The proportion of time that the generation available is unable to meet the system load (kilowatt). The loss-of-load probability is normally expressed in terms of days when the load is not met in the years studied (e.g. a LOLP of one day in 10 years means the load is not met one day in a period of 10 years).

Lot—A group of items manufactured or processed under substantially the same conditions.

Machine language—The numeric form of specifying instructions, ready for loading into memory and execution by the machine. This is the lowest-level language in which to write programs. The value of every bit in every instruction in the program must be specified (e.g. by giving a string of binary, octal or hexadecimal digits for the contents of each word in the memory).

Main memory—The internal memory of the computer contained in its circuitry, as opposed to peripheral memory (tapes, disks).

Mainframe—The box that houses the computer's main memory and logic components—its CPU, RAM, ROM, input/output (I/O) interface circuitry, and so on. The word is also used to distinguish the very large computer from the minicomputer and microcomputer. Usually uses 32-bit word.

Maintainability—A measure of the ease and rapidity with which a system or equipment can be restored to operational status following a failure, expressed as the probability that an item will be retained in or restored to a specified condition within a given period of time when the maintenance is performed in accordance with precribed procedures and resources.

Maintainability engineering—The engineering discipline which formulates an acceptable combination of design features, repair policies and maintenance resources, to achieve a specified level of maintainability, as an operational requirement, at optimum life cycle costs.

Maintenance—All actions necessary for retaining an item in a specified condition before failure or breakdown (preventive maintenance) or the process of restoring an item to return it to a workable condition (corrective maintenance).

Maintenance analysis—The process of identifying required maintenance functions by analysis of the design, to determine the most effective means to accomplish these functions.

Maintenance capabilities—The facilities, tools, test equipment, drawings, technical publications, trained maintenance personnel, engineering support and spare parts required to restore a system to serviceable condition.

Maintenance concept—A description of the planned general scheme for maintenance and support of an item in the operational environment. The maintenance concept provides the practical basis for design, layout and packaging of the system and its test equipment and establishes the scope of maintenance responsibility for each level (echelon) of maintenance and the personnel resources (maintenance manning and skill levels) required to maintain the system.

Maintenance Downtime Rate (MDT)—Equipment downtime per operating hour, comprised of downtime due to corrective maintenance and downtime required for preventive maintenance.

Maintenance engineering analyses—An analytical process in which the quantitative requirements, support resources, cost, operational objectives, and safety considerations that affect each preventive and anticipated corrective maintenance action are estimated documented.

Maintenance task—Actions required to preclude the occurrence of a malfunction or restore an equipment to satisfactory operating condition.

Man-function—The function allocated to the human component of a system.

Manufacturing defect—A flaw cased by in-process errors or uncontrolled conditions during assembly, test, inspection or handling.

Maximum time to repair (M_{maxct})—The maximum time required to complete a specified percentage of all maintenance actions.

Mean corrective maintenance time (M_{ct})—The mean time required to complete a maintenance action, i.e. total maintenance downtime divided by total maintenance actions, over a given period of time. Mean time to repair (often denoted as MTTR) is the sum of all maintenance downtime during a given period divided by the number of maintenance actions during the same period of time.

Mean Downtime (MDT)—The average time an equipment is down during a maintenance action and during which the system is not in a condition to perform its intended function. Downtime is subdivided into the following categories: active repair time and supply delay time.

Mean maintenance manhours—Average total maintenance manhours required to perform preventive maintenance (servicing) and corrective maintenance (repairs or replacements of failed items). This measure is important in evaluating required maintenance staffing and projecting future maintenance costs.

Mean maintenance time—The total preventive and corrective maintenance time divided by the number of preventive and corrective maintenance actions during a specified period of time.

Mean preventive maintenance time (M_{PT})—The mean (or average) equipment downtime required to perform scheduled preventive maintenance on the item, excluding any preventive maintenance time expended on the equipment during operation and excluding administrative and supply delay downtime.

Mean Time Between Failure (MTBF)—Total operating time (frequently stated in hours) divided by the total number of failures.

Mean Time Between Maintenance (MTBM)—The mean of the distribution of the time intervals between maintenance actions (either preventive, corrective or both).

Mean Time Between Outage (MTBO)—Operating time divided by the number of outages experienced.

Mean Time To Outage (MTTO)—Operating time divided by the number of outages experienced. This measure can be calculated for full, partial or all outages (i.e. planned, forced) and is most applicable to mature technologies.

Mean Time To Repair (MTTR)—The mean time required to complete a maintenance action, i.e. total active maintenance downtime (i.e. fault isolation, fault correction, calibration, and checkout) divided by the total number of maintenance actions, over a given period of time, excluding those time elements which are related to preparation and delay, administrative and supply delay downtime.

Mean Time to Restore (MTR)—That time associated with reinitiation of the system's functional capabilities. For non-redundant systems, this time is usually equivalent to MTTR. In the case of standby redundant systems, or systems where a different hardware type can provide backup service, system restoration time is equal to the time required to switch operation to the backup unit. It is computed by dividing the total system outage time by the number of system outages, over a given period of time.

Memory—That part of a computer which holds data and instructions. Each instruction or datum is assigned a unique address which is used by the CPU when fetching or storing the information. The memory is usually arranged in a rectangular matrix. Each cell is capable of storing information in either of the two binary states, a 1 or a 0, commonly termed 1 bit. A square matrix with 64 rows and 64 columns is capable of storing $64 \times 64 = 4096$ bits, and is designated as having a 4K bit capacity.

Memory device—An article of hardware capable of storing machine readable computer programs and associated data. May be an integrated circuit (IC) chip, a circuit card containing ICs, a core memory module, a disk, a magnetic tape, etc.

Microcomputer—A small but complete computer system, including CPU, memory, input/output (I/O) interfaces and power supply. Generally smaller, slower and less sophisticated than a 'minicomputer', and uses an 8 bit word.

Microprocessor—An IC (or set of a few ICs) that can be programmed with stored instructions to perform a wide variety of functions, consisting at least of a controller, some registers and an arithmetic and logic unit. Contained usually in an IC package with 18–64 leads.

Minicomputer—A computer in a certain range or size and speed, generally smaller, slower and less sophisticated than a 'computer'.

Model—A computer reproduction (or simulation) or a real or imaginary person, process, place or thing. Models can be simple or complex; artistic, educational, or entertaining; serious or part of a game.

Module—A uniquely identified element of a computer program which performs a specific function or set of related functions.

Monolithic semiconductor—Two main types: Bipolar—TTL, ECL, IIL; and MOS—PMOS, NMOS, CMOS, HMOS.

Non-destructive testing—A test that is neither functional nor potentially destructive. It is performed to establish acceptability; e.g. X-ray analysis, leak tests, ultrasonic tests, etc.

Non-volatile memory—A form of computer memory that will store information for an indefinite period of time with no power applied. Magnetic core is a non-volatile memory, as are ROMs and EPROMs.

On-line—Directly connected to the computer system in a performance ready condition.

On–off cycling—Switching of an item's primary power on and off at specified intervals.

Operational Characteristics (OC) curve—The quality curve which shows for a particular sampling plan the relation between (1) the fraction defective in a lot and (2) the probability that the sampling plan will accept the lot.

Operational software—Operational software (or flight software or resident software) is all software resident in the system and in use while installed in its operating environment. It includes executive software, functional (or applications) software, database software and BITE software.

Operating System (OS)—Software required to manage the hardware and logical resources of a system, including scheduling and file management.

Outage, forced—The failure of a system resulting in loss of all or part of the output. A full outage results in complete loss of output; a partial outage results in degraded system output.

Outage, planned—The period a unit is unavailable due to inspection, testing, nuclear refueling or overhaul. A planned outage is scheduled well in advance and is of a predetermined duration.

Output device—A machine that transfers programs or information from the computer to some other medium. Examples of output devices include tape, disk and bubble memory drives; computer printers, typewriters and plotters; the computer picture screen (video monitor); robots and sound synthesis devices that enable the computer to talk and play music.

Part—An element of a subassembly, or an assembly, of such construction that it is not practical to disassemble the element for maintenance purposes.

PCB—Printed circuit board.

Percent defective—That proportion of a lot which is defective.

Percentage assurance—A method for the systematic treatment of reliability, maintainability, availability, life cycle cost, standardization, configuration management and quality assurance in the design, construction and operation of a system.

Percentage reserve—The margin of installed capacity in excess of the expected peak load.

Performance indices—Completely describe the performance of an electrical power operating unit. Included are capacity factor, availability (operating and equivalent) and outage rates (forced and planned).

Peripheral—Any human interface device connected to a computer.

Predicted—That which is expected at some future date, postulated on analysis of past experience.

Predicted reliability—The reliability of an equipment computed from its design considerations and from the reliability of its parts in the intended conditions of use.

Prediction techniques—Methods for estimating the future behaviour of a system on the basis of a knowledge of its parts, functions and operating environments, and of their interrelationships.

Preliminary Design Review (PDR)—A review to determine the compatibility of the selected design approach with the performance and functional requirements of the system requirements document, to formalize the allocation baseline and to obtain approval for commencement of the detailed design phase.

Preventive Maintenance—A procedure in which the system is periodically checked and/or reconditioned in order to prevent or reduce the probability of failure or deterioration in subsequent service.

Program—A collection of instructions properly ordered to perform some particular task.

Quality—A measure of the degree to which an item conforms to applicable specification and workmanship standards (measured by percent defective).

Quality assurance—The system of engineering activities that assures quality by performing and preparing implementation documents for quality control. It includes analyzing all quality-related considerations for the development, implementation and continuing evaluation of a quality control system.

Quality, average outgoing—The ultimate average quality of products shipped to the customer which are the result of the composite techniques of sampling and screening.

Quality control—The system of inspection and testing activities that are performed, documented and used to measure, monitor and control quality, as well as to initiate corrective and/or preventive action in controlling selected characteristics of an item. It also performs the acceptance or rejection function at key points in the evaluation and/or use of a product and implements the quality control system developed by quality assurance.

Random failure—A failure whose failure rate is constant and whose occurrence within any given interval of time is, therefore, unpredictable.

Random sample—A sample in which each item in the lot has an equal chance of being selected in the sample.

Random vibration—Vibration excitation where magnitude and frequency are specified by a probability distribution function.

Redundancy—The existence of more than one means for accomplishing a given task, where all means must fail before there is an overall failure to the system.

Redundancy, active—That redundancy wherein all redundant items are operating simultaneously rather than being switched on when needed.

Redundancy, parallel—The existence of two systems working at the same time to accomplish the task, where either system can handle the job itself in case the other system fails.

Redundancy, standby—That redundancy wherein the alternative means of performing the function is inoperative until needed and is switched on upon failure of the primary means of performing the function.

Reliability—The characteristic of an item expressed by the probability that it will perform a required function under a stated condition for a stated period of time.

Reliability, dynamic—The ability to withstand a sudden outage in its first few seconds or minutes without causing additional loss of facilities (i.e. preventing a cascading effect that may lead to widespread blackout).

Reliability engineering—The engineering discipline which formulates an acceptable combination of design features, repair philosophy and maintenance resources, to achieve a specified level of reliability as an operational requirement, at optimum life cycle costs.

Reliability growth—The improvement in a reliability parameter caused by the successful correction of deficiencies in design or manufacture.

Reliability growth testing—The improvement process during which hardware reliability increases to an acceptable level.

Reliability, inherent—The potential reliability of an item as defined by its design configuration.

Reliability, operational—The assessed reliability of an item based on operational data.

Reliability, predicted—The reliability of an equipment computed from its design considerations and from the reliability of its parts in the intended conditions of use.

Reliability requirement—A level of reliability expressed in an equipment specification as a design requirement and supported with a reliability acceptance test.

Reliability, starting—The ratio of starting successes to total number of starting attempts.

Reliability, steady-state—The system's ability to meet demand within specified voltage limits and the ratings of transmission lines during outages of some generating units and transmission lines.

Reliability, system—The probability that the complete system will perform its intended function for the prescribed period in the specified environment without failure.

Risk—The probability of occurrence of a specific deleterious consequence with a specific dimension, e.g. number of fatalities.

Safety—The quality of being devoid of whatever exposes one to danger or harm.

Screening—A process or combination of processes for the purpose of identifying and eliminating defective, abnormal or marginal parts and manufacturing defects.

Screening attrition costs—Costs associated with replacing and rescreening parts that fail during part level screening or repairing higher indenture items that fail during higher level screening.

Screening cost-effectiveness—The dollar difference between the cost of screening and the benefits derived.

Screening effectiveness—The percentage of defective items which can be identified through the use of a screening process.

Security—System reliability in the steady-state and dynamic sense during actual operation, in contrast to its assessment (used by utility operating personnel).

Sensor—Any device that acts as 'eyes' or 'ears' for a small computer. Types of sensors include photoelectric sensors that are sensitive to light; image sensor cameras that record visual images and transform the images into digital signals; pressure sensors that are sensitive to any kind of pressure; contact sensors that record infrared information; and ultrasonic transducers that produce a high frequency sound wave that bounces off objects and lets the computer calculate the distance between itself and those objects.

Service life—The period of time during which a material, component, equipment, system or process is expected to perform in a satisfactory manner under specified operational conditions prior to wearout or obsolescence and consequent removal from service.

Servicing—The performance of any act (other than preventive or corrective maintenance) required to keep an item of equipment in operating condition, such as lubricating, fueling, oiling, cleaning, etc., but does not include periodic replacement of parts or any corrective maintenance tasks.

Single failure point—A single item of hardware, the failure of which would lead directly to total loss of the hardware system performance.

Software error—A conceptual, syntactic or clerical discrepancy which causes one or more faults in the software.

Software failure—A software failure occurs when a fault in the computer program is evoked by some input data, resulting in the computer program not correctly computing the required function.

Software fault—A specific manifestation of an error. A discrepancy in the software which impairs its ability to function as intended. An error may be the cause of several faults.

Software maintainability—The probability that the software can be retained in or restored to a specific status in a prescribed period compatible with mission requirements.

Software quality—Attributes of a software package other than performance requirements that indicate the character of the software: usually defined in terms of quality factors, e.g. correctness, reliability, acceptability, flexibility, efficiency, human factors engineering, integrity and testability.

Software reliability—From a system, user or 'macroscopic' viewpoint, the probability that the use of the software does not result in failure of the system to perform as expected by more than a specified frequency. From a subsystem, developer or 'microscopic' viewpoint, the probability that the software is fault-free.

Software reliability prediction model—Mathematical model that could include appropriate parameters such as code complexity, branching numerics, structured/modular format utilization, execution rate, timing restrictions and data complexity, predictability and variability, as may be verified by test data.

Specification—A document intended primarily for use in procurement which describes the essential technical requirements for items, including the procedures by which it will be determined that the requirements will be met. A detailed description of the characteristics of a product and of the criteria which must be used to determine whether the product is in conformity with the description.

Standard—A prescribed set of rules, conditions, or requirements established by standards setting bodies, concerning definition of terms, classification of components, specification of materials, performance of operations, delineation of procedures, or measurement of quantity and quality in describing materials, products, systems, services or practices.

Step stress test—A test consisting of several stress levels applied sequentially for periods of equal duration to a sample. During each period, a stated stress level is applied and the stress level is increased from one step to the next.

Storage life (shelf life)—The length of time an item can be stored under specified conditions and still meet specified requirements.

Stress analysis—The evaluation of stress conditions (electrical, thermal, vibration, shock, humidity, etc.) under which parts are applied in the design of a system or equipment. On the basis of a stress analysis, failure rates are appropriately adjusted to reflect the deleterious effects of the stresses on the reliability of the parts involved.

Stress component—The stresses on component parts during testing or usage which affect the failure rate and, hence, the reliability of the parts. Voltage, power, temperature and thermal environmental stress are included.

Structured programming—In a program proceeding in a systematic way from section to section rather than branching widely on GOTO instructions.

Subassembly—Two or more parts which form a portion of an assembly, or form a unit replaceable as a whole, but having a part or parts which are replaceable as individuals.

Subsystem—A major subdivision of a system that performs a specified function in the overall operation of a system.

Support equipment—Items that are necessary for the operation and/or maintenance of the system but are not physically part of the system.

Support software—All software that is used in the development, verification, validation and modification of the operational software. Support software includes: compilers, assemblers, emulators, simulators, editors, linking loaders, debugging programs, operator training programs and document generation and control programs.

Survivability—The measure of the degree to which an item will withstand a hostile manmade environment and not suffer abortive impairment of its ability to accomplish its designated function.

System—A combination of complete operating subsystems, equipments, assemblies, subassemblies, components, parts or accessories interconnected to perform a specific operational function.

System availability—The probability (or proportion of operational time) that the hardware and software are in the required operable and committable state when the function is required with a specified data environment.

System capability—The probability that the hardware and software can achieve the required objectives given the operational conditions, including data environment.

System dependability—The probability that the hardware and software will perform successfully during one or more of the required sequences, given the hardware status at the start.

System effectiveness—The probability that a system can successfully meet its specified operational requirements within a given period of time when operated under specified conditions.

System safety engineering—The activities identified with the analysis of system design and operation for the timely identification and elimination of hazards. System safety activities closely parallel those of reliability to ensure that system safety is achieved early in the design phase and maintained throughout the system life cycle.

Temperature cycle—Temperature testing where an item is repeatedly tested within high or low temperature limits. The transition between temperature limits is a moderate rate of change.

Temperature stabilization period—Time required to reach thermal equilibrium starting from the end of the prior stabilization point.

Test specification—A document defining tests to be performed and the specified parameter limits.

Test-to-failure—The practice of inducing increased electrical and mechanical stresses in order to determine the maximum capability of an item so that conservative use in subsequent applications will thereby increase its life through the derating determined by these tests.

Thermal shock—Temperature testing where an item is repeatedly tested within high and low temperature limits. The transition between temperature limits is at a high rate of change.

Throughput—A measure of the computing capability of a processor, normally expressed in thousands of operations per second (KOPS) of a specified instruction mix.

Time, adjustment or calibration—That element of maintenance time during which the needed adjustments of calibrations are made.

Time, checkout—That element of maintenance time during which performance of an item is verified to be in a specified condition.

Time, down (downtime)—That element of time during which the item is not in a condition to perform its intended function.

Time, fault correction—That element of maintenance time during which a failure is corrected by (a) repairing in place; (b) removing, repairing and replacing; or (c) removing and replacing with a like serviceable item.

Time, up (uptime)—That element of active time during which an item is either alert, reacting or performing a function.

Tradeoff—The process by which a designer can evaluate one or more proposed design considerations in terms of possible effects in other areas and make an intelligent decision based upon these evaluations.

Tradeoff analyses—Studies performed to optimize design in which interrelationships among performance, technical risk, cost, schedule and safety are established and the effects of variations in these factors are determined.

Unit—An assembly or any combination of parts, subassemblies and assemblies mounted together and normally capable of independent operation in a variety of situations.

Uptime ratio—The quotient of uptime divided by uptime plus downtime.

Useful life—The length of time an item operates with an acceptable failure rate.

Validation—The process of establishing that the delivered product, of which the software is a part, complies with aircraft level requirements; the higher-level follow-up to verification. Validation objectives are to demonstrate compliance with aircraft requirements under operational conditions and the absence of undesired effects. Validation procedures may include use of simulated aircraft characteristics and/or flight test.

Verification—The process of establishing that the developed software satisfies both the system requirements and the software requirements.

Volatile—Pertaining to the characteristic of certain types of computer memory which lose their contents when power is removed.

Wearout—The process of attrition which results in an increase in hazard rate with increasing age (cycles, time, miles, events, etc., as applicable for the item).

Word—The basic group of bits which is manipulated (read in, stored, added, read out, etc.) by the computer in a single step. Two types of words are used in every computer: data words and instruction words. Data words contain the information to be manipulated. Instruction words cause the computer to execute a particular operation.

Appendix B: Bibliography

STANDARDS AND SPECIFICATIONS

AFR 800.18: US Air Force Reliability and Maintainability Program. [This regulation outlines the policy for implementing and managing the US Air Force R & M program for systems, subsystems, equipment and munitions.]

British Standard, BS4778: Glossary of Terms Used in Quality Assurance. [Includes reliability and maintainability.]

British Standard, BS5179 (1, 2 & 3): Guide to the Operations and Evaluation of Quality Assurance System.

British Standard, BS5760: Reliability of Systems, Equipments and Components.

British Standard, BS9000: Electronic Components of Assessed Quality.

IEEE, Definitions for Use in Reporting Electric Generating Unit Reliability, Availability and Productivity, Power Plant Productivity Definitions Task Force, Application of Probability Methods Subcommittee, Power System Engineering Committee, New York. [This standard was developed to overcome present difficulties in the interpretation of electric generating unit performance data from various systems and to facilitate comparisons among different systems. It should also make possible the future exchange of meaningful data among systems in the US, Canada and throughout the world. This document standardizes terminology and indices for reporting and evaluating electric generating unit performance. Performance measures can generally be categorized into three groups—reliability, availability, and productivity.]

IEEE STD-500-1977: Nuclear Reliability Data Manual. [Includes failure rates, failure rate ranges, failure modes and environmental factor information on generic components actually or potentially in use in nuclear power generating stations.]

IEEE STD-730-1981: Standard for Software Quality Assurance Plans. [Provides uniform, minimum acceptable requirements for the preparation and content of software quality assurance plans (SQAP). Applies to the development and maintenance of critical software, e.g. where failure could impact safety or cause large financial or social losses. For non-critical software or for software already developed, a subset of the requirements may be applied.]

US DoD Directive 5000.40: Reliability and Maintainability. [This directive establishes policies and responsibilities for the R & M of defense systems, subsystems and equipments. It requires each DoD component to establish R & M programs consistent with its roles and missions.]

US DoD-H-108: Sampling Procedures and Tables for Life and Reliability Testing. [Provides standard sampling procedures and tables for life and reliability testing in US Government procurement, supply and maintenance quality control operations.]

US DoD-STD-1686: Electrostatic Discharge Control Program for Protection of Electrical and Electronic Parts, Assemblies and Equipment (Excluding Electrically Initiated Explosive Devices). [Covers the establishment and implementation of an electrostatic discharge (ESD) control program for any activity that design, tests, inspects, services, manufactures, processes, assembles, installs, packages, labels or otherwise handles electrical or electronic parts susceptible to damage from personnel discharges of up to 4000 V. This standard does not apply to electrically initiated explosive devices. *DoD-HDBK-263* provides information for implementing the requirements of this standard.]

US MIL-M-38510: Microcircuits, General Specification for. [This specification establishes the general requirements for monolithic, multichip, and hybrid microcircuits and the quality and reliability assurance requirements which must be met in the procurement of microcircuits. Detailed requirements, specific characteristics of microcircuits and other provisions that are sensitive to the particular circuits, and other provisions which are sensitive to the particular use intended shall be specified in the applicable procurement document or details specification. Multiple levels of product assurance requirements and control are provided for in this specification.]

US MIL-Q-9858: Quality Program Requirements. [Most widely referenced 'system' specification concerning the actions required of the contractor in establishing a quality program. Requires the establishment of a quality program by the contractor covering all areas of contract performance, including design, development, fabrication, procuring, assembly, inspection, test, maintenance, packaging, shipping, storage and site installation. Prescribes that personnel performing the quality function must have sufficient well-defined responsibility, authority and organizational freedom to identify and evaluate quality problems and to initiate, recommend or provide solutions.]

US MIL-S-750: Test Methods for Semiconductor Devices. [This standard is intended to apply only to semiconductor devices (i.e. transistors, diodes, voltage regulators, rectifiers and tunnel diodes). The test methods described have been prepared to serve several purposes, including control of laboratory conditions, uniform methods and format or results.]

US MIL-S-781: Reliability Tests: Exponential Distribution. [Outlines test levels and test plans for reliability qualification (demonstration), reliability production acceptance (sampling) tests and for longevity tests. (The test plans are based upon the exponential, or Poisson distribution, and are intended for the testing of equipment.) Provides uniformity in reliability testing by: (a) facilitating the preparation of military specifications and standards through the establishment of standard test levels and test plans; (b) restricting the variety of reliability tests so that those conducting the test can establish facilities; (c) facilitating the determination of more realistic correlation factors between test and operational reliability; and (d) facilitating the direct comparison of MTBF test results through the establishment of uniform test levels and plans. Includes graphic examples and examples of records and reports.]

US MIL-S-883: Test Methods and Procedures for Microelectronics. [This standard establishes uniform methods and procedures for testing microelectronic devices, including basic environmental tests to determine resistance to deleterious effects of natural elements and conditions surrounding military and space operations, and physical and electrical tests. For the purpose of the standard, the term 'devices' includes such items as monolithic, multichip, film and hybrid microcircuits, microcircuit arrays, and the elements from which the circuits and arrays are formed.]

US MIL-S-19500: Semiconductor Devices, General Specification. [This specification covers the general requirements for semiconductor devices used in military equipment. Specific requirements for a particular type of semiconductor device are listed in the applicable detail specification which is included within the specification as slash sheets.]

US MIL-S-52779: Software Quality Assurance Program Requirements. [This specification requires the establishment and implementation of a software quality assurance (QA) program by the contractor. The purpose of the software QA program is to assure that software delivered under the contract complies with the requirements of the contract. The purpose of this specification is to assure that the program is effective, economical and planned and developed in consonance with the contractor's other administrative and technical programs. The term 'software QA program' as used herein identifies the collective requirements of the specification. The software QA program may be an extension of the contractor's existing QA program. The software QA program shall provide for periodic assessment and, where necessary, realignment of the QA program to conform to changes in the acquisition program.]

US MIL-STD-105: Sampling Procedures and Tables for Inspection by Attributes. [Contains the sampling plans most widely used by government and industry for inspection and testing. Provides approximately 150 sampling plans for attribute inspection and testing. Attribute inspection can be used where the results can be classified as good or bad, passing or failing, etc.]

US MIL-STD-109: Resistors, Selection and Use of. [Establishes the requirements for the selection of resistors used in the design and manufacture of military equipment. Lists those parts considered to be standard for military applications.]

US MIL-STD-198: Capacitors, Selection and Use of. [Establishes the requirements for the selection of capacitors used in the design and manufacture of military equipment. Lists those parts considered to be standard for military applications.]

US MIL-STD-199: Resistors, Selection and Use of. [The purpose of this standard is to provide the equipment designer with a selection of standard resistors for use in most military applications; to control and minimize the variety of resistors used in military equipment in order to facilitate logistic support of the equipment in the field; and to outline criteria pertaining to the use, choice and application of resistors in military equipment.]

US MIL-STD-202: Test Methods for Electronics and Electrical Component Parts. [Establishes uniform methods for testing electronic and electrical component parts (capacitors, resistors, switches, relays, etc.). Includes environmental, physical and electrical test methods which attempt to simulate or provide equivalent-to-field-test conditions in the laboratory. Does not include sample sizes, lot acceptance or evaluation methods.]

US MIL-STD-210: Climatic Extremes for Military Equipment. [Indicates the probable extreme climatic conditions of the natural environment to which military equipment may be exposed. Establishes uniform limits not to be exceeded in normal design requirements. In addition, environments induced because of worldwide short-term storage and transportation are presented to indicate the extreme conditions to which any military item might be subjected in storage or during transportation.]

US MIL-STD-414: Sampling Procedures and Tables for Inspection by Variables for Percent Defective. [Contains approximately 150 sampling plans by variables. Intended originally to be a companion document to *MIL-STD-105*. Advantages in the use of variables inspection are the saving in sample size over that of attributes inspection: usually only half or less of the attributes sample is required for the same statistical assurance.]

US MIL-STD-454: Standard General Requirements for Electronic Equipment. [Provides the technical baseline for the design and construction of electronic equipment for the US DoD. Presents in one document fundamental design requirements for thirteen general electronic specifications.]

US MIL-STD-470: Maintainability Program Requirements. [This standard provides requirements for establishing a maintainability program and guidelines for the preparation of a maintainability program plan, that encompasses: (a) analysis; (b) design criteria; (c) design tradeoffs; (d) parameter values; (e) subcontractor and vendor contract specifications; (f) design reviews; (g) data collection, analysis and corrective action systems; and (h) inputs and status reports.]

US MIL-STD-471: Maintainability Demonstration. [This standard provides methods for demonstrating various maintainability parameters, MTTR, M_{CT}, M_{PT}, etc. Identifies statistical methods, number of trials, confidence intervals and other information for demonstration of maintainability.]

US MIL-STD-480: Configuration Control—Engineering Changes, Deviations and Waivers. [Provides specific guidance on configuration control of systems during acquisition in terms of how to handle engineering changes, waivers and deviations.]

US MIL-STD-483: Configuration Management Practices for Systems, Equipment, Munitions and Computer Programs. [Provides general guidance and criteria for configuration management of equipment, systems and computer programs during acquisition.]

US MIL-STD-499: Engineering Management. [Describes recommended engineering management practices to be applied during equipment/system acquisition.]

US MIL-STD-690: Failure Rate Sampling Plans for Procedures. [Provides uniform requirements and procedures for establishing a quantitative reliability statement for parts produced from a qualified process. The standard is concerned primarily with process qualification and, thus, differs in concept from the lot acceptance plans for *US MIL-STD-105* or *US MIL-STD-414*.]

US MIL-STD-701: Lists of Standard Semiconductor Devices. [This standard provides device characteristics, ratings and other parameters of standard semiconductors. It is intended to guide designers in the selection process for non-critical applications where an established device can be used.]

US MIL-STD-721: Definitions of Terms for Reliability and Maintainability. [Defines words and terms most commonly used which are associated with R & M. Intended for use as a common base for R & M definitions and to reduce the possibility of conflicts, duplications and incorrect interpretations either expressed or implied elsewhere in documentation. Statistical and mathematical terms which have gained wide acceptance are not defined in this standard.]

US MIL-STD-750: Test Methods for Semiconductor Devices. [Establishes uniform methods for testing semiconductor devices, including basic environmental tests to determine resistance to deleterious effects of natural elements and conditions surrounding military operations and physical and electrical tests. For the purpose of this standard, the term 'devices' includes such items as transistors, diodes, voltage regulators, rectifiers and tunnel diodes. This standard is intended to apply only to semiconductor devices.]

US MIL-STD-756: Reliability Prediction. [Establishes uniform procedures for predicting the quantitative reliability of aircraft, missiles, satellites, and electronic equipment. Graphically portrays the effects of system complexity on reliability to permit the ready prediction of tolerance and interaction problems. Provides appropriate k factors by which to adjust *MIL-HDBK-217* predictions for airborne, missile and space environments.]

US MIL-STD-757: Reliability Evaluation from Demonstrated Data. [Discusses procedures for evaluating reliability of equipment/systems from test data.]

US MIL-STD-781: Reliability Qualification and Production Acceptance Test—Exponential Distribution. [This standard outlines test levels and test plans for reliability qualification (demonstration), reliability production acceptance (sampling) tests and for longevity tests. The test plans are based upon the exponential, or Poisson distribution, and are intended for the testing of equipment. These tests do not replace design, performance, environmental, preproduction, individual, or other required tests (i.e. all functional and environmental tests) specified for the equipment.]

US MIL-STD-785: Reliability Program for Systems and Equipment—Development and Production. [Establishes uniform criteria for reliability programs and provides guidelines for the preparation of reliability program plans. Lists detailed requirements as program elements including: (a) reliability management (reliability organization, management and control, subcontractor and supplier reliability program, program review); (b) reliability design and evaluation (design techniques, reliability analysis, parts reliability, failure mode and effect analysis, reliability, critical items, effects of storage, design reviews); (c) reliability testing and demonstration (reliability test plans, development testing, reliability demonstration); (d) failure data (failure data collection analysis and corrective action, failure summaries); (e) production reliability (transition from development reprocurement); and (f) status reports.]

US MIL-STD-790: Reliability Assurance Program for Electronic Parts Specification. [Provides guidelines to assure uniform evaluation of manufacturers' reliability assurance programs. Establishes the controls and procedures that a manufacturer must provide and maintain in order to qualify parts to an established reliability level.]

US MIL-STD-803 (A-1, A-2, A-3): Human Engineering Design Criteria for Aerospace Systems and Equipment—Part 1, Aerospace System Ground Equipment; Part 2, Aerospace System Facilities and Facility Equipment; Part 3, Aerospace Vehicles and Vehicle Equipment. [Provide design criteria, requirements and definitions for human engineering in military systems. Standardization, automation, visual and auditory displays, controls, workspace design, maintainability, remote handling devices, safety hazards and environmental requirements are some of the subjects treated.]

US MIL-STD-810: Environmental Test Methods. [Describes environmental test procedures and criteria for military equipment/systems.]

US MIL-STD-839: Parts with Established Reliability Levels, Selection and Use of. [Provides guidance and criteria for the procurement of ER level electronic component parts.]

US MIL-STD-882: System Safety Program Requirements. [Defines those elements of a system safety program which are required during the development, production and initial deployment of systems and equipment.]

US MIL-STD-891: Contractor Parts Control and Standardization Program. [This standard establishes the criteria and guidelines for the preparation and implementation of a planned contractor parts control and standardization program. Includes (a) reference documents; (b) definitions; (c) general requirements; (d) detail requirements; (e) equipment performance; (f) data and graphics designating parts selected for proposed and additional program preferred parts lists.]

US MIL-STD-965: Parts Control Program. [Establishes the guidelines and requirements for the implementation of a parts control program. Describes two procedures covering the submission, review and approval of program parts selection lists and changes thereto. Procedure I is applicable to those contracts that do not ascribe to the 'parts control board' concept. Procedure II is applicable to contracts that include a parts control board. Both procedures contain provisions for processing of requests for approval to use parts both within and external to the Military Parts Control Advisory Group assigned commodity classes.]

US MIL-STD-977: Test Methods and Procedures for Microcircuit Line Specification. [This standard sets forth test methods and procedures applicable to the control of materials and processes used in the manufacture of microcircuits. This standard is intended to cover only silicon wafer fabrication. The tests listed and the various procedures may or may not apply to other materials. These methods and procedures are intended to complement those of *US MIL-STD-883* and, therefore, cover the operations required during wafer processing and inspection, starting with the raw material and ending with the finished wafer.]

US MIL-STD-1132: Switches and Associated Hardware, Selection and Use of. [Similar to *MIL-STD-1346*, this standard provides data and information relative to switches. Application guidelines, data and notes are provided to aid designers in proper selection.]

US MIL-STD-1286: Transformers, Inductors and Coils, Selection and Use of. [This standard provides guidelines for selection and application of inductive devices. Data and information covering classes of insulators, power ratings, etc., are provided to aid designers in selecting suitable devices.]

US MIL-STD-1304: Reports—Reliability and Maintainability Engineering Data. [Defines reporting requirements necessary to show compliance with reliability and maintainability design and test requirements applicable to Naval Air Systems Command procurements.]

US MIL-STD-1346: Relays, Selection and Use of. [Establishes the requirements for selection of relays for use in military equipment. Lists those parts considered to be standard for military applications.]

US MIL-STD-1353: Connectors, Selection and Use of. [Establishes standard criteria for all DoD departments and agencies and their contractors governing performance of a logistic support analysis (LSA), integral to the engineering process to define support system requirements and inject support criteria into system/equipment design and acquisition.]

US MIL-STD-1547: Parts, Materials, and Processes for Space and

Launch Vehicles, Technical Requirements for. [Establishes the criteria and minimum technical requirements for electronic parts in the design, development and fabrication of space and launch vehicles.]

US MIL-STD-1562: Lists of Standard Microcircuits. [The purpose of this standard is: to provide equipment designers and manufacturers with lists of microcircuits considered to be most acceptable for military applications; to control and minimize the variety of microcircuits used by military activities in order to facilitate effective logistic support of equipment in the field; to maximize economic support of and to concentrate improvement on production of the microcircuits listed in this standard.]

US MIL-STD-1629: Procedures for Performing a Failure Mode, Effects and Criticality Analysis. [Establishes requirements and procedures for performing an FMECA to systematically evaluate and document by item failure mode analysis the potential impact of each functional or hardware failure on mission success, personnel and system safety, system performance, maintainability and maintenance requirements. Each potential failure is ranked by the severity of its effect in order that appropriate corrective actions may be taken to eliminate or control the high-risk items.]

US MIL-STD-1635: Reliability Growth Testing. [Establishes the requirements and procedures for reliability development (growth) tests. These tests are conducted during the hardware development phase on samples which have completed environmental tests prior to production commitment. These tests provide engineering information on the failure modes and mechanisms of a test item under natural and induced environmental conditions of military operations. Reliability improvement (growth) results when failure modes and mechanisms are identified and their recurrence prevented through implementation of corrective action.]

US MIL-STD-1679 (Navy): Weapon System Software Development. [This standard contains requirements for the design and development of weapon system software which are applicable in US Government contracts. A standard specifically addressing weapon system software is necessary because of factors concerning this software which are not common to general software or which carry a significantly different degree of emphasis. Major factors are:

(a) *Criticality of perfomance.* The combat capability of weapon systems and the combat survivability of combatant units of the operating forces depend in part upon the effective operation of the weapons system software. Therefore, extraordinary efforts are justified in the development phase to ensure maximum R & M. Special emphasis shall be placed on the accuracy and effective operation of the software.

(b) *Changing operational requirement.* Weapon system software implements weapon system operations and doctrine in areas susceptible to many changes of performance requirements. These changes often impact the software and need expeditious implementation. This demands that weapon system software be designed to facilitate efficient change, sometimes at the expense of technical design efficiency. Continuation of an efficient change capability over the operational life of the weapon system also requires detailed documentation describing the software. Proposed changes and their total impact must be easily discernible and capable of being implemented by personnel not associated with the original development effort.

(c) *Life cycle cost.* Development and implementation of changes to weapon system software over the operational life of the weapon system are costly. The design of the software during development must be strongly influenced by factors which will reduce LCC. Among these are various standardization requirements, such as those imposed upon program design, languages and intersystem and intrasystem interfaces. An additional benefit of these standardization requirements is to ensure that changes developed and implemented in one system will have applicability in other systems.]

US MIL-STD-2068: Reliability Development Tests. [This standard establishes requirements and procedures for a reliability development test to implement the *US MIL-STD-785* requirement for such a test. The purpose of the reliability development test is reliability growth and assessment to promote reliability improvement of systems and equipments in an orderly and standardized manner.

US RDT-F2-9T: Reliability Assurance. [This standard contributes toward safe and reliable nuclear facilities and associated systems and components through the application of reliability principles and techniques. The standard provides for the establishment of reliability and maintainability requirements, design techniques and verification methods, including analysis, review, testing and data collection and feedback.]

USPS-170: Reliability and Maintainability Requirements for Postal Hardware Systems. [This standard prescribes uniform R & M program requirements, specific assurance engineering and test and evaluation tasks, and essential management and reporting provisions applicable to the design, development, production and installation of postal hardware systems and equipments. Requirements and provisions of this standard are applicable to all US Postal Service (USPS) material acquisition contracts, subcontracts and purchase orders to the extent specified in requests for proposals and contract work statements.]

HANDBOOKS, GUIDEBOOKS AND NOTEBOOKS

AD-A066979: Reliability Centered Maintenance (Nowland & Heap), 29 December 1978.
AFSCP 800: Reliability and Maintainability Management Guide. [This pamphlet explains how to insure appropriate levels of R & M over the life cycle of systems and equipments through effective management actions by staff, the program office and contractor personnel.]
AMCP 706-132: Maintenance Engineering Techniques. [The fundamental purpose of this US Army handbook is to provide authoritative information requisite to the planning discussion of maintenance engineering functions that must be accomplished in order to insure cost-effective acquisition, operation and support of US Army material.
AMCP 706-133: Maintainability Engineering Theory and Practice. [This US Army handbook is concerned with the theory and practice of maintainability as an engineering discipline which influences design.]
AMCP 706-134: Maintainability Guide for Design. [The objective of this US Army handbook is to influence design so that equipment can be (1) serviced efficiently and effectively if servicing is required and repaired efficiently and effectively if it should fail, or (2) operable for the period of intended life without failing and without servicing, if possible. The handbook embraces information on the extent and nature of the maintenance problem as it exists today and the principles and techniques that, if included in future design, will reduce this problem. Part 1 describes the extent of the maintenance problem in terms of the expenditure of money, men and material. Part 2 presents maintainability objectives, principles and procedures. Part 3 describes the nature of the maintenance problem in terms of the conditions under which weapon systems must be operated and maintained from the logistical, human and environmental points of view. Part 4 deals with design considerations that have general applicability to all types of army material. Design considerations

applicable to specific types of army material are presented in Part 5. Specific references are listed after chapters or sections. A glossary of maintainability terms and a bibliography are included near the end of the handbook. An appendix presents a tabulation of applicable military specifications, standards and publications.]

Anderson, R. T., Kos, D. and Schiller, J., *Reliability and Maintainability Planning Guide for Army Aviation Systems and Components*, R & M Division, Directorate for Product Assurance, US Army Aviation Systems Command, St Louis, Missouri, 1974. [This guidebook serves as a management tool to use in planning, managing and monitoring R & M programs for aviation systems. It provides specific guidelines for structuring work efforts, allocating resources and evaluating all life cycle R & M activities.]

DA-P-750-40: Guide to Reliability Centered Maintenance for Fielded Equipment (*DARCOM Final Report DAG39-77-C-0169*). [Illustrates how the elements of reliability centered maintenance (RCM) are planned, developed and incorporated into maintenance plans/programs for material systems. Individual material developers are expected to tailor the techniques to fit their particular item/system needs.]

Earles, M. E., *Factors, Formulas and Structures for Life Cycle Costing*, Eddins-Earles, Concord, Massachusetts, 1981. [This handbook is a compilation of factors, formulas and structures useful in the conduct of life cycle cost estimates and analyses. It provides the user with a quick reference to standard and representative information.]

IECCA(P) 1/78: A Guide to the Management of Software-Base Systems for Defence, Procurement Executive of the Ministry of Defence. [Addresses the problems associated with the whole life coverage of software-based systems and presents the concensus of IECCA experience in the management of a wide variety of such systems. The aim of the guide is to assist those responsible for the whole-life management of software-based MOD projects by discussing relevant techniques and making recommendations.]

IES, ESSEH, Environmental Stress Screening Guidelines. [Environmental Stress Screening of Electronic Hardware (ESSEH) is a process performed on all 'items' at various levels of assembly. It is intended to identify, force and/or segregate those items (part, module, unit or system) defined as defective. This document provides guidelines to plan and implement a ESSEH program. It is a compendium of information relative to the state-of-the-art of environmental stress screening of electronic hardware. This information has been derived from hard data solicited from companies which have developed and implemented successful stress screening programs in support of their product lines. All levels of assembly have been addressed. In addition, cost–benefit models have been included which are easily understood, and which may be used to assess the cost-effectiveness of various stress screens being considered for implementation. The document supports the decision-making process, as it relates to the technical and economic aspects of environmental stress screening of electronic hardware.]

NAVAIR 00-65-502/NAVORD OR-41146: Reliability Engineering Handbook. [This handbook provides step-by-step procedures for the definition, pursuit and acquisition of required reliability in naval weapon systems, equipments and components. The methods presented are generally applicable to all categories of weapon system elements: electronic, electromechanical, mechanical, hydraulic, chemical, etc. Although this is primarily a 'reliability' handbook, considerable attention has been given to maintainability as a closely related system effectiveness parameter. Procedures are therefore included for the computation, assessment, measurement and specification of maintainability as a design-controlled characteristic essential to overall system operational effectiveness.]

NAVAIR 01-1A-31: Reliability and Maintainability Management Handbook. [Presents guidelines for management and control of reliability and maintainability in *NAVAIR* systems and equipment throughout its entire life cycle.]

NAVAIR 01-1A-32: Reliability Engineering Handbook. [Provides reliability procedures and practices particularly suited to naval aeronautical systems and equipments. Enumerates engineering and management procedures for reliability analysis, prediction, allocation, design, demonstration, contract specification, and control to achieve a desired level of reliability and to reduce costs by reducting failures.]

Naval Electronics Systems Command, *Software Configuration Management, Software Management Guidebooks*, Vol. 2. [Configuration Management (CM) is required in varying degrees throughout the entire system acquisition life cycle, generally beginning at a low level and increasing in stages as the product evolves. The major tasks required to implement CM are described in this volume against the background of the system acquisition life cycle.]

NAVORD OD 39223: Maintainability Engineering Handbook. [This handbook has been developed for the guidance of project managers, system engineers and equipment designers in the search for ways of applying the division of labor concept to the maintenance of complex equipments—i.e.

of adapting design techniques and fleet maintenance concepts to available human resources. The handbook provides engineering and management procedures for maintainability analysis, design, demonstration, contract specification and management control, to achieve the necessary level of maintainability and to reduce the cost of keeping naval ordnance systems 'up and ready'. The procedures may be applied to other systems as well and are equally useful in both government and industrial activities.]

NAVSHIPS 0967-316-8010: Reliability Design Handbook. [Provides guidelines for the design of electronic equipment for ship and shore, including a discussion of reliability concepts and design criteria that form the basis for the handbook.]

NPRD-2: Nonelectronic Parts Data, Reliability Analysis Center, Griffiss AFB, New York. [This handbook contains failure rate and failure mode information on electrochemical and nonelectronic components and assemblies not presently covered in *US MIL-HDBK-217*.

NRC, NUREG-0492: Fault Tree Handbook. [This handbook has been developed not only to serve as text for a system safety and reliability course, but also to make available to others a set of otherwise undocumented material on fault tree construction and evaluation. The publication of this handbook is in accordance with the recommendations of the Risk Assessment Review Group Report (*NUREG/CR-0400*) in which it was stated that the fault/event tree methodology both can and should be used more widely by the NRC. It is hoped that this document will help to codify and systematize the fault tree approach to systems analysis.]

RADC-TR-67-108 (Vol. 2), Hughes Aircraft Co., September 1967.

RADC-TR-74-235: Study of Reliability Prediction Techniques for Conceptual Phases of Development. [This report presents the results of a study to develop a reliability prediction technique to estimate system complexity for application during the early conceptual phases of system development. The prediction technique is based on system performance data derived from design specifications, detailed parts summaries, and detailed handbook predictions using *US MIL-HDBK-217A* on existing systems.]

RADC-TR-74-308 (Vols I, II and III): Maintainability Engineering Design Notebook, Revision II, Cost of Maintainability. [The RADC Maintainability Engineering Design Notebook brings together currently available knowledge of maintainability engineering and treats such knowledge from a practical rather than theoretical viewpoint. The notebook provides both quantitative and qualitative information and techniques which can serve as guidelines for those personnel who are directly responsible for establishing maintainability requirements and maintainability design and for the acceptance of the maintainability of air force ground electronic systems and equipments. Although the notebook is directed at ground electronic systems, the majority of the material is applicable to a much broader class of hardware. Specifically, the notebook includes a description of the time phasing of the maintainability program tasks, a breakdown of maintainability into its roots with a detailed description, guidelines and methodology, procedures, and an example of each maintainability task, as applicable. Since maintainability covers a wide range of disciplines ranging through electronic and mechanical design, instrumentation requirements, logistic support, personnel requirements and statistics, it is not anticipated that any single group will find all of its responsibilities completely described in this notebook. It should, however, contribute significantly to improved maintainability programs and subsequent improved system/equipment maintainability.]

RADC-TR-75-22: Nonelectronic Reliability Notebook. [The purpose of this notebook is to select those useful reliability analysis methods that have been developed over the years and that are applicable for the types of parts of concern and collect them under one cover for use in the design employing these classes of parts. The aim is to present the methods in their most direct and useful form by step-by-step instruction accompanied by appropriate examples. The notebook is divided into sections containing the failure rate information, applicable statistical methods, reliability prediction, demonstration and specification.]

RADC-TR-78-155 (Vol. I): Bayesian Software Prediction Models, an Imperfect Debugging Model for Reliability and other Quantitative Measures of Software Systems. [Develops a stochastic model for software failure phenomena when errors are not corrected with certainty. Derives expressions for several quantities of interest to establish quantitative measures of performance assessment. Discusses approximations for large-scale software systems using a gamma distribution. Illustrates the computation with numerical examples.]

RADC-TR-79-200: Reliability and Maintainability Management Manual. [The R & M management guidelines have wide application. Even though the technical approach to R & M design and demonstration may be quite different in various technologies, the R & M management issues and considerations are quite similar. Therefore, whether working on electronic, hydraulic or mechanical systems, the management guidelines explained should be largely applicable to the program, even though technical details will be different.]

RDH-376: Anderson, R. T., *Reliability Design Handbook*, Reliability Analysis Center, USAF/RADC, 1976. [The purpose of the handbook is to provide information and direction to the designer which will help him engineer reliability into an equipment during its basic design stage. To this end, it provides design data and guidelines for those safety, mission, maintenance and cost factors which together form the working elements of reliability engineering, system engineering and cost-effectiveness. This handbook is primarily intended for use in the design of new equipment or systems which are largely composed of electronic parts and components. However, it can also be used for the design of systems which encompass both nonelectronic and electronic parts, as well as for the modification of existing systems. The handbook embodies a preventive approach to reliability.]

Swain, A. D. and Guttmann, H. E., *Handbook of Human Reliability with Emphasis on Nuclear Power Plant Applications*. [This handbook aids qualified persons in evaluating the effects of human error on the availability of engineered safety features and systems in nuclear power plants. The handbook expands the human error analysis presented in *WASH-1400* and includes principles of human behavior and ergonomics, analytical procedures, mathematical models, and human error probabilities derived from related performance measures and experience. The derived probabilities should be adequate to determine the relative merits of different configurations of equipment, procedures, and operating practices within a plant, and for gross comparisons among plants. Limitations of the handbook and cautions to be observed in its use are explicitly stated.]

US DoD-HDBK-263: Electrostatic Discharge Handbook for Protection of Electrical and Electronic Parts, Assemblies and Equipment (excluding Electrically Initiated Explosive Devices). [Provides guidance for developing, implementing and monitoring elements of an electrostatic discharge (ESD) control program, including: identification of causes and effects of ESD on electrical and electronic parts, assemblies and equipment; development of ESD control program controls; selection and application considerations for ESD protective materials and equipment; design and construction of ESD-protected materials and equipment; design and construction of ESD-protected areas and and grounded work benches; the preparation of ESD operating, handling, packaging and marking procedures; development of ESD personnel training program; and certification of ESD-protected areas and grounded work benches.]

US DOE No. 2634-03-29 (*Draft*): The Fossil Performance Assurance Procedures Handbook, Washington DC. [This handbook is to aid fossil energy contractors in preparing and aiding the DOE Project Managers in their review of performance assurance (PA) program plans and procedures. The handbook provides a systematic approach for the contractor to use in preparing his PA program plans and procedures which, when implemented, are expected to strengthen the fossil energy programs. The handbook contains self-standing procedures to aid in writing program plans for reliability, maintainability and availability, system safety, life-cycle costing, configuration management, and quality assurance. In addition, each program plan procedure is accompanied by additional self-standing procedures, as required, to implement each program plan. Use of these procedures will assure that the contractor will, when generating or modifying his supplied requirements, satisfy his basic obligation to provide an effective and economical performance assurance program.]

US DOE, Performance Assurance Contractual Specification Guide (Draft), Washington DC, 1979. [This guide contains standardized performance assurance-related specifications for use in all major fossil energy projects. The specifications are general in nature and the format has been designed so the DOE project manager may select and/or tailor any element of these specifications to fit the unique needs of any project. The Performance Assurance Office of the Planning and Systems Engineering Division also is available for assistance in tailoring these specifications. A separate 'stand alone' specification has been provided for each performance assurance discipline.]

US DOE, Performance Assurance Project Managers Manuals, Fossil Energy Programs, Washington DC, 1979. [This document contains a description of performance assurance (PA) elements, functions and disciplines. It describes what PA is, why it is needed, when it should be applied and how it should be applied to various types of fossil energy projects. It also provides illustrations of the application of PA disciplines during the typical phases of a fossil energy project. It is addressed to the fossil energy project managers, their staffs and appropriate contractors to aid them in ensuring the design, development and implementation of performance assurance programs applicable to particular projects, thereby enhancing the potential for success of these projects.]

US MIL-E-4158: Electronic Equipment, Ground, General Requirements for. [Specific requirements and criteria for development of ground electronic equipment.]

US MIL-E-5400: Electronic Equipment, Aircraft, General Specifications for. [Similar to *US MIL-E-4158* but unique to aircraft electronic equipment.]

US MIL-E-8198: Electronic Equipment, Missiles, Boosters and Allied Vehicles, General Specification for. [Similar to *US MIL-E-4158* but unique to missile systems and components thereof.]

US MIL-E-16400: Electronic, Interior Communication and Navigation Equipment, Naval Ship and Shore, General Specification for. [Similar to *US MIL-E-4158* but unique to navy equipment.]

US MIL-E-19600: Electronic Modules, Aircraft, General Requirements for. [Provides guidance and criteria for recommended electronic modules for aircraft applications.]

US MIL-E-25366: Electrical and Electronic Equipment, Guided Missile, Installation of, General Specification for. [Guidance and criteria for installation of electrical and electronic equipment in guided missiles.]

US MIL-H-46855: Human Engineering Requirements for Military Systems, Equipment, and Facilities. [Provides design criteria, requirements, and definitions for human engineering in military systems.]

US MIL-HDBK-175: Microelectronic Device Data Handbook. [Provides guidance for the selection and application of microelectronic devices in military systems. Emphasis is placed upon considerations affecting reliability of systems employing such devices. This handbook comprises six sections of user-oriented technical discussion, ranging from design, specification, manufacture and use of the devices in subsystems to reliability and failure physics.]

US MIL-HDBK-189: Reliability Growth Management. [Provides an understanding of the concepts and principles of reliability growth, advantages of managing reliability growth and guidelines and procedures to be used in managing reliability growth.]

US MIL-HDBK-217: Reliability Prediction of Electronic Equipment. [Establishes uniform methods for predicting the reliability of military electronic equipment and systems. Provides a common base for reliability prediction during acquisition programs of military equipment and serves as a means of comparing the reliability of related or competitive designs. This document provides two methods of reliability prediction: (a) parts stress analysis and (b) parts count. Mathematical expressions for part failure rates are provided for use in computer programming. Tables, rather than curves, are used for base failure rates to improve ease of manual application of the prediction methods. These prediction methods will be continually updated as new information becomes available. This handbook includes information relating to part stress analysis prediction in the areas of: (a) microelectronic devices; (b) discrete semiconductors: (c) tubes, electronic vacuum; (d) lasers; (e) resistors; (f) inductive devices; (g) rotating devices; (h) relays; (i) switches; (j) connectors; (k) wire and printed wiring boards; and (l) miscellaneous parts. Also covered is parts count reliability prediction. Appendices dealing with system reliability modelling and approximation for reliability calculation, and a comprehensive bibliography are also included.]

US MIL-HDBK-251: Reliability/Design Thermal Application. [Recommends and describes electronic parts stress analysis methods which lead to the selection and control of maximum safe temperatures for parts so that the ensuing thermal design is consistent with the required equipment reliability.

US MIL-HDBK-338: Electronic Reliability Design Handbook (ERDH). [The ERDH is a comprehensive update of *RDH-376*. It provides comprehensive coverage of all aspects of electronic reliability as an integrated methodology from both a system and a component engineering standpoint. The approach taken was to emphasize the practical aspects of R & M design and management techniques and to concentrate on 'real world' examples which will give the reader insight into how the techniques are applied. The intent was to provide sufficient theoretical and practical information to solve those reliability and maintainability problems frequently encountered. In addition, any excellent standard textbooks which treat the theoretical and mathematical areas in great depth are referenced. Also, through a comprehensive list of reference material the reader will be able to explore for himself aspects of the techniques required by those special problems which inevitably appear. The handbook, containing over 1500 pages, is organized into two volumes. Volume 1— System Reliability—describes a comprehensive methodology covering all aspects of electronic system R & M design engineering and cost analysis as they relate to the design, acquisition and deployment of DoD equipment/systems. Volume 2—Component Reliability—contains detailed procedures for the selection, control and application of electronic components to insure a reliable end product.]

US MIL-HDBK-472: Maintainability Prediction. [Establishes uniform methods for predicting the maintainability of military equipment and systems. Provides a common basis for maintainability prediction during acquisition programs of military equipment and serves as a means of comparing the maintainability of related or competitive designs. Includes procedures dependent on the use of recorded R & M data and experience which have been obtained from comparable systems and components under similar conditions of use and operation. Prescribes four maintainability prediction procedures. Procedure I—system downtime of airborne electronic and electromechanical systems involving modular replacement at the flight-line. Procedure II—methods and techniques used to predict corrective, preventive and active maintenance parameters. Procedure III— method of performing a maintainability prediction of ground electronic systems and equipment by utilizing the basic principles of random sampling. Procedure IV—historical experience, subjective evaluation, expert judgment and selective measurements for predicting the downtime of a system/equipment; uses existing data to the extent available; provides an orderly process by which the prediction can be made and integrated preventive and corrective maintenance; task time to perform various maintenance actions are estimated and then combined to predict overall system/equipment maintainability. Procedures I and III are solely applicable to electronic systems and equipment. Procedures II and IV can be used for all systems and equipments.]

US MIL-HDBK-xxx (Draft): Guidelines for Derating of Electronic Products. [Establishes guidelines and criteria for derating of electronic components and products.]

US MIL-HDBK-xxx (Draft): Human Reliability Models. [Addresses human reliability for: (1) successful man–machine system interface; (2) efficient allocation of man–machine functions; and (3) optimization of field performance. The document includes reliability prediction terminology to address the human elements and R & M program activities to account for the human in the system. It also addresses assessment and evaluation of human reliability design through developmental and operational testing, identifying responsibilities and procedures which are applicable to those tests and different test phases.]

OTHER PUBLICATIONS

Anderson, R. T. and Bass, S., How to control reliability from design through burn-in, *Evaluation Engineering*, March 1980. [The increasing number of products which incorporate microprocessors and other complex and sensitive electronic components present a particularly difficult reliability problem. Special handling techniques are required during assembly, and manufacturing-induced defects are often difficult to detect with ordinary inspection techniques. This paper describes an overall cost-effective program to control reliability based on screening and burn-in procedures, coupled with a well-designed system of failure reporting and analysis as well as periodic process modification.]

Anderson, J. E. and Macri, F. J., *Multiple Redundancy, Applications in a Computer*. [This paper presents a computer design in which the selected form of redundancy was a result of various tradeoff studies performed to yield the optimum overall design. The units are the launch vehicle digital computer and data adapter (LVDC/LVDA).]

Angus, J. and James, L., Combined hardware/software reliability models, *Proc. Reliability and Maintainability Symposium*, Los Angeles, 1982. [A theory for combining well-known hardware and software reliability models is developed around the principles of Markow processes. The use of steady-state availability as a reliability/maintainability measure is shown to be misleading for systems exhibiting both hardware and software faults. This work is the results of a study performed for Rome Air Force Base, New York under contract F30602-80-C-0085.]

Anon., A matrix technique for reliability evaluations, *National Electronics Conference*, Chicago, October 1978. [This paper presents a new approach for evaluating reliability models based on digraphs and related matrix methods. The new results involve matrices that contain fewer non-zero entries than are used by other methods. These are termed sparse matrices and the main advantage gained from their use is a saving in computation effort.]

Anon., Electronic equipment screening and debugging techniques, *Technical Report*, Defense Technical Information Center, 1978. [The objective of this study was to develop techniques which permit cost-effectiveness tradeoff analyses among various screening test approaches that may be used during the development and production of electronic equipment. A computerized optimization model was developed for use as a decision tool in selecting the most cost-effective test sequence from among the many possible alternatives that are presented in a given program. Although the model is rather complex, the computer program allows for optional user input and relative ease in performing sensitivity and tradeoff analyses. Use of the model is recommended as an aid in selecting optimum test sequences in terms of minimum expected test and rework costs and maximum screening effectiveness.]

Anon., Methodology in the establishment and implementation of cost-effective reliability and maintainability (R & M) parameters for complex systems. *Proceedings National Reliability Conference*, 1977. [This paper describes the R & M and cost methodology, defines the R & M program structures for the FAA, and outlines the engineering tasks that are being carried out to assure its implementation.]

Anon., Paper from *National Conference of Reliability a View of the*

Reliability Scene, 1977. [This paper presents an historical view of reliability, the applications of reliability techniques, problem areas experienced in the reliability field. It also points out some of the organizations engaged in reliability studies.]

Anon., Performance and reliability: a many faceted component improvement program, *Reliability Conference for the Electric Power Industry*, Madison, 1980. [This paper discusses EPRI's reliability and performance improvement program of both new and existing fossil-fueled steam power plants. The program addresses every major problem area that has been identified to ultimately improve reliability. The program, as currently established, deals with five separate but interrelated areas. They are turbine/generators, steam generators, plant auxiliaries, plant chemistry and integrated plant. As currently designed, this program will cost approximately $50 million for the five-year period 1980–1984. Conservatively, the utility industry should accrue benefits of at least $20 billion over the following 20-year period as a result of these R & D efforts.]

Anon., *Reliability and Maintainability of Naval Material; Policy for SECNAVINST 3900.36*. [This document establishes policy for the guidance of efforts to increase the R & M of navy and marine corps systems and equipment and to assign responsibility for its achievement.]

Anon., *Research Study of Radar Reliability and its Impact on Life Cycle Costs for the APQ-113, -114, -120 and -144 Radar Systems*, 1972. [The purpose of this study was to provide insight into reliability worth through quantifying the relative values of reliability activities and their impact on life cycle costs. The study is based on data obtained and analyzed for the APQ-120 and the APQ-113, -114 and -144 radar systems. In-service reliability performance data was gathered and analyzed for both radar families, the objective being to correlate differences in performance with the equipment reliability requirements and programs structured. The reliability disciplines and methodologies applied to these radar programs were analyzed with emphasis placed on providing measurable quantified analyses and conclusions. Recommendations are provided, based on conclusions derived from study findings, relative to reliability contracting practices, prerelease disciplines and testing programs.]

Anon., Some precepts of reliable electronic design, *Proceedings, Reliability and Maintainability Symposium* 1966. [Presents precepts for reliable design. The guidelines given are practical and are derived from actual experience.]

Anon., The economics of power transformer reliability-improvement by profit incentive, *Reliability Conference for the Electric Power Industry*, Madison, 1980. [The cost of buying a step-up power transformer is only the first cost incurred in owning the transformer. Utilities have long recognized that transformer losses are an important economic factor. Therefore, methods have been developed to calculate the value of losses over the life of a transformer, and these are added to the purchase cost. A major forced outage on a large power transformer can also have a large economic impact. The cost of repairing or replacing failed transformers with long lead time is an important element of cost. Frequent reports of transformer failure suggest the problem may be increasing. While the transformer has a forced outage, replacement energy must be obtained. Cascading outages represent a greater risk when a transformer is out of service. These factors also contribute to the additional transmission system and generation reserves needed to compensate for the increased system forced outage rate resulting from the transformer failure rate.]

Anon., Tradeoff of thermal cycling versus life cycle costs. *Reliability and Maintainability Symposium*, 1976. [This paper presents a practical example of how and why total life cycle costs are affected by environmental testing. The example illustrates graphically why it is good management to spend more money initially during the production phase to avoid a larger number of field failures and their associated repair costs that would have occurred months or years after hardware delivery to the government. We relate predicted (*US MIL-HDBK-217B*) and field reliability, the ratio of dormant to operating reliability, the costs of thermal cycling at the factory, and the effectiveness of thermal cycling in catching weak or marginal parts to arrive at an optimum number of thermal cycles during acceptance testing.]

Arsenault, J. E. and Roberts, J. A. (eds.) *Reliability and Maintainability of Electronic System*, Computer Sciences Press Inc., Rockville, Maryland, 1980. [The book is intended for engineers, managers and academics engaged in system engineering and concerned with reliability and maintainability. Accordingly it takes a broad approach to the subject. The editors have tried to select sufficient theoretical and practical information to solve those reliability and maintainability problems frequently encountered. In addition, through a comprehensive set of reference material, the reader will be able to explore, for himself, aspects of the techniques required for those special problems which inevitably appear.]

Barton, S. P. and Tapper, D. N., Designing turbine generators for high availability power plants, *Westinghouse Electric Corporation Steam Turbine and Generator Technology Symposium*, Philadelphia, 1978. [This paper describes the availability and design assurance techniques Westinghouse uses for the availabilities in excess of 90%. The application of these techniques is discussed for: making preliminary availability estimates; designing new generic subsystems; developing turbine-generator systems for contract applications; and performing service related activities.]

Baumgardner, J. S., Determination of a generation expansion planning LOLP design criterion, *Reliability Conference for the Electric Power Industry*, Madison, 1980. [This paper presents the results of a comprehensive generation system reliability study to determine a loss-of-load probability (LOLP) design criterion for the Florida Power Corp. system. This paper also presents the results of the investigation into the relationship between a daily LOLP and an hourly LOLP for the Florida Power generation system.]

Bazovsky, I., *Reliability Theory and Practice*, Prentice-Hall, Englewood Cliffs, 1961. [The objective of this book is to develop reliability concepts and methods in a logical way, from simple components to complex systems, to give the reader a thorough understanding of the subject and show him how to solve reliability problems by analysis, design and testing. There is an abundance of useful reliability formulas in the book which will help the reader predict system reliability, establish reliability goals and determine the procedures necessary to achieve them. Also included is a quantitative treatment of system maintainability, availability and safety and outlined methods which have to be followed.]

Blanchard, B. S. and Lowery, E. E., *Maintainability Principles and Practices*, McGraw-Hill Book Co., New York, 1969. [This book is an introduction to maintainability engineering. Its focus throughout is on the principles and practices of organization, planning, actuation and control of a company maintainability program. This book is primarily designed for use in courses at either the undergraduate or graduate level.]

Bonetti, A. F., An effective thermal cycle screen for electronic subassemblies, *IES National Conference and Workshop*, 1979. [IBM studied the effects of a thermal cycle screen for electronic subassemblies manufactured at its Federal System Division (FSD) in Oswego, New York. The screen consists of 55 temperature cycles from -55 to $+80\,°C$ with the hardware not operating. Test data were collected, collated and analyzed. This screen caused failures in 6.8% of the subassemblies tested, thereby providing early detection of many temperature-sensitive failure mechanisms. The process also reduces manufacturing costs.]

Brinkmiller, F. J. and Coogan, F. C., A corporate approach to reliability in an electrical industry, *Reliability Conference for the Electric Power Industry*, Madison, 1980. [This paper presents a unique approach to corporate reliability, availability, maintainability and safety. Traditionally, such programs are developed by corporate staffs and passed down to the division level. The approach described here develops the program at the division level and reports results to the corporation. The paper discusses the development of the program, program description and expected results of the program.]

British Aerospace (Dynamics Group), *Reliability Aspects of Microprocessor Systems* (*U*), prepared for Ministry of Defense Procurement Executive, 1981. [The result of a study carried out to identify and evaluate past and present research in reliability aspects of microprocessor systems and to identify key areas affecting microprocessor system reliability. Emphasis is placed on a system viewpoint, with rather more attention given to software than to hardware. After defining various basic concepts and stating the problem, the report summarises reliability improvement methods and reliability evaluation methods. An extensive annotated bibliography is included as an appendix.]

Chamow, M. F., Directed graph techniques for the analysis of fault trees, *IEEE Transactions on Reliability*, **R-27**(1), April 1978. [Some of the difficulties encountered with previous fault tree valuation methods are avoided by using the methods described in this paper. The new methods involve using directed graphs (digraphs) and related matrix methods and solutions for paths in a manner similar to that for conventional digraphs. Most of the attractiveness stems from the fundamental philosophy of speedily transforming the graphics into corresponding matrices. This puts the bulk of the solution effort into the mathematics where it belongs. The major benefit arises because the mathematical solutions are readily performed by standard matrix techniques, which can be implemented either manually or with the aid of a computer. The new methods have been used on various hypothetical logic combinations and actual fault trees of typical sizes.]

Conrad, J. D., Jr., *Steam Turbine-Generator Quality Assurance: A Management Approach*, Westinghouse Power Generation, Philadelphia, 1979. [New management concepts are being used to coalesce advanced turbine-generator technology and innovative quality assurance concepts. This paper describes the system synthesized by Westinghouse to provide for disciplined achievement of goals during the entire turbine-generator life cycle; i.e. from contract negotiation through design, manufacture and operation to unit retirement.]

Cooper, J. D. and Fisher, M. J. (eds.), *Software Quality Management*, Petrocelli Books, New York, 1979. [A collection of all papers presented at

the *Software Reliability, Availability and Maintainability* (*RAM*) *Conference*. It treats reliability, availability and maintainability merely as subsets or factors of software quality. Many of our software development problems in the past were brought about by the blind or uninformed application of traditional hardware RAM techniques and/or procedures. A major purpose of this book is to take that issue head-on. It is pointed out that hardware and software really are different in many important ways. It is also shown where and why many of these hardware-oriented techniques and procedures are inappropriate for software. Finally, some of latest proven and effective software techniques and procedures are presented.]
Cunningham, C. E. and Cox, W., *Applied Maintainability Engineering* (Wiley Series in Human Factors), John Wiley & Sons Inc., New York, 1972. [This book, backed by 10 years of implementation experience using *US MIL-STD-470*, describes every facet of developing and implementing a maintainability engineering program for electronic systems using the noted standard. Specific examples and methodology are given for each maintainability task, with a complete chapter for each numbered paragraph of the standard. This book will provide implementation guidance for human factors specialists, maintainability engineers, systems engineers and equipment design engineers. It will also be valuable to students and professors in human factors and systems engineering courses. Contents include: maintainability program management; maintainability program plan; the maintainability analysis; the maintenance concept; maintainability design criteria; maintainability design tradeoffs; maintainability prediction; vendor and subcontractor maintainability efforts; the integration of other items; design review, data collection, analysis and corrective action system; maintainability demonstration; maintainability status reports; and human factors in maintainability.]
Cutting, J. C., DelBueno, R. P., Eckels, R., Maruvada, S. N., Chamow, M. F. and Lynch, J. J., The influence of component redundancy on the availability of open-cycle MHD power plants, *Fourth US/USSR Colloquium on Magnetohydrodynamic Electrical Power Generation*, Washington DC, 1978. [This paper describes a continuing effort to identify methods of improving availability/reliability of open-cycle magnetohydrodynamic (MHD) power plants. Analytical and design studies have been performed that indicate that the use of redundant MHD topping cycle equipment can result in significant improvement in plant availability and/or reduced MTBF requirements for channels and combustors. Design considerations for rapid component replacement/repair have been incorporated into the study and a safety analysis is presently underway.

Final results of the study will include an assessment of the impact of these cycle or equipment arrangements on the commercialization of open-cycle MHD.]

Deger, E. and Jobe, T., For the real cost of a design factor in reliability, *Electronics*, 30 August, 1973. [Presents a method of figuring total reliability costs into design tradeoffs. Includes information on the economics of reliability, reliability data sources, a guide to transistor reliability, adding components for lower total cost, guidelines in actual practice, reliability compared to modular design and presently used reliability data.]

Duvall, L., Design of a software analysis center, *COMPSAC '77: First International Computer Software and Applications Conference*, Chicago, 1977. [Summarizes the results of a study to design a software analysis center. This analysis center will serve the government/university/industrial community as a focal point for the acquisition, analysis and dissemination of software experience information. A functional definition of the center, including a discussion of the inputs, processes and outputs is presented. Discussed are the textual information and production/development data that are the main inputs to the center, two databases including summarized and detailed software experience data, data subsets to assist on software development research and output services provided by the center.]

Ekings, J. C., Profit and customer satisfaction equals the specification for commercial reliability programs, *Reliability and Maintainability Symposium*, 1976. [This paper describes the manufacturing reliability assurance program successfully applied by Zerox Corp., a producer of leased business machines. The program covers all the reliability tasks which are accomplished by each function of the manufacturing department. The degree of application of each task depends on the amount of improvement in service cost which is a direct influence on profit. The relevancy of reliability methods developed by aerospace industries to a commercial organization are illustrated.]

Fasano, R. M. and Lemack, A. G., A QUAD configuration—reliability and design aspects, *Proceedings 8th National Symposium on Reliability and Quality Control*, 1962. [Primarily discusses the reliability design aspects and limitations of electronic circuits in which a particular type of component redundancy called QUADing is used. Proposes that a nonredundant and a QUAD redundant circuit are analyzed to establish the magnitude of the increased reliability, effects on loading, power, terminal supplies, transient aspects, etc.]

Foster, R. C., How to avoid getting burned with burn-in, *Circuits*

Manufacturing, August 1976. [Appropriately selected screens and testing at high and low temperatures will weed out the vast majority of infant mortalities caused, for instance, by process defects, marginal design and testing errors. On the other hand, determining the inherent failure rate may require exotic and expensive test conditions (high humidity, high temperatures and expensive fixturing for biasing). This paper discusses how Xerox Business Products Group examines the pros, cons and costs of burn-in.]

Franklin Research Center, *A Review of Equipment Aging Theory and Technology* (*EPRI NP-1558*), Philadelphia, PA, 1980. [The theory and technology of equipment aging is reviewed, particularly as it relates to the qualification of safety-system equipment for nuclear power generating stations. A fundamental degradation model is developed, and its relation to more restricted models (e.g. Arrhenius and inverse stress models) is shown. The most common theoretical and empirical models of aging are introduced, and limitations on their practical application are analyzed. Reliability theory and its application to the acceleration of aging are also discussed. The difficulty of accelerating the aging of an assembly of materials and components in a scientifically rigorous manner, through the application of aging models, is demonstrated. A compendium of aging data for materials and components, including degradation mechanisms, failure modes and activation energies, is included.]

Gansler, J. S. and Sutherland, G. W., *Defense Management Journal: a Design to Cost Overview*, 1974. [This article presents a philosophy of design-to-cost, applicable to the DoD. The need for principles, commercial practice, life cycle cost considerations, characteristic features, application and the relationship to the DoD decision process are covered, as well as the challenge to DoD and the defense industry. Also presented is a hypothetical design to cost program.]

Garver, L., Electric utility planning models, *Joint National ORSA/TIMS Meeting*, Chicago, 1975. [This paper introduces several electric utility planning models currently being used in long-range generation and transmission expansion studies.]

Green, A. E. and Bourne, A. J., *Reliability Technology*, John Wiley and Sons, New York, 1972. [The purpose of this book is to examine the problems of reliability against a background of cost, efficiency and safety. It describes the techniques for solving them, and deals with applications over a wide range of technological products. The opening chapter formulates a definition of reliability to make it a measurable quantity and the following chapters show how reliability criteria based on this definition can be applied to items in many branches of technology. Contents include: reliability concepts; an approach to reliability assessment; the performance requirements; the performance achievement; variations in the performance achievement; the transfer characteristic; properties of distribution; sampling, estimation and confidence; reliability considerations for systems; synthesis of system reliability; synthesis of complex systems; and the application of reliability assessment.]

Haynes, R. and Thompson, W. E., Combined hardware and software availability, *Annual Reliability and Maintainability Symposium*, Philadelphia, 1981. [Presents a Bayesian availability model for combined hardware and software systems. Assumes that all system malfunctions are assumed to be random events generated by one or more independent, stationary, discrete-valued random processes with unknown rate parameters that have known prior distributions. Presents techniques that yield exact expressions for the posterior probability density function for each rate parameter and for the total system availability. Discusses approximate methods and presents an example. Results are applicable to the availability of embedded computer systems. The procedures presented can also serve as the basis for system specifications, warranty provision, or to other contractual agreements related to combined hardware and software system availability.]

Homes & Narver Inc., *Power Plant Data Systems*, EPRI, Palo Alto, 1978. [Several data systems now exist that collect and report certain kinds of data from operating units. This study was undertaken to determine what kind of power plant data are needed and how the information can best be provided. More than 150 people in 35 organizations were interviewed. The results of several industry/government meetings on the subject of data were reviewed. All known existing data systems were examined. Several methods were useful to the collection and dissemination of generating unit data. This final report summarizes the data needs of the power industry and the requirements of government agencies. The existing data gathering activities of the power industry and government agencies are described. Two plans for acquiring a single, national generating unit data system that would satisfy the needs of the power industry and the requirements of government agencies are presented.]

Honeywell (Aerospace Division), *Unit Product Cost System*, St. Petersburg, Florida. [This pamphlet provides informative data related to a unit product cost system. The design is broken down into four phases, those being: Phase I—preliminary design; Phase II—design; Phase III—pilot production; and Phase IV—production. The data are presented in the form of diagrams, flowcharts and worksheets.]

Hsiao, B. M. Y., Hardware error detection and failure isolation design evaluation technique, *IFIP Working Conference on Reliable Computing and Fault-Tolerance in the 1980's*, London, 1979. [A design evaluation technique to evaluate/project digital system error detection and failure isolation capability has been developed. The projection methodology goes hand in hand with a maintenance philosophy which relies primarily on instantaneous detection and isolation of errors in the operational environment, rather than on conventional error recreation diagnostics. Gives the basic definitions of error detection percentage and isolation percentage. Shows how failure likelihood and error detection are calculated from the circuit count and failure rate together with the particular error checker being considered. Also establishes definitions and procedures to evaluate the automatic failure isolation probability of a digital system based on the set of detection syndromes.]

IEC Publications 605: Equipment Reliability Testing, International Electrotechnical Commission, Geneva, 1978.

IEEE, Recommended Practice for Design of Reliable Industrial and Commercial Power Systems, Standards Project No. 493, New York. [The objective of this book is to present the fundamentals of reliability analysis as it applies to the planning and design of industrial and commercial electrical power distribution systems. The test material is primarily directed toward consulting and plant electrical engineers. The material in this book should enable engineers to make more use of quantitative cost versus reliability tradeoff studies during the design of industrial and commercial power systems. Included are: basic concepts of reliability analysis by probability methods, fundamentals of power system reliability evaluation, economic evaluation of reliability, cost of power outage data, equipment reliability data and examples of reliability analysis. In addition, discussions and information are provided on: emergency and standby power, electrical preventive maintenance, and evaluation and improvement in the reliability of existing plants.]

Joksimovic, V. and Vesely, W. E., Use of PRA in evaluating safety of nuclear power, *Reliability Engineering*, 1(1), July–September 1980. [This paper offers an overview of uses and limitations of probabilistic risk assessment techniques in evaluating safety of nuclear power in the US as seen and experienced by the authors. It discusses subjects like PRA contributions to the defense-in-depth philosophy adopted in the US many years ago for licensing nuclear power plants, needs to expand its present role, present pitfalls, the Three Mile Island (TMI) accident sequence, retrospective and prospective safety improvements, quantitative safety goals and the nuclear fuel cycle.]

Karam, D., *Burn-in: Which Environmental Stress Screens Should be used*, RADC, 1981. [This report is based on a literature survey of stress screening studies and concludes that thermal and random vibration are the two most powerful screens. Reports dealing with thermal cycling and random vibration tests are discussed and some conclusions and areas that need further research are drawn from them. Recommendations are also given for a military standard on burn-in based on the findings in this report.]

Klass, P. J., USAF weighing standardized modules, *Aviation Week and Space Technology*, 16 September, 1974. [Discusses the success of the US Navy's attempt at circuit-module standardization through their standard hardware program (SHP), and how it might affect the USAF to consider a similar program for future airborne avionics.]

Klein, W. E., Modified aerospace reliability and quality assurance method for wind turbines, *Annual Reliability and Maintainability Symposium*, 1980. [This paper describes the safety, reliability and quality assurance (SR & QA) approach developed for the first large wind turbine generator project, MOD-QA. This SR & QA approach had to ensure that the machine would not be hazardous to the public or operating personnel, would operate unattended on a utility grid, would demonstrate reliable operation and would help establish the quality assurance and maintainability requirements for future wind turbine projects. Since the ultimate objective of the wind energy program is to provide wind power at a cost competitive with other energy sources, the final SR & QA activities were to be accomplished at a minimum of cost and manpower. The final approach consisted of a modified failure mode and effects analysis (FMEA) during the design phase, minimal hardware inspections during parts fabrication, and three simple documents to control activities during machine construction and operation. This low cost approach has worked well enough that it should be considered by others for similar projects.]

Konakovsky, R., Reliability and safety issues in diversified parallel microcomputer systems, *IFIP Working Conference on Reliable Computing and Fault Tolerances in the 1980's*, London, 1979. [Discusses the use of parallel diversified hardware and software units in order to enhance the failure detection capability of a computing system. Such capability can then be used to meet reliability and safety requirements in many areas of application.]

Krasnodebski, J. and Bartko, I., How to manage design to assure quality operations and construction feedback to design, *Fourth Annual National Conference on Nuclear Power*, Washington DC, 1977. [The nuclear power industry needs to design more timely and comprehensive feedback of operational and construction experiences. This paper reviews nuclear industry information systems in the United States. The organization of the design and construction and operations branches of Ontario Hydro, the quality engineering program, and responsibilities for information feedback are described. Application of information feedback to various design tasks is discussed. Prerequisites for component data collection systems that will provide the required information efficiently are described. The limitations of existing systems are briefly reviewed and suggestions for improvements made.]

Krasnodebski, J. and Christians, J., Reliability and maintainability in design of power stations, *Availability Engineering Workshop*, Albuquerque, 1977. [This paper describes the development of a reliability and maintainability (R & M) engineering program and its application to the design of Ontario Hydro thermal power stations. The effect of the unavailability of these stations on the reliability of the power system and resulting cost are described. The R & M program and application of the various R & M design tasks are outlined. Allocation of availability goals, application of reliability analysis, design reviews, maintainability program, data collection, R & M activities in equipment procurement and R & M training are discussed.]

Krasnodebski, J. and Ravishanker, T. J., Reliability through effective specifications and programs, *Reliability Conference for the Electric Power Industry*, Miami Beach, 1979. [This paper briefly reviews Ontario Hydro's reliability and maintainability (R & M) program in designing generating stations and develops the need for adequate specification of equipment R & M requirements. The R & M program and tasks required of architect–engineers (AE) and suppliers are outlined. Experience in carrying out the program is reviewed and approaches to increase effectiveness of both supplier and AE effort are suggested. R & M costs and their effect on equipment life cycle cost and supplier income are also discussed.]

Lakner, A. A. and Anderson, R. T., An analytical approach to determining optimum reliability and maintainability requirements, *The Radio and Electronic Engineer*, **48**, July/August 1978. [This paper describes the methodology, its approach and the specific life cycle cost (LCC) models used in the computation of optimal levels of R & M. It provides criteria, guidelines, rationale and formulas which the Federal Aviation Authority

(FAA) is applying to its procurements in order to determine optimum mean time between failures (MTBF) and mean time to repair (MTTR) design values for use in its hardware specification. The methodology is based upon LCC principles which are fully in accord with FAA needs.]

Lakner, A. A. and Anderson, R. T., A methodology in the establishment and implementation of cost-effective reliability and maintainability parameters for complex systems, *Second National Reliability Conference*, Birmingham, 1979. [The development of an overall Federal Aviation Administration (FAA) reliability and maintainability (R & M) methodology has included the establishment of a complete R & M planning and implementation process. This paper discusses the process as it has been formulated and can be applied for various procurement types. Included is a discussion of the FAA tailoring approach to R & M activities, R & M specification, R & M program evaluation, as well as actual FAA R & M requirement implementation efforts. It is through the implementation of such a methodology that the FAA hopes to obtain equipment with high field R & M levels in a cost-effective manner.]

Lakner, A. A. and Anderson, R. T., Cost-effective reliability testing, *Proceedings Annual Reliability and Maintainability Symposium*, New York, 1978. [This paper addresses the design and implementation of cost-effective reliability tests and is based on work performed by the airways facilities of the Federal Aviation Administration in their development of a total methodology and database for reliability and maintainability and cost.

Lawson, J. W. and Anderson, D. R., Reliability cost benefits: a utility operations viewpoint, *Reliability Conference for the Electric Power Industry*, Miami Beach, 1979. [This paper describes an approach to measuring and evaluating the performance of fossil- and nuclear-generating units and components with a view to instituting improvements based on relative economics. Examples of cost–benefit studies and their application are given.]

Lennox, C. R., *Experimental Results of Testing Resistors under Pulse Conditions*, Sandia Laboratories, Albuquerque, 1967. [Presents results of tests performed on three types of resistors; those being wirewound, metal film and carbon composition; also describes testing equipment.]

Lipow, M. and Thayer, T. A., Prediction of software failures, *Proc. 1977 Annual Reliability and Maintainability Symposium*, IEEE, 1977. [Computer program size appears to be a good predictor of the number of errors, but an accurate estimate of size is available only when most of the coding has been performed. A prediction available earlier than size of program is the number of design problem reports (DPRs) generated during design reviews prior to coding. The data of one project showed that the number of software problem reports (SPRs) obtained during software testing was proportional to the number of DPRs. Once the programs are produced, and testing begins, it is desirable to refine error predictions. Although size was considered a good predictor, other parameters were investigated for improved capability of error prediction. A detailed statistical analysis suggests that numbers of errors could be best predicted by numbers of branches, measures of program complexity. The number of software errors found in operation by the user show a good correlation with the number of SPRs. This observation indicates that the consequences of design errors continue to surface as software failures in operational use.]

Lloyd, D. and Lipow, M., *Reliability: Management, Methods and Mathematics*, 2nd edn, Redondo Beach, 1977. [This book describes the management, methods and mathematics of reliability with a comprehensive treatment suitable for both graduate engineering students and practicing engineers. All the activities of a reliability department are described, including failure and operating time reporting systems and a thorough treatment of the mathematics of reliability is given. This includes basic probability theory and statistics, reliability point and confidence limit estimation, reliability demonstration methods based on various discrete and continuous statistical distributions, reliability growth models and various forms of system models. Binomial and exponential sampling plans and the latest methods of industrial experimentation are described. Added in this second edition is a chapter on computer software describing techniques for design and production of reliable computer programs and methods for measuring their reliability. A feature of the book is that in the mathematical chapters an introductory discussion is presented in non-mathematical language. Many examples are given from the authors' technical and management experience. Also included are numerous exercises in the statistical methods contained within the text and useful charts and tables of sample sizes and confidence levels.]

Lockheed Missiles and Space Co., *Reliability Design and Application Considerations for Classical and Current Redundancy Schemes*, Sunnyvale, California, 1973. [This report presents a summarization of the reliability, application and design aspects of both classical and state-of-the-art redundancy techniques currently being investigated or developed.]

Locks, M. O., *Reliability, Maintainability & Availability Assessment*, Hayden Book Company Inc., Rochelle Park, New Jersey, 1973. [The material is organized around the subject of confidence assessment, the measure of the quality of estimated reliability. The assumption is made that the component success or failure data used for reliability estimation are governed by some parametric probability distribution. The distribution groups treated include: binomial and other Bernoulli-type distributions; exponential, Poisson and gamma; normal and lognormal; and Weibull. Simplified graphical goodness-of-fit analysis is included as well as the use of optimum (maximum likelihood or linear) or Bayesian models for point and interval estimation and Monte Carlo simulation. The book covers reliability, maintainability, and availability analysis of both repairable and non-repairable systems and components. This book is written for quality, reliability and safety analysts, engineers, and operations researchers. This volume features a textbook style, with problems, to facilitate classroom use as well as self-study. Models include both point and interval values (confidence levels) for assessing the reliability, maintainability or availability using attributes data, and time-to-failure or time-to-repair data.]

Longbottom, R., *Computer System Reliability*, Wiley-Interscience, New York, 1980. [Explains the complications of reliability assessment of computer systems and is based on the author's 18 years experience in analyzing reliability statistics, evaluating new systems, appraising maintainability techniques, specifying and supervising acceptance trials and providing comparative assessments for projects requiring computer systems. Chapter 5 examines software reliability illustrating reliability characteristics with particular emphasis on software errors which give rise to overall system failures. Chapter 6 examines hardware and software fault symptoms as observed by the user.]

Loranger, J. A., The case for component burn-in: the gain is well worth the price. *Electronics*, January 1975. [This paper discusses how burn-in significantly increases systems reliability and how burn-in can give substantial cost savings to the systems builder, as it greatly reduces the need for reworking and field repair.]

Maruvada, S. N., Ideise, K. E. and Charnow, M. F., Failure rate as a design parameter: possibilities and limitations, *Annual Reliability and Maintainability Symposium*, Los Angeles, 1978. [This paper examines the availability performance of a fossil-fired generating unit taking into account several derated states. The probabilities of being in the up, down, and various derated states are computed using state space techniques and a derived measure of unit performance. The sensitivity of this measure of performance to the failure and repair rates of major components is obtained. The results can be used to aid decision-making during the design of a power generating unit to obtain the most cost-effective reliability improvement.]

Milow, R. *Microprocessors—Meeting the Reliability Challenge*, Rome Air Development Center, 1982. [Addresses some of the work to be done to assure that microprocessors can be procured which will have adequate reliability for use in the various military environments. Most of the challenges associated with microprocessors can also be extended to include any of the complex LSI devices.]

Mueller, C. J. (ed.), *Quantitative Software Reliability Analysis of Computer Codes Relevant to Nuclear Safety*, Nuclear Regulatory Commission, 1981. [Presents the results of the first year of an ongoing research program to determine the probability of failure characteristics of computer codes relevant to nuclear safety. An introduction to both qualitative and quantitative aspects of nuclear software is given. A mathematical framework is presented which will enable the *a priori* prediction of the probability of failure characteristics of a code given (1) the proper classification system for software errors and code failures; (2) probabilistic modeling for selected reliability characteristics; (3) multivariate regression analyses to establish predictive relationships among reliability character- istics and generic code property and development parameters; and (4) the associated information base. Preliminary data of the type needed to support the modeling and the predictions of this program are described. Illustrations of the use of the modeling are given but the results so obtained, as well as all 'results' of code failure probabilities presented herein, are based on data which at this point are preliminary, incomplete and possibly non-representative of codes relevant to nuclear safety.]

Multhaup, H. A., Design for reliability and maintainability: life cycle cost minimization, *Reliability Conference for the Electric Power Industry*, Madison, 1980. [The basic objective of the life cycle cost analysis in this paper is to integrate all aspects of the reliability engineering function into a decision-making math model that is closely aligned with the end-use goals of gas turbine users. It allows a more rational global view relative to the establishment of R&M goals and requirements. This total economic analysis, however, requires quantitative inputs from many organizations on both sides of the manufacturer/user interface.]

Myers, G., *Software Reliability Principles and Practices*, John Wiley, New York, 1976. [This book is divided into four parts: Part 1 defines software reliability and analyzes the major causes of unreliability; Part 2 discusses design of reliable software; Part 3 covers the broad area of software testing;

and Part 4 deals with important factors related to development of reliable software, such as project organization, programming languages, predictive models and software support systems. In addition, it covers: structured programming, step-wise refinement, composite design, design walk-throughs, management of programming teams and librarians, programming style and language design, and mathematical proofs of program correctness and reliability of software.]

Myers, R., Wong, K. and Gordy, H., *Reliability Engineering for Electronic Systems*, John Wiley and Sons, New York, 1964. [This book represents an introductory treatment of reliability engineering as applied to electronic systems. Suitable for use as a text book for students or research workers, the subject matter covers fundamental concepts drawn from probability and statistics and applies them to reliability engineering. Simple problems and references are provided at the end of each chapter.]

NAVMAT P-9492: Navy Manufacturing Screening Program, Department of the Navy, 1979. [This report outlines, primarily for US Navy contractors, an adapted and effective manufacturing screening program consisting of temperature cycling and random vibration. With the recognition that test facility cost has been a major obstacle to the use of random vibration, a technical report, which describes in detail a proven means to generate random vibration at low cost, is included as an appendix. Together, temperature cycling and random vibration provide a most effective means of decreasing corporate costs and increasing fleet readiness.]

Niebo, R. J., Transition of equipment availability data system to NERC, *Reliability Conference for the Electric Power Industry*, Miami Beach, 1979. [This paper reviews the current status of the database since its transition to NERC on 1 January 1979. The database rebuilding program initiated by NERC, the new data reporting procedures manual and anticipated changes to the database are outlined. The initial reactions of the utilities and the Department of Energy regarding the transition and subsequent revisions to the database are also discussed.]

O'Connor, P. D. T., *Practical Reliability Engineering*, Heyden, London, 1981. [The mathematical concepts described are limited to those necessary for the solution of the problems covered. Practical approaches to problem solving such as probability plotting techniques and computer programs are stressed throughout. Full coverage is given to major national and international standards and specifications on reliability engineering. This is a vital aspect of the practical approach since so much engineering development is now governed by such documents. The effects of current engineering, commercial and legislative developments, such as microelectronics, software-based systems, consumerism and product liability, are covered in detail.]

Randell, B., Software fault tolerance, *IFIP Working Conference on Reliable Computing and Fault Tolerance in the 1980's*, London, 1979. [Software design fault tolerance is discussed as a means of obtaining improved system reliability. Approaches being investigated usually include structuring a system to separate provisions for automatic backward error recovery. Such comparatively simple approaches are complicated by the presence of parallelism. Distributed systems further complicate the problem when there is no centralized control of parallel activities. It is concluded that software fault tolerance shows promise of enabling increased levels of reliability from complex systems but is still only a substitute for software fault avoidance.]

Ravishanker, T. J., R&M program for gas turbine-generators in nuclear generating stations, *Reliability Conference for the Electric Power Industry*, Madison, 1980. [This paper describes Ontario Hydro's R&M program in procuring gas turbine-generator (GT) sets. A brief description of the functions of the on-site backup power supplies and their GT is given. A description of the approach used in determining the number of GT constituting the on-site backup power supplies and identification of their R&M requirements is then presented. The essential elements of the supplier's R&M program, communicated through the R&M clause in technical specifications, are described. This paper also describes Ontario Hydro's experience in contract monitoring and follow-up with three manufacturers on four contracts. Areas where difficulties were experienced and areas that require careful consideration are identified.]

Read, M. R., Brown, S. G. and Menze, K. L., High temperature burn-in and its effects on reliability, *Reliability and Maintainability Symposium*, 1976. [Reliability of electronic calculators is a substantial aspect of their production economics. Repairs are usually at the factory and a one year parts and labor warranty is standard. Hence, any calculator failure during a year is done at the expense of the manufacturer. Burn-in prior to shipment has been used to induce early life failures and to reduce costly customer returns and the subsequent loss of good will. Deciding on a suitable burn-in period has always been a compromise between the quantity of latent early life failure that can be permitted in delivered calculators and time and money that could be allocated to the burn-in operations.]

Reifer, D. J., Software failure mode and effects analysis, *IEEE Transactions on Reliability*, August 1979. [Discusses the possible use of failure mode and effects analysis (FMEA) as a means to produce more reliable software. FMEA is a fault avoidance technique whose objective is to identify hazards in requirements that have the potential to either endanger mission success or significantly impact life cycle costs. As hazards are identified, software defences can be developed using fault-tolerant or self-checking techniques to reduce the probability of their occurrence once the program is implemented. Critical design features can also be demonstrated *a priori* analytically using proof-of-correctness techniques prior to their implementation if warranted by cost and criticality.]

Retter, B. and Thompson, W., *Computer System Software Acquisition Management*, ARNIC Research Corp., 1976. [Comprises the management support required for the planning, specification, design, production and testing of computer software, with the purpose of realizing a software procurement within all performance, cost and schedule constraints. Successful management of computer software acquisition has been difficult to achieve, as evidenced by many examples of cost overruns and missed schedules in recent development programs. Further, there has been the recurring problem of discovering errors on software after systems have been tested and accepted as ready for operation. Formulation of computer software has been treated in the past as more of an art than a science. A general trend now perceived within the industry is to apply, to software acquisitions, engineering and management techniques similar to those used in hardware development. It is the objective of this perspective to review some of the background issues in software acquisition, and the means that may be used to achieve desired levels of quality.]

Reynolds, J. Q., Use of aerospace technology in consumer products, *Reliability and Maintainability Symposium*, 1976. [Increased consumer awareness and government agency activity will continue to create a demand for improved reliability in consumer products. Technology transfer from aerospace and military developments offers significant opportunities in consumer products, but application of these techniques must be carefully controlled to avoid unrewarding payoffs which will negatively offset profitability.]

Rickers, H. C., *Microcircuit Screening Effectiveness* (IRS-1), Analysis Center, 1978. [The purpose of this report is to broaden the readers awareness of the factors affecting the reliability of microelectronics through the application of screening techniques. The information compiled within this document should provide a valuable resource to those responsible for design, specification, testing and failure analysis of integrated circuits.]

Rogovin, M., *Three Mile Island: A Report to the Commissioners and the Public* (*NUREG/CR-1250*), NTIS, Nuclear Regulatory Commission, 1980.

Selby, J. and Miller, S., Reliability planning and management—RPM, *Symposium for Reliability and Maintainability Technology for Mechanical Systems*, Washington AOA, 1982. [This paper presents a new approach to the reliability planning and management of complex weapon systems. RPM is essentially a management tool for bridging the gap between stated reliability requirements and implementation planning. The RPM methodology, equally usable by buyer and contractor, is applicable to establishing plans, projecting effort, evaluating proposals and monitoring contract performance.]

Shooeman, M. L., *Software Engineering: Design, Reliability, Management*, McGraw-Hill Book Co., New York, 1981. [There is a need for software engineering methods which are essential to the timely development of quality, cost-effective software. This book introduces the subject and develops the techniques to service the needs of a diverse audience. The book serves both as a text for university and professional courses, and as a reference for the practitioner.]

Singer Co. (Kearfoor Division), *Standard Printed Wiring Practices* (ESP18), 1971. [The purpose of this document is to insure that all printed wiring designs represent the highest quality consistent with current economically sound manufacturing procedures.]

Smith, C. O., Introduction to Reliability in Design, McGraw-Hill Inc., New York, 1976. [This book has two parts, differing substantially in approach and format. Part I consists of ten chapters covering definition of reliability, the general reliability function and basic concepts. The fundamental aspects of series, parallel, standby, and mixed systems are treated. Conditional probability and multimode function are used as additional methods of approach to problem solution. Consideration is given to derating and reliability testing. Part I concludes with a set of broad guidelines for designing reliability into a given situation and a discussion of factors involved in deciding how much reliability is appropriate. A number of relatively simple problems are included, both as part of the text material and at the end of each chapter to permit development of some rudimentary skill for dealing with reliability problems. Part II consists of a series of specific examples taken from published literature. It must be recognized that inclusion of these examples does not imply the solutions are the best, or only, ones, neither does inclusion imply use of the latest techniques. The examples only show how real people dealt with real problems. This book is primarily intended for use in a first course in reliability at the senior-graduate level, but can also be of substantial value and use to any practicing engineer who desires to gain a basic understanding of reliability, especially with relation to design.]

S. M. Stoller Corp., *Consolidating Power Plant Data System*, EPRI, Palo Alto, 1978. [This report describes the experience of several large industries with collecting and using equipment performance data in large complex systems and assessing the feasibility of consolidating power plant data-bases. It discusses the results of interviews conducted with two utilities and six organizations outside the utility industry. It concludes that consolidation of data collection in a single organization would reduce the burden of reporting by the utilities and would improve the quality and availability of the data. Several potential improvements in the quality and scope of plant outage course data are recommended.]

S. M. Stoller Corp., *Power Plant Early Alert Reporting System*, EPRI, Palo Alto, 1979. [The early alert reporting system (EARS) as presented in this report collects information on significant generic failures in power plants and broadcasts the details of these failures in a timely manner to other utilities, architect engineers and manufacturers in an attempt to avert similar failures in other plants. The need for such an alert system was identified in an industry study of power plant data systems sponsored by the American National Standards Institute (ANSI). As a result of this identified need, the executive committee of the Edison Electric Institute's engineering and operations committee requested that EPRI undertake a study to define the scope and operating method for an alert system. This report contains the results of that study.]

Starr, M. K., *Product Design and Decision Theory*, Prentice-Hall Inc., Englewood Cliffs, 1963. [This book attempts to explain the nature of decision theory and how it can be utilized to improve design decisions. Many different kinds of design situations are presented, but the surface of possibilities has only been scratched. Extensive detail has been avoided in presenting the examples for lack of space. This book has been planned so that all members of the design team, as well as students of engineering, production, marketing, and management, can obtain a different point of view about the product design problem.]

Stone & Webster Engineering, *Analysis of Utility Industry Data Systems*, EPRI, Palo Alto, 1979. [This project examines the usefulness of the three most prominent availability and reliability related data systems currently being applied and supported by the electric power industry: (1) Equipment Availability Data System (Edison Electric Institute); (2) Operating Units Status Reports (Gray Books, US NRC); and (3) the Nuclear Plant Reliability Data System (NPRDS, American National Standards Institute). The aim was to test completeness, accuracy and utility of these data reporting systems by attempting to apply the data to power plant reliability and availability analyses. Results of some of the computations that can be made with existing data are provided together with cautions concerning their interpretation. The study indicates limitations and deficiencies in each of the systems. The basis for judging limitations and deficiencies was not necessarily what the data systems were designed to do, but their ability to supply data necessary to perform basic reliability and availability analyses applicable to power plants and their equipment. Recommendations are offered concerning actions necessary to enhance existing systems and actions that should be considered in developing any new data reporting system.]

Sukert, A. N., An analysis of software reliability prediction models, *Proc. NATO Advisory Group on Aerospace Research and Development Avionics Panel Symposium on Avionics Reliability, its Techniques and Related Disciplines*, Ankara, 1979. [A study to validate several mathematical models for predicting the reliability and error content of a software package against error data extracted from the formalized testing of four large DoD software projects, was undertaken by the RADC. Describes the results of this empirical study for three models, the Jelinski–Moranda, Schick–Wolverton and a modified Schick–Wolverton, using both maximum likelihood and least spares methods for estimating model parameters. Model predictions are compared on a total project, functional and error severity basis and on an errors per day and errors per week basis for defining model time intervals. Draws conclusions concerning the application for these models, the main conclusion being that model predictions should be begun when all modules in a system are ready for testing.]

Sukert, A. N., An investigation of software reliability models, *Proceedings Annual Reliability and Maintainability Symposium*, IEEE, 1977. [Reports on the initial phase of a software modeling study, in which nine software reliability models were applied against software error data detailing the complete error history from the start of formal testing through delivery of a large command and central software development project with over 100 000 lines of Jovial code. Describes the models considered and the procedures used to prepare the data for model input. Model predictions are then compared and analyzed against the actual post-delivery error data for this project.]

System Development Corp., *Procedures for the Operation and Use of the Fossil Energy Equipment Data System (FEEDS)*, 1978. [This document provides operating procedures for the first phase of DOE's Fossil Energy Equipment Data System (FEEDS). The system is being established to provide for the collection, analysis, storage and dissemination of equipment and material failure, availability, operability and maintainability information generated through the operation of fossil energy plants.]

TM-38-750 and 750-1: The Army Maintenance Management System (TAMMS). [The US Army maintenance data collection system; similar to the US Navy's 3M system and the US Air Force's 66-1 system.]

Tracor Sciences & Systems, *Final Report Engineering Study for Optimum Burn-in Determination*, 1979. [This report presents the results of a multi task program to develop guidelines for test planning during development and for continuing reliability assessment and trend analysis during production. Efforts included developing an analytical model to analyze the data which will be generated during the performance of a burn-in program. The report discusses a method (and the contractual language) for having the cost and/or cost-saving of changes in the burn-in requirement shared by the equipment developer and the government.]

US Air Force, *Investigation of Secondary Effects for the Checkout of Nonelectronic Systems*, Aero Propulsion Laboratories Research and Technology Division (AFAPL-TR-65-57), August 1965. [The main objective of this investigation was to demonstrate the feasibility of sensing secondary phenomena of nonelectronic system operation and to obtain information concerning the status of the system or its components. The primary emphasis has been on secondary effects not requiring the physical dismantling of the system under test and not currently employed in existing checkout operations.]

US Air Force, *Maintenance Data Collection System* (ARM66-1). [This system is the US Air Force's counterpart of the Navy's 3M field data collection system. Useful for comparing relative R&M experience on hardware items.]

US Army, *Secondary Effect Technique Study* (Armament Command FCF-13-74), 1974. [The object of this program was to identify and develop secondary effect detection technique concepts for use with the MAID-TECH program. The purpose of these techniques is to simply checkout procedures and to provide information that cannot be easily obtained through primary measurements.]

US Army, *Systems Reliability, Availability and Maintainability* (*AR702-3*). [This document is the US Army's equivalent of *AFR 800-18*, as applied to Army equipment/systems.]

US Report WASH-1400 (*NUREG-75/104*): Reactor Safety Study: An Assessment of Accident Risks in US Commercial Nuclear Power Plants.

Vessley, J. E. and Cowdery, J. W., RAM—A management challenge, *Reliability Conference for the Electric Power Industry*, Madison, 1980. [The presentation examines the current needs and benefits of RAM (reliability, availability, maintainability) programs from three different viewpoints, i.e. general public, customers and stockholders. The FPL organization and approach in response to these needs is described. Specific activities, including data use, illustrate the effectiveness of this organization. The paper concludes with expected future activities.]

VonAlven, W. H. (ed.), *Reliability Engineering*, Prentice-Hall Inc., Englewood Cliffs, 1964. [Recognition of reliability and maintainability as vital factors in the development, production, operation and maintenance of today's complex commercial, military and space systems has placed greater emphasis on the training of management, engineering procurement and administrative personnel in the application of these concepts. This book offers a valuable contribution to this important educational requirement by presenting a unified introduction to the concept of system effectiveness—relating mission reliability, operational readiness and performance capability (design adequacy). Prediction techniques associated with R&M and successful reliability management concepts are discussed in detail. Outstanding features of this book are: emphasis on concepts of system effectiveness, including multimodal systems; predesign reliability prediction for system feasibility evaluation; methods for determining confidence limits on reliability predictions; prediction techniques for the reliability of propulsion systems for space vehicles (includes failure effects analyses); recently developed maintainability prediction methods; and an appendix listing government documents establishing R&M requirements. This text has been addressed to practicing engineers and senior level engineering students.]

Walters, G. F. and McCall, J. A., Development of metrics for software R&M, *Proceedings Annual Reliability and Maintainability Symposium*, IEEE, 1978. [Describes the derivation and validation of software metrics which provide a means for quantitatively specifying and measuring software quality. The work was part of a study of the factors on software quality conducted on support of the US Air Force Systems Command Electronic Systems Division and RADC. The software metrics discussed are those which relate specifically to the quality factors of reliability and maintainability.]

Weiss, G. H. and Dishon, M., Some economic problems related to burn-in programs, *IEEE Transactions on Reliability*, **R-20**(3), August 1971. [The problems raised by the requirement of *N* components, some or all of which are subjected to a burn-in program lasting a time *T*, were considered. A model is first analyzed in which *N* components are initially subjected to burn-in and failures are replaced at the end of the program by components that have not been burnt-in. Then generalizations of this model are analyzed. These include one in which *M* > *N* components are subjected to burn-in initially and one in which intermediate replacements can be made at times before *T*.]

Welsh, J. and McKeag, M., *Structured System Programming*, Prentice Hall Inc., Englewood Cliffs, 1980. [Demonstrates the application of structured programming to the construction of system programs—in particular compilers and operating systems. Summarizes the structured programming style and notations used and then presents the development of a complete compiler and complete operating system, with working code for each, in a suitable high-level language (PASCAL PLUS).]

Zagusky, G. P. and Pillar, C. S., A systems approach to reliability program development and implementation, *Reliability Conference for the Electric Power Industry*, Madison, 1980. [This paper discusses the techniques used by Florida Power & Light Co. to develop and implement a practical and effective power generation reliability program. Although the overall reliability objectives were clear, the division of responsibility was not. In an attempt to manage this monumental effort, a systems approach was used to identify and separate the task into its basic elements for analysis and program development. The results produced a reliability team, complete with a program charter and functional department guideline documents for implementation.]

Index

Accelerated aging techniques,
  developed by NSAC, 6–7
Acceptance tests, 339
  post-manufacturing, 141–2
  reliability acceptance, 150, 318, 322
Acquisition
  audit of, 184–9
  costs affected by reliability, 164, 165,
    171, 172
  Management Systems and Data
    Requirements Control List
    (AMSDL), 91
Active parallel (AP) mode
  compared with standby mode, 46–7
  reliability calculations of, 46
Aging
  accelerated testing of, 6–7
  failure mechanisms, 55–7
  forced outage affected by, 39
  premature, causes of, 27
  reliability
    affected by, 277
    degradation affected by, 22, 113,
      114
Airlines, reliability-centered
  maintenance used by, 272,
  274, 277
Analysis techniques, 19, 21–79
AND/OR gates, fault tree analysis, 73,
  291, 352
Architect–engineer (A–E) firms, listed,
  6

Attributes
  and control, minimum requirement
    for, 17
  demonstration testing of, 134
Auditing
  reliability program, 182–93
  supplier, 84, 109–10, 171
Automatic test equipment (ATE), 235
Availability
  analysis, 36–47
    formulae for, 37–8
    methodology of, 39, 40
    procedures for, 40, 41
  definition of, 36, 340
  engineering, 19
  factors
    inherent, 42, 340
    total system, 44
  improvement, 47–51, 87, 158–9

Backfitting, cost of, 82
Bayesian statistics, 74, 155, 206, 266–7
Bibliography, 367–411
  handbooks, etc., 377–86
  other publications, 386–411
  standards and specifications, 367–77
Binomial distribution function, 72
Block diagrams, reliability analysis, 29,
  30, 31, 76, 119, 286
  fault tree diagrams compared with,
    73–4

Boiling water reactors (BWRs),
    suppliers of, 4, 6
British Standards, 211, 367
Built-in test (BIT), 49
Burn-in tests, 22, 48, 142, 148
    costs of, 229
    definition of, 341
    examples of, 229
    in audits, 189

Capacitors
    burn-in tests for, 229
    component quality levels for, 169
    cost trends for, 222
    derating of, 241
    failure
        modes for, 57
        rates for, 262
    reliability information for, 221
    selection guidelines for, 219–22
    specifications for, 220
    temperature–reduction schemes for,
        252
    types of, 220
Capacity factor, 37, 342
Catastrophic failures, 305
Certification, definition of, 342
Change control functions, software
    development, 333–4, 342
Checklists
    design review, 202, 315
    maintainability design, 246
    reliability specification, 175, 176
Chi-square statistical test, 71
Closed-loop corrective action systems,
    158, 160–1
    see also Failure reporting...
        (FRACA)
Coils, *see* Inductive devices...
Commercial equipment, reliability
    aspects of, 177–9
Commercial procurements, 174, 175
Common cause failure (CCF) analysis,
    85, 129–30
Common mode failure (CMF)
    analysis, 85, 129–30

Component
    availability, program elements, 86,
        137–47
    control criteria, monitoring of,
        199–200
    derating, 86, 138–9, 240–4
        in audits, 188–9
    failure rate modeling and estimation,
        86, 140–1
    quality levels, 86, 139–40, 169
    reliability
        database, INPO, 8
        maintained by NRC, 11
    selection, 86, 143–4
    test methods and compliance, 86,
        141–3
Computer(s)
    communications network, INPO,
        7–8
    programs
        definition of, 343
        development of, 61, 334–5
    terminology, 60, 61, 343–4
        algorithm, 340
        architecture, 61
        assembly language, 340
        BAUD rate, 341
        byte, 341
        compiler, 343
        erasable programmable ROM
            (EPROM), 61
        file, 349
        firmware, 349
        high-level language, 350
        input device, 351
        instruction, 61, 351
        interface, 351
        interpreter, 351
        kilobyte, 351
        language, 352
        machine code, 353
        mainframe, 353
        memory, 356
        microcomputer, 59, 356
        microprocessors, 59, 356
        output device, 357
        peripheral, 358

Computer(s)—*contd.*
terminology—*contd.*
random access memory (RAM), 61
read only memory (ROM), 61
software, 61, 361–2
volatile memory, 61, 365
word, 366
Computerized systems, reliability
program tasks for, 329–30
Conceptual design phase
management tasks in, 100–1
reliability engineering tasks in, 2, 13,
15
Condition monitoring, 277
Configuration(s)
management, 84, 108–9, 344
audit of, 193
by NRC, 13
importance of, 18
software, 70, 332–4
reliability model, 29–31, 45–6
Connectors
derating of, 243
failure
modes for, 57
rates for, 262
Contacts, electrical, deterioration of,
56
Controlled screening and burn-in tests,
148–9
Corrective action, 308
request (CAR) forms, 308, 310, 314
Corrective maintenance (CM), 32,
269–71, 344
cost of, 167, 168
worksheet for, 270
Correlation analysis, 72
Cost-effective maintenance, 36, 146,
179, 272, 273–82
Cost-effectiveness, screening test, 223
Cost–reliability relationships, 165–9,
171, 172
Costs, downtime, 1, 82
Critical component lists, 281
Critical design reviews (CDRs), 135,
194, 197, 201, 202, 316, 317,
344

Critical items, 344
control, 50
list, 84, 110–11
monitoring of criteria, 199
Criticality
analysis, 77, 291–2
definition of, 344
levels, 89, 140
reliability program requirements
affected by, 92–3
software, 331, 332
Cyclic inspection, 87, 153–4

Data
collection criteria, monitoring of,
205
item descriptions (DIDs), 91, 94–9
recording, operational reliability, 18,
20, 79
requirements for, 78–9, 87, 162–3
sources listed, 9, 79, 160, 281
Databases, information available from,
79, 162, 254
Defects
definition of, 298, 345
process-induced, 52–3, 54, 354
examples of, 297
reliability affected by, 24, 55,
113, 114
Definitions, in alphabetic listing,
339–66
Degradation
failures, 305
reliability
causes of, 22–3, 24, 51–3, 113–15,
297
control of, 53–8, 87, 147–52
in audits, 189
Delivery/packaging specifications, 173
Demonstration tests, 85, 133–5, 318,
321, 345
costs of, 324
Derating, 240–4
definition of, 138, 240, 345
in audits, 188–9

Derating—*contd.*
  reliability improved by, 48, 138–9,
    169–70
  values quoted, 241–3
Design
  assurance elements, 84–5, 112–37
  integrity, assured by NRC, 11
  parameter, reliability as, 2, 16
  phase, screening tests applied
    during, 223
  reliability activities, listed, 207–9
  reliability program starts in, 114
  reliability in, 2, 3, 21, 207–53
  review(s), 85, 135–7, 314–17, 346
    audit of, 187–8
    checklists for, 202, 315
    criteria, monitoring of, 201
  simplification, 249
Development
  combined hardware/software, 327
  management tasks in, 101
  phase, software, 333–5
  procurements, 173–4
  reliability in, 2, 3, 17, 21
  screening tests applied during, 223,
    224
Diagnostics, 49, 124–5, 244–5, 248,
    346
Distribution functions, reliability
    analysis, 72
Documentation
  listed, 194
  procurement specification, 173
  R & M, 9, 11, 91, 367–76
  requirements for
    common mode/cause failure
      analysis, 129
    component
      derating, 139
      failure rate modeling, 141
      quality levels, 140
      selection/specification/QA, 144
      test methods, 143
    configuration management, 109,
      333
    controlled screening tests, 149
    critical item list, 111

Documentation—*contd.*
  requirements for—*contd.*
    cyclic inspection, 154
    data recording, 163
    design reviews, 136–7
    diagnostics, 125
    fault tree analysis, 128
    FMEA/FMECA, 128
    FRACA, 161
    human factors reliability, 126
    life cycle cost analysis, 131
    maintainability prediction and
      allocation, 124
    maintenance planning, 146
    mathematical modeling, 120
    modularity, 125
    monitoring reports, 194
    organization, 107
    PRA/reliability analysis, 157
    program
      planning, 103, 105
      review, 108
    R & M
      assessment, 156–7
      specifications, 118
    reliability
      acceptance, 150
      degradation control, 149, 150,
        152
      demonstration, 134–5
      growth, 132–3
      improvement, 159
      prediction and allocation, 122
    single failure analysis, 130
    spare parts logistics analysis, 147
    storage effects, 145
    supplier audit, 110
    tradeoff analysis, 116–17
    training, 112
  software, 334–5
DoD, *see* US Department of Defense
    (DoD)
DOE (US Department of Energy), 6
Dormant periods, effects of, 86,
    144–5
Downtime, 32, 33, 346, 364
  costs of, 1, 82

Duane reliability growth model, 132,
170, 319

Effectiveness, reliability program, audit
of, 191–2
Efficiency factors, production process,
301–2, 303
Electric Power Research Institute
(EPRI), 6
Electric utilities, nuclear power plant
operated by, 4, 5
Electronic components
control quality levels for, 169
derating of, 240–4
effect of environmental stresses on,
56–7
failure
modes for, 56–7
rate data for, 260, 262; *see also*
US MIL-HDBK-217 . . .
selection guidelines for, 209–22
temperature-reduction schemes for,
252–3
see also Capacitors . . .;
Microcircuits . . .;
Resistors . . .;
Semiconductors
Electronic equipment
design simplification of, 249
effect of environmental stresses on,
56–7, 250–2
reliability improvement techniques
for, 250–3
screening test guidelines for, 226–8
Environmental stresses
effects of, 48, 56–7, 250–2
reliability improvement techniques
for, 250–2
Equipment characteristics, nuclear
plant, 81
Errors, software, 61–3, 335–6, 347, 361
Essential (E) criticality level, 89, 140
Evaluation
of reliability programs, 180–206
by audit, 182–93
by monitoring, 193–206

Evaluation—*contd.*
of reliability programs—*contd.*
criteria for, 181, 182
reports, INPO, 7
Experience feedback, 87, 159–63
Exponential failure models, 28–9,
348
process-induced defects not
accounted-for in, 53

Failure(s)
definition of, 304, 305, 348, 349
during test, action upon, 305, 308
Failure analysis, 308, 347
report (FAR) forms, 308, 309,
313–14
see also Failure reporting, analysis
and corrective action
(FRACA)
Failure distribution curves, 25–7, 296
new hardware, 53–4
software/hardware combined, 64–5
Failure mode(s)
analysis, 282–93
combined hardware/software
systems, 282
concepts, 282–93
monitoring of, 197
purposes of, 283
techniques for, 284
and effects analysis (FMEA), 85,
126–8, 348
used in audits, 186, 187
effects and criticality analysis
(FMECA), 284–6
applied to computer systems, 68
availability improved by, 50
development application of, 27
information sources for, 285
reliability program element, 85,
126–8
software/hardware, 329
used in audits, 186, 187
used in RCM, 281
worksheet for, 286, 287
electronic components, 56–7

Failure rates
  definition of, 28, 348
  in reliability calculations, 45, 46,
    289, 291
  MTBF related to, 29, 31
  quoted for electronic components,
    260, 262
Failure report (FR) forms, 305, 307,
    311–13
Failure reporting, analysis and
    corrective action (FRACA),
    87, 160–2, 304–14, 348
  activities flowchart of, 306
  applications of, 304
  forms for, 307, 308–14
  monitoring of, 204–5
Fault
  avoidance techniques, software,
    328–9
  compensating methods, 283;
    see also Modularity;
    Structured programming
  isolation test (FIT), 49
  matrix, 292, 293
  tree
    analysis (FTA), 73–4, 127, 128,
      286, 288–92, 349
    as reliability program element,
      85, 126–8
    corrective suggestions from, 292
    used in audits, 186, 187
  diagrams (FTDs), 76, 119, 289, 290
    compared with reliability block
      diagrams, 73–4, 76
Federal Aviation Administration
    (FAA), R & M documen-
    tation by, 11
Feedback of information, 87,
    159–63
  by INPO/NRC, 7–8, 13, 162
Figures of merit
  failure, 28–9, 31, 47
  maintainability, 34, 35
Final assembly testing, 226–8, 237–9
Fisher F-test, 72
Flowcharts
  information processing, 328

Flowcharts—*contd.*
  reliability
    analysis methodology, 75
    planning, 14, 15
Forced outage
  causes of, 38–9
  rate (FOR), 37–8, 39, 350
Forms, FRACA, 307, 308–14
Function reliability prediction
    technique, 121
Functional-level diagrams, 123–4,
    268–9

Gas-cooled reactors (HTGRs),
    suppliers of, 6
Generating Availability Data Systems
    (GADS), 8–9, 162, 163
Glossary, 339–66

Hardtime replacement policies, 276–7
Hazard rate, 25, 350
  curves, 64–5; *see also* Failure
    distribution curves
High-temperature gas-cooled reactors
    (HTGRs), suppliers of, 6
Human error rates, 291
Human factors reliability analysis,
    125–6, 351

IEEE Standards, 7, 367–8
Inductive devices
  derating of, 242
  failure
    modes for, 57
    rates for, 262
  temperature-reduction schemes for,
    252
Infant mortality period
  failure distributions in, 25–6, 53–4,
    296, 351
  hazard rate distribution in, 65
  time-reduction of, 148
Inherent availability, definition of, 42,
    340

Inherent reliability, 113, 147, 151, 360
Inspectability, 22
Inspection
  efficiency, 303
  procedures, 58
Installation errors, prevented by NRC, 13
Institute of Nuclear Power Operations (INPO), 6, 7–8, 162
Instruction set architecture (ISA), computer, 61
Integrated circuits (ICs), 59, 351
Integration testing, software, 70
Intermediate testing, 226–8, 234

Kolmogorov–Smirnov statistical test, 71

Latent defects, 52–3, 298
Licensee event reports (LERs), 8, 162, 163
Life curves
  failure rate, 25–7, 53–4, 296
  hazard rate, 65
Life cycle
  cost (LCC)
    assessment and analysis of, 85, 130–1
    components of, 165–6
    definition of, 130, 352
    factors affecting, 166–7
    minimization of, 15–16, 42, 82–3, 115, 117, 164–5
    versus reliability, 165
  failure distributions, 25–7
  reliability
    activities in, 11, 12–13, 18–19
    model, 22, 24
    programs, 80–91
      elements of, 84–7
      purpose of, 80–1
    summarized, 2–3
Lifespan, nuclear plant, 83
Log normal distribution function, 72

Logistics analysis, spare parts, 86, 146–7
Loss-of-load probability (LOLP), 38, 352
Lot acceptance testing, 142, 200, 296
Lubricants, electronic component, 56

Maintainability
  allocation criteria, monitoring of, 198
  concepts of, 31–6, 267
  definition of, 353
  demonstration test criteria, monitoring of, 203–4
  design checklist, 246
  prediction
    and allocation, 84, 123–4, 267–82
    criteria, monitoring of, 198–9
    techniques, 123, 267–9
    audit of, 185–6
Maintenance
  definition of, 353
  plan criteria, monitoring of, 197–8
  planning, 86, 145–6
  policy, audit of, 190
  procedures, checked by NRC, 13
  significant items (MSI), 273
  time elements, 33–4, 354–5
Management tasks, reliability, 84, 88–9, 91, 100–12
Manufacturing
  defects introduced during, 52–3, 54, 297, 345
    reliability affected by, 24, 55, 113, 114
  errors prevented by NRC, 13
  reliability in, 21–2
Mathematical modeling, 84, 119–20
Mean time between failures (MTBF)
  definition of, 28–9, 31, 47, 355
  improvement of, 47–9
  in reliability specification, 16, 323
  life cycle costs versus, 16
  post-production estimate of, 299–303
  relationship to acquisition cost, 164, 165, 171, 172

Mean time to repair (MTTR)
  allocation of, 198
  definition of, 32, 355
  determination of, 34, 123, 267–8
Methodology, reliability analysis, 74–7
Microcircuits
  burn-in tests for, 229
  component quality levels for, 169
  cost-effectiveness trends for, 213
  derating of, 241
  failure
    modes for, 57
    rates of, 262
  production costs for, 212
  reliability levels for, 212
  screening requirements for, 211,
    212–13
  selection guidelines for, 211–13
Microcomputers, 59, 356
Microprocessors
  described, 59, 356
  listed, 60
  specification for, 61
MIL-STDs, see US MIL-STD...
Military systems, nuclear plant
    compared with, 83, 88
Models
  failure rate prediction, 61, 140–1
  reliability growth process, 132, 170,
    319
Modularity, 124–5, 170, 244–8
  computer software, 69
  maintainability improved by, 48,
    244–5
Monitoring guidelines, 193–206
  component control and
    standardization criteria,
    199–200
  critical item control criteria, 199
  design review criteria, 201
  failure mode analysis criteria, 197
  FRACA criteria, 204–5
  maintainability
    allocation criteria, 198–9
    demonstration test criteria, 203–4
    prediction criteria, 198–9
  maintenance plan criteria, 197–8

Monitoring guidelines—*contd.*
  prediction criteria, 196–7
  production reliability assurance
    criteria, 205–6
  project organization, 195
  reliability
    allocation criteria, 196
    assessment criteria, 206
    data collection criteria, 205
    demonstration test criteria, 203
    growth tests criteria, 201–3
  supplier reliability control criteria,
    200–1
Motors
  failure modes for, 57
  failure rates of, 262

National Aeronautics and Space
    Administration (NASA), 9, 11
National Electric Research Council
    (NERC), 39
NAVMAT P-9492 (US Navy screening
    program), 238
Non-destructive testing (NDT), 298,
    308, 356
Non-essential (NE) criticality level, 89,
    140
Normal-distribution function, 72
NOTEPAD computer communications
    network, 7–8
NRC (US Nuclear Regulatory
    Commission), 4, 9, 11, 13
Nuclear Plant Reliability Data
    System (NPRDS), 8, 162, 163,
    281
Nuclear Safety Analysis Center
    (NSAC), 6–7
Nuclear steam supply system (NSSS)
    vendors, listed, 4, 6

Off-line repairs, 34
Off-the-shelf procurements, 174,
    175, 177–9
Operating availability, definition of,
    37, 341

Operational and support (O & S) costs, effect of reliability on, 165, 167
Operational phase
management tasks in, 101
reliability in, 2, 3, 18
Operational practices, audit of, 189–91
Operational procedures, checked by NRC, 13
Operational reliability assurance, 87, 153–9
Organization, 84, 106–7
audit of, 183–4
documentation requirements for, 107
monitoring of, 195
responsibility for FRACA system, 305, 308, 312
specifications applicable to, 107
Outage
analysis, 19, 37, 38
types of, 38–9
Overview, US nuclear industry, 3–11

Parallel model, reliability calculations on, 46
Parts
control and standardization, 294–7
count predictive techniques, 73, 121, 178, 260–3
screening, 225, 229, 234; *see also* Electronic components
selection, 209–22
for capacitors, 219–22
for microcircuits, 211–13
for resistors, 216–19
for semiconductors, 213–16
Planned outage, 38
Plant logs, component reliability data from, 281–2
Plastics, deterioration of, 56, 57
Prediction techniques
maintainability, 123, 267–9
reliability, 73–8, 121–2, 178, 260–7
Preferred component lists (PCL), 144
Preliminary design reviews (PDRs), 135, 194, 197, 201, 202, 316, 358

Pressurized water reactors (PWRs), suppliers of, 4, 6
Preventive maintenance (PM), 34–6, 168, 267, 271, 273, 358
cost of, 167, 168
worksheet for, 271
Printed circuit boards (PCBs)
screening test guidelines for, 226–8
temperature-reduction schemes for, 253
Probabilistic nature of reliability, 25
Probabilistic risk analysis (PRA), 9, 82, 87, 157
sensitivity analysis of, 157
Process flow diagram, MTBF-estimating process, 301, 302
Process-and-inspection analysis, 298–9
Procurement
classification affecting test levels, 324
classification of, 173–4, 324
specification
components of, 172–3
importance of, 15
Production reliability assurance, 297–303
monitoring of, 205–6
Production-induced reliability degradation, 24, 55, 113, 114, 148
Program
elements, 84–7, 176
audit of, 185–8
component availability, 86, 137–47
design assurance, 84–5, 112–37
experience feedback, 87, 159–63
management tasks, 84, 91, 100–12
operational reliability assurance, 87, 153–9
reliability degradation control, 87, 147–52
planning, 84, 101–6
basic requirements for, 102–3
documentation requirements for, 103, 105
specifications applicable to, 106
review, 84, 108
Proof tests, 142

Qualification testing, 142, 200, 296
Quality
  assurance (QA), 173, 296, 358
    monitoring of, 200
    software, 331
  designators, multi-level
    specification, 265
  levels, 86, 139–40, 169, 265

Random failures, caused by
    stress/strength factors, 26–7,
    359
Random vibration screen tests, 238,
    239, 359
Redundancy
  reliability affected by, 30, 31, 359
  used to achieve required
    availability, 44, 248–9
Redundant systems, reliability
    calculations affected by, 44–6
Regression analysis techniques, 71
Reject rates, 149, 239
Relays
  derating of, 242
  failure
    modes for, 56, 57
    rates for, 262
Reliability
  as design parameter, 2, 16
  as performance parameter, 11
  definition of, 25, 359
Reliability acceptance tests, 150, 318,
    322
Reliability allocation
  criteria, monitoring of, 196
  definition of, 120–1
  prediction and assessment, 256–67
  techniques
    compared with prediction
      techniques, 256–7
    weighting factors for, 259
Reliability analysis
  methodology flowchart, 74, 75
  software, 59–70
  techniques, 70–9, 157
    audit of, 186

Reliability and maintainability
    (R & M)
  assessment, 87, 154–7
  documentation, 9, 11, 122, 124,
    367–77
  growth, 49–50, 113
  improvement affecting availability,
    49–51
  specifications, 84, 117–19
Reliability assessment, 266
  monitoring of, 206
Reliability assurance
  production factors, 297–303
  programs (RAPs), elements of, 11–20
Reliability block diagrams (RBDs), 29,
    30, 31, 76, 119
  compared with fault tree diagrams,
    73–4, 76
Reliability centered maintenance
    (RCM), 36, 146, 272, 273–82
  decision logic for, 274, 275, 278
  default decisions in, 278
  information management system
    for, 280–1
  maintenance plan for, 278–9
Reliability concepts, 25–31
Reliability data, purpose of, 160
Reliability degradation
  causes of, 22–3, 24, 51–3, 113–15, 297
  control of, 53–8, 87, 147–52
Reliability demonstration tests, 85,
    133–5, 318, 321, 324
  cost of, 324
  monitoring of, 203
Reliability engineering
  procedures, 255, 359
    design reviews, 85, 135–7, 314–17
    failure mode analysis, 282–93
    FRACA, 87, 160–2, 304–14
    maintainability prediction, 267–82
    parts control and standardization,
      294–7
    production reliability assurance,
      297–303
    reliability allocation, prediction
      and assessment, 256–67
    reliability testing, 317–26

Reliability engineering—*contd.*
 procedures—*contd.*
  software, 326–37
 programs, objectives of, 23
Reliability growth, 49–50, 85, 113, 359
 models for, 132, 170, 319
 test programs, 131–3, 170, 318–26, 360
  audit of, 191
  monitoring of, 201–3
Reliability improvement, 87, 158–9
 warranty (RIW), 178
Reliability levels, definition of, 243
Reliability math models (RMMs), 76
Reliability mathematical modeling, 84, 119–20
Reliability organization, *see* Organization
Reliability prediction
 and allocation, 84, 120–3
 criteria, monitoring of, 196–7
 techniques, 73–8, 121–2, 178, 260–7
  audit of, 185
  purpose of, 258
Reliability program
 audit of, 181, 182–93
 effectiveness, audit of, 191–2
 elements of, *see* Program elements
 monitoring of, *see* Monitoring guidelines
 planning of, 84, 101–6
Reliability specification requirements
 components of, 16, 163–4
 determination of, 163–4
Reliability testing, 317–26
Repair
 costs, throwaway module costs compared with, 246–7
 effect on reliability calculations of, 47
Replacement
 policies, 276–6
 strategy, audit of, 190–1
Resistors
 burn-in tests for, 229
 component quality levels for, 169
 cost trends for, 219

Resistors—*contd.*
 derating of, 241
 failure
  modes for, 57
  rates for, 262
 screening requirements for, 217
 selection guidelines for, 216–19
 temperature-reduction schemes for, 252
 typical specification for, 218
Return on investment (ROI), 165

Safety Assurance and Failure Elimination (SAFE) program, 9, 10, 11
Safety requirements, nuclear plant, 1, 81, 82, 360
Screening
 strength, definition of, 301
 tests, 22, 53, 142, 148–9, 170–1, 222–40, 321–2, 360
  audits, in, 189
  flowchart for, 224
  insufficient safety margins detected by, 27
  MTBF improved by, 48
  risks associated with, 234
SEE-IN feedback program, 8
Semiconductors
 burn-in tests for, 229
 component quality levels for, 169
 cost ratios for, 215–16
 derating of, 241
 failure
  modes for, 57
  rates for, 262
 manufacturing processes for, 295, 296
 reliability levels for, 215
 screening requirements for, 214–15
 selection guidelines for, 213–16
 temperature-reduction schemes for, 252–3
Sensitivity analysis, 77, 157
Series model, reliability calculations on, 45

Series/parallel concepts of reliability,
    29–31, 45–7
Shelf life, 86, 144–5, 362
Similar-complexity predictive
    techniques, 121
Similar-equipment predictive
    techniques, 73, 121
Single point failure (SPF), 361
    analysis of, 85, 130, 187
Software
    definitions for, 61, 361–2
    error(s), 61–3, 361
        correction costs, 63
        modes in, 66, 361
        reports, 335–6
    interaction with hardware, 329
    planning, 331
    program matrix, 331, 332
    reliability
        analysis, 59–70
        engineering and control, 326–37
        techniques, 66, 67
    reporting and correction procedure
        for, 337
Spare parts logistics analysis, 86, 146–7
Specification(s)
    definition of, 362
    quoted for each program element
        common mode/cause failure
            analysis, 130
        component
            derating, 139
            failure modeling, 141
            quality levels, 140
            selection/specification/QA, 144
            test methods, 142, 143
        configuration management, 109
        critical items list, 111
        data recording, 163
        design review, 137
        fault tree analysis, 128–9
        FMEA/FMECA, 128–9
        FRACA, 162
        human factors reliability, 126
        life cycle cost analysis, 131
        maintainability prediction, 124
        maintenance planning, 146

Specification(s)—*contd.*
    quoted for each program element—
        contd.
        mathematical modeling, 120
        organization, 107
        PRA/reliability analysis, 157
        program planning, 106
        R & M
            assessment, 157
            specifications, 118–19
        reliability
            degradation control, 149, 150–1,
                152–3
            demonstration, 135
            growth, 133
            improvement, 159
            prediction, 122–3
            supplier audit, 110
        reliability included in, 2, 3
        requirements, reliability,
            components of, 16, 163–4
        see also US MIL-STD...
Standard(s)
    components, definition of, 138, 294
    definition of, 362
    listed, 366–77
Standby parallel (SP) mode, reliability
    calculations of, 46–7
State-of-the-art (SDA), reliability
    testing affected by, 324,
        325
Statistical methods, reliability analysis,
    71–2, 90, 155
Status accounting function, software
    development, 334
Storage life, 362
    effects of, 86, 144–5
Stress-analysis tests
    as reliability prediction techniques,
        73, 122, 178, 263–7
    worksheets for, 263–4
    definition of, 151, 362
    latent defects revealed by, 298, 301,
        321–2
    purpose of, 53, 142, 244
Structured computer-programming,
    69, 283, 363

Supplier(s)
  audit, 84, 109–10, 171
  of nuclear reactors, listed, 4, 6
  reliability control criteria,
    monitoring of, 200–1
Switches
  failure modes with, 56, 57
  failure rates for, 262
System
  availability factor, 44, 363
  definition of, 363
  integration, reliability in, 2, 3
  reliability, 29–30
  restoration times, definition of, 34

Temperature reduction schemes,
    electronic components, 252–3
Temperature/time screening tests,
    235–7, 364
  final assembly, 238–9
  parts failure rate improvement due
    to, 234
Test
  analyze and fix (TAAF) approach,
    90, 132, 319, 323
  methods, listed, 142
  plans, 134, 322–3
  programs, components of, 326
Throwaway maintenance modules,
    246–8
Time-of-onset (Tos) concept, 276
Times to repair (TTRs), 32
  elements of, 33–4
Tradeoff analysis, 15–16, 84, 115–17,
    365
  audit of, 186–7
  availability studies, 42–4
  reliability growth, 322–3
Training
  cost of, 167, 168
  reliability, 84, 111–12
Transformers, *see* Inductive devices ...

US Army handbooks, listed, 377

US Department of Defense (DoD)
  R & M documentation, 9, 91, 94–9,
    368
  reliability-centered maintenance used
    by, 146, 273
US Department of Energy (DOE), 6
US MIL-HDBK-217, 175, 384–5
  factors affecting failure rates in, 213,
    216, 222
  failure rate
    data from, 73, 178, 211, 260
    versus stress rate data in, 139, 241
  reliability prediction models in, 61,
    121, 141, 152, 212, 263,
    265–6
US MIL-HDBK-472, 175, 385–6
  maintainability prediction
    techniques in, 123
  maintenance time elements in, 271
US military specifications, listed, 169,
    174, 294, 368–76
US MIL-M-38510, 211, 368
US MIL-R-390088, 218
US MIL-S-19500, 214, 369
US MIL-S-52779, 330–1, 369–70
US MIL-STD-198, 169, 294, 370
US MIL-STD-199. 169, 217, 294, 370
US MIL-STD-202, 142, 294, 370
US MIL-STD-471, 203, 371
US MIL-STD-701, 169, 294, 372
US MIL-STD-750, 142, 294, 369, 372
US MIL-STD-756, 174, 196, 372
US MIL-STD-781, 134, 150, 203, 321,
    322, 324, 369, 372
US MIL-STD-785, 102–5, 128, 161,
    174, 200, 372–3
  applications matrix for, 104
US MIL-STD-883, 142, 211, 212, 225,
    230–3, 294, 369
US MIL-STD-1562, 169, 294, 375
US MIL-STD-1629, 127, 375
US Navy handbooks, 379–80
US nuclear industry, overview of, 3–11
US Nuclear Regulatory Commission
    (NRC), 4, 9
  system reliability activities, listed,
    11, 13

Useful life period
   hazard rate distribution in, 65
   reliability in, 26–7, 53, 65

Validation, definition of, 365
Vendor control program, availability
   improved by, 50–1
Verification, definition of, 365
Vibration screening tests, 237–8,
   239

Visual inspection, importance of, 58

Warranty, 178
Wearout
   concept, 276–7, 365
   period
     failure rate distributions in, 26, 27,
     54, 296
     hazard rate distribution in, 65
Weibull probability distribution
     functions, 72

T - #0457 - 101024 - C0 - 220/142/24 - PB - 9781138561731 - Gloss Lamination